# Renewable Energies in the Built Environment

# Renewable Energies in the Built Environment

Guest Editors

**Elena Lucchi**
**Tianyi Chen**
**Wen Zhang**

MDPI

Basel • Beijing • Wuhan • Barcelona • Belgrade • Novi Sad • Cluj • Manchester

*Guest Editors*

Elena Lucchi
Dipartimento di Architettura, Ingegneria delle Costruzione e Ambiente Costruito (ABC)
Politecnico di Milano
Milano
Italy

Tianyi Chen
Solar Energy Research Institute of Singapore (SERIS)
National University of Singapore
Singapore
Singapore

Wen Zhang
School of Architecture and Design
Beijing Jiaotong University
Beijing
China

*Editorial Office*

MDPI AG
Grosspeteranlage 5
4052 Basel, Switzerland

This is a reprint of the Special Issue, published open access by the journal *Sustainability* (ISSN 2071-1050), freely accessible at: www.mdpi.com/journal/sustainability/special_issues/0FBN75O7XL.

For citation purposes, cite each article independently as indicated on the article page online and using the guide below:

| Lastname, A.A.; Lastname, B.B. Article Title. *Journal Name* **Year**, *Volume Number*, Page Range. |
|---|

**ISBN 978-3-7258-3858-5 (Hbk)**
**ISBN 978-3-7258-3857-8 (PDF)**
**https://doi.org/10.3390/books978-3-7258-3857-8**

# Contents

# About the Editors

**Elena Lucchi**

Elena Lucchi, Getty Conservation Guest Scholar 2024/2025. She is an Architect with a PhD in building physics from the Politecnico di Milano, complemented by a post-Master's degree in environmental education from the Università Cattolica. Her research addresses the connections between energy, environment, and architecture, with a focus on energy efficiency, climate mitigation strategies, and the integration of renewable energies in historic buildings and protected landscapes.

**Tianyi Chen**

Dr. Tianyi Chen, originally from China, is a Research Fellow at the Solar Energy Research Institute of Singapore (SERIS) and Co-Founder of Power Facade Pte. Ltd. His research focuses on Building-Integrated Photovoltaics (BIPV), with a particular emphasis on prefabricated unitized BIPV wall systems, heritage building solar integration, and BIM for BIPV implementation. After completing his architectural education at Hunan University, ETH Zurich, and the Architectural Association, Dr. Chen earned his Ph.D. from the National University of Singapore where he developed innovative approaches to merging solar energy technologies with modern construction methods. His work bridges the disciplines of architecture, renewable energy, and digital innovation to advance sustainable building solutions for dense urban environments in tropical climates.

**Wen Zhang**

Dr. Wen Zhang has long been committed to research and education in the field of low-carbon urban planning, with a particular focus on photovoltaics applied in the built environment and urban microclimate research. He is currently an associate professor in the Department of Architecture, School of Architecture and Design, Beijing Jiaotong University, and was formerly a research fellow in the Department of buildings, National University of Singapore. In 2023, he was selected for the Beijing Youth Talent Support Program. He presided over and participated in 5 National Natural Science Foundation of China projects, obtained 6 invention patents, and published more than 20 journal papers.

# Preface

It is with great pleasure that we present this comprehensive volume, "Renewable Energies in the Built Environment," which brings together significant contributions from a Special Issue of the *Sustainability* journal.

We stand at a transformative moment in human history, where our built environment is undergoing profound change driven by the imperative for sustainable development and energy efficiency. This reprint explores the frontier where renewable energy technologies intersect with architectural and urban design, offering innovative approaches to creating sustainable, resilient communities.

The research presented herein spans multiple disciplines—engineering, architecture, urban planning, environmental sciences, policymaking, and heritage conservation—providing a holistic perspective on integrating renewable energy systems into our buildings and cities. From solar photovoltaics that reshape architectural aesthetics to wind technologies that redefine our landscapes, from geothermal applications that tap the earth's energy to ocean systems that harness the power of waves, this collection examines both the technical challenges and creative solutions emerging in this rapidly evolving field.

A particular focus of this volume is the delicate balance between innovation and conservation, especially in cultural heritage contexts. We explore how cutting-edge renewable technologies can be sensitively integrated into historically significant buildings and areas, preserving their cultural value while enhancing their sustainability.

The contributions in this reprint reflect the dedication and expertise of researchers and practitioners worldwide who are committed to advancing sustainable energy solutions. We extend our sincere gratitude to all authors for sharing their valuable insights and to the reviewers whose rigorous evaluation has ensured the high quality of this work.

It is our hope that this collection will inspire further research, inform policy development, and guide the practical applications of renewable energy in the built environment. As we face the urgent challenges of climate change and resource depletion, the knowledge shared in these pages offers pathways toward a more sustainable future for all.

**Elena Lucchi, Tianyi Chen, and Wen Zhang**
*Guest Editors*

*sustainability*

MDPI

*Editorial*

# Renewable Energies in the Built Environment

**Elena Lucchi [1,2,*], Tianyi Chen [3] and Wen Zhang [4]**

1 Department of Architecture, Built Environment and Construction Engineering (DABC), Politecnico di Milano, 20133 Milan, Italy
2 Getty Conservation Institute, Los Angeles, CA 90049, USA
3 Solar Energy Research Institute of Singapore (SERIS), National University of Singapore, Singapore 117574, Singapore; tianyi.chen@u.nus.edu
4 Department of Architecture, School of Architecture and Design, Beijing Jiaotong University, Beijing 100044, China; zhwen@bjtu.edu.cn
* Correspondence: elena.lucchi@polimi.it or elucchi@getty.edu

## 1. Introduction

The attention to environmental sustainability and energy has progressively increased in recent years, driven by the global energy crisis and mounting climate change concerns. These fundamental challenges, combined with the COVID-19 pandemic, have transformed the worldwide energy situation, impacting energy security, climate response, and lifestyles (e.g., smart working, home–school, online shopping). Our built environment is undergoing a metamorphosis, driven by the need for sustainable development and energy efficiency. This revolution brings with it an opportunity to reimagine and redefine the way we harness and utilize energy. This situation opens new possibilities for the energy retrofit of buildings, as well as for the adaptive reuse of existing buildings, asking for a conscious use of passive and active systems.

In this context, the idea of sustainable transition arose, defined as "[...] long-term, multidimensional, and fundamental transformation processed through which established socio-technical systems shift to more sustainable alternatives" [1]. Inside it, the energy transition asks for the replacement of fossil fuels with low or zero-carbon energy sources, such as renewable energy sources (RESs) like active solar energy, wind technologies, biomass and bioenergy, geothermal energy, hydropower, and ocean energy.

The energy transition is suggested at the international level to face the effects of climate changes and global warming through the "*decarbonization*" of the energy system, especially for building, transportation, and industrial sectors. Several policies have been developed internationally for decreasing greenhouse gas (GHG) emissions and for improving the energy efficiency of the built environment, with a particular emphasis on the design, construction, operation, and integration of RESs. For example, the Sustainable Development Goals (SDGs) adopted at the international level aim at reaching a sustainable development balancing social, economic, and environmental aspects [2]. Key goals supporting RES integration include Goal 7 "*Affordable and clean energy*" that boosts energy efficiency and renewable energy production in the infrastructural sector; Goal 11 "*Sustainable cities and communities*" advancing sustainable urbanization through green infrastructure and RES integrated housing; and Goal 13 "*Climate action*" supporting fossil-free adoption [3]. Specific legislative frameworks have been developed in Europe, the United States, Asia, and Australia.

RES development has reached advanced solutions, both on technological and building integration issues. Special attention is devoted to the evolving area of solar energy technologies, as they are shaping a new paradigm in architectural and urban design. This

Received: 26 February 2025
Accepted: 4 March 2025
Published: 17 March 2025

**Citation:** Lucchi, E.; Chen, T.; Zhang, W. Renewable Energies in the Built Environment. *Sustainability* **2025**, *17*, 2661. https://doi.org/10.3390/su17062661

includes both pioneering works on module design, photovoltaics, or solar thermal integration in retrofitted buildings to foster the harmonious coexistence of esthetic appeal, technologic development, and energy production [4]. In this direction, the study of practical applications and tests of energy, esthetic, and sustainable performances (e.g., life cycle assessment, building and landscape integration, grid connection, energy communities) of groundbreaking photovoltaic (PV) and solar thermal (ST) systems is an important field to be investigated. At the same time, solar energies' potential applied in the built environment is also deeply studied, including a comparison of traditional and innovative evaluation potential methods [5,6]. In line with this, the potential of Building Information Modeling (BIM) to aid the integration of solar technologies into the design and construction processes deserves further exploration, both for implementing the visual appearance and for calculating the energy production from the early design to the construction phase [7]. Also, the policy environment influencing their implementation is scrutinized, shedding light on how these policies can further propel the adoption of renewable energy in the built environment [8].

Presently, special attention is given to wind energy technologies, considering wind farms, landscape integration, visibility mapping, policy, and acceptability issues. The integration of wind technologies is studied mainly at the territorial level for offshore and wind farms in natural areas, while the integration of wind technologies in the built environment, especially in heritage or sensitive contexts, is less studied due to the presence of strict regulation constraints. Applications and studies in this direction can demonstrate the feasibility of these systems, showing good and wrong practices.

Furthermore, the integration of ground source heat pumps in existing buildings requires adequate design and construction to overcome administrative and technical problems. This topic is less studied in the literature, delineating problems and opportunities for energy production [9]. Finally, hydropower and ocean energy integration in the built environment or in existing landscapes is rarely addressed [10].

At the intersection of built environment and RES lies the challenge of balancing conservation and innovation, especially when it comes to cultural heritage [9]. In addition, the uncritical application of RESs in the built environment could generate serious conservation issues, especially in culturally significant built environments and heritage-built environments as well as in protected landscapes. Here, it can compromise heritage values, biodiversity, traditional visual appearance, and materiality. Here, RES integration presents a delicate interplay of tradition and modernity, requiring a balance between heritage preservation and energy production with the support of clear rules, policies, criteria, and heritage-compatible technologies [11]. Furthermore, the acceptability (and acceptance) of these technologies, especially of solar and wind systems, deserves to be deeply analyzed for the presence of technical, economic, informative, and legislative barriers [9]. Furthermore, the importance of the energy efficiency of the built environment requires a focus on energy auditing and simulation [12]. These tools enable the efficient use of energy resources for improving the sustainability in the built environment [13].

## 2. Contributions

According to this background, the Special Issue on "*Renewable Energies in the Built Environment*" aims at presenting a holistic view of the opportunities and challenges associated with integrating renewable energy into the fabric of our built environment. The development of these technologies from planning, design and construction perspectives has been investigated. This Special Issue provides a platform for thought-provoking discussions, innovative ideas, and the sharing of best practices, heralding a sustainable future for our cities and communities. Its scope spans across several domains of research—engineering, architecture, urban planning, environmental sciences, policymaking, finance, and heritage conservation—aiming to foster innovative and sustainable solutions to energy use in the context of the built environment.

Through various studies, innovative approaches have been revealed to optimize energy consumption, reduce environmental impacts, and foster energy self-sufficiency. Research areas include the following:

- Building components, where emerging technologies in RES integration are analyzed (e.g., innovative PV technologies, such as tensioned membranes and agrivoltaics building envelopes).
- Building applications, with a particular focus on historic and protected buildings.
- Urban areas, discussing energy communities, renewable energy company development and RES applications at the large scale.

Complementing these technological advances, Yu et al. provide a comprehensive bibliometric analysis of research trends in low-carbon rural areas through systematic examination of 583 papers published between 2013 and 2023. Their findings demonstrate the evolution of rural carbon emission research through three distinct stages: from initial conceptual development (2003–2010) to technical implementation (2011–2018), and finally to the current deepening stage focused on comprehensive carbon reduction strategies. The study reveals that while rural carbon emissions have gained increasing global attention, research remains concentrated in China, the United States, and the United Kingdom, with limited cross-regional collaboration. The authors make an important contribution by identifying critical research gaps and proposing future directions, including the need for standardized emission inventories, multi-scale statistical methods, and region-specific reduction pathways that balance environmental goals with economic development. Their analysis provides valuable guidance for researchers and policymakers working to develop effective, locally adapted solutions for rural carbon emission challenges.

## 3. Building Components

Milošević, Marchwiński and Lucchi examine PV-integrated hyperbolic paraboloid membranes to optimize their design, minimizing strain and enhancing energy efficiency. A major technical hurdle lies in managing the substantial strains caused by external forces, which can negatively impact both the PV system's efficiency and the structural stability of the membrane. While current design methods focus on stress and deflection, overall strain behavior is often overlooked. Thus, the study evaluates the impact of load types, geometry, material properties, and patterning direction using the Finite Element Method (FEM) with Sofistik software 2023. The results show that wind loads induce higher strains than snow loads, larger structures and increased curvature amplify strain, and diagonal patterning helps reduce it. High tensile-strength materials and optimized prestress further mitigate strain, while edge type has minimal influence. The findings provide design guidelines to improve the structural performance and efficiency of PV-integrated tensioned membranes in sustainable construction.

Zhang, Chen, Gasparri and Lucchi present an innovative agrivoltaics building envelope (ABE) system that integrates thin-film photovoltaics with hydroponic vertical farming in a modular design approach. The system incorporates transparent thin-film PV glass modules alongside low-E glass modules in a "*checkerboard*" pattern, achieving PV conversion efficiencies of 18% while maintaining 20–30% transparency to support plant growth. The modular design employs recycled aluminum frames as primary structural support, offering both sustainability benefits and flexibility for new construction or retrofitting applications. Their experimental investigation demonstrates that this integrated approach can generate approximately 80–100 $W/m^2$ under standard test conditions while providing adequate natural light transmission for agricultural modules through multiple pathways: diffused light through semi-transparent PV panels, direct sunlight through clear glass sections, and reflected light from adjacent surfaces. The system's adaptable tilt angles (5° to 20° C) and ventilation cavities (65–100 mm) help optimize both energy generation and growing conditions while managing heat accumulation. This research represents a significant advance in multi-functional building envelope technology, demonstrating how careful integration of renewable energy and food production systems can enhance building sustainability while maintaining architectural flexibility.

## 4. Building Applications

Han, Qu, Wu, Su, Qiu and Zhang examine the spatial patterns of carbon emissions and differentiated peak paths at the county level in Shandong Province, China. Through a comprehensive analysis utilizing principal component analysis and k-means clustering, they identify five key driving factors of carbon peak: green transformation, urbanization, industrial construction, energy consumption, and environmental constraints. Their findings reveal distinct spatial patterns characterized by an east–west gradient, with higher emissions concentrated in coastal areas. The study identifies five cluster areas with different carbon peak characteristics: agricultural transformation pending areas, low-carbon lagging areas, industrial transformation areas, low-carbon potential areas, and low-carbon demonstration areas. Each cluster represents different stages of development and faces unique challenges in achieving carbon peak targets. The research demonstrates how county-level analysis can provide valuable insights for formulating targeted low-carbon development strategies, highlighting the importance of considering local conditions and development stages when designing carbon reduction policies. Their findings underscore the critical role that county-level regions play in achieving China's carbon neutrality objectives while revealing the complex interplay between economic development, industrial structure, and environmental constraints at the local level.

Xie et al. examine passive energy conservation strategies for traditional dwellings in the Peking area through a systematic investigation combining field experiments and numerical simulations. Their findings demonstrate that optimizing building envelope components can significantly reduce energy consumption, including an up to 744.68 kWh reduction through optimal wall thickness of 800 mm, 2514.58 kWh through roof insulation of 15 mm thickness, and 1997 kWh through external wall insulation of 50 mm thickness. Additionally, installing photovoltaic tiles achieved annual electricity generation of 12,057 kWh while reducing $CO_2$ emissions by approximately 6631.35 kg annually. This comprehensive analysis provides valuable guidance for retrofitting traditional brick dwellings to improve energy efficiency while preserving architectural heritage, with implications for similar building types across northern China.

Ziozas et al. analyzed the potential for renewable energy application and design direction from the perspective of energy consumption analysis of an Italian cultural heritage building. They utilized a dynamic simulation tool named *INTEMA.building* to simulate the energy-saving retrofit performance of a cultural heritage building in Italy. Their findings demonstrated that by renovating the building envelope, applying the PV panels, using ground-source heat pumps, and installing LED lighting systems, the project can reduce the primary energy demand by 30.49%, the final energy demand by 36.74%, and the net electricity demand by 8.72%. These included a 48% reduction in natural gas consumption due to the using of a ground-source heat pump and a 22.8% reduction in electricity demand attributed to PV applications in the case of a cultural heritage building in Italy.

## 5. Urban Areas

At the urban level, studies focus on protected areas, such as high-value natural and architectural marine landscapes, rural areas, or new urban developments. Romano, Baiani and Mancini explores the opportunities and challenges of integrating renewable energy into the built environment, focusing on less than 3000 residential buildings on Procida, a small island in southern Italy. Through territorial, urban, historical, structural, and environmental analyses, it develops a classification system using in-house software to assess building conditions and propose intervention scenarios. These scenarios combine technological and environmental upgrades (e.g., improved building envelopes and bioclimatic strategies) with energy retrofitting measures like system replacements and solar panel installations. The findings highlight strategies for enhancing building performance and achieving self-sufficiency energy. With a different approach, Abouaiana and Battisti examine rural energy consumption in Egypt. They explore its relationship with the built environment and socio-economic factors, also assessing the feasibility of PV systems. Using analytical, field, and statistical methods, the research focuses on a typical agricultural village in the Delta region, evaluating its built environment, building types, and socio-economic conditions. They discovered that rooftop PV panels could generate substantial energy, but they remain financially unviable under current regulations. Thus, the research offers policy recommendations for rural energy transition. In addition, Ahern et al. proposed an Electrical Water Heater Aggregation (EWHA) scheme with wind energy for fuel-poor households in Ireland, using electricity generated from surplus wind-generated electricity to provide domestic hot water (GHW) to address the hot water supply issues of fuel-poor families. They assessed the feasibility and cost-effectiveness of the solution using a developed wind-generated electricity allocation model and calculated that the scheme can meet the needs of poor users, providing a full tank of hot water every three weeks, while reducing 33 Mkg $CO_2$ annually. It is expected to save Ireland about EUR 4 million in carbon costs by 2030, increasing to EUR 11 million by 2050.

Another key topic of interest is Renewable Energy Communities (RECs), which enable collective RES generation by actively involving citizens in the energy transition. The primary goals of RECs are to maximize energy sharing among members and optimize RES use by reducing supply-demand imbalances. One strategy to align energy demand and supply in RECs is the use of Thermal Energy Storage (TES) systems that, through power-to-heat technology, offer a cost-effective and versatile solution compared to battery storage. Brunelli et al. examine the role of TES in RECs through a bibliometric analysis focused on TES applications in buildings and districts. The research reveals that TES and REC studies have gained momentum in recent years, particularly in Europe, due to energy efficiency goals and European directives. Despite the increasing interest, the findings highlight a lack of direct focus on TES in the existing literature on RECs, while energy storage is generally discussed within broader RES studies. However, terms like "*district heating*", "*smart grid*", and "*micro grid*" suggest a potential overlap in collective self-consumption research, which includes thermal energy. The merging of these keywords points to potential future research in applying TES technologies to RECs, particularly in Europe, where research on both topics is concentrated. In addition, another key topic of interest that has been studied is the development trends of the renewable energy industry. Osiichuk conducted a comparative analysis of the financial conditions of renewable energy companies and conventional energy companies in the European Union and explored the factors affecting the relationship between energy price fluctuations, financial conditions, and the external financing needs of renewable energy companies. The research shows that the financial conditions of renewable energy companies are currently improving, and the return on assets generated by them is significantly higher than that of their peers in conventional energy companies. He also pointed out that the biggest challenge for the operating performance of renewable energy companies may lie in the restrictive revenue cap, price regulation, and the exposure of renewable energy companies to spot market price fluctuations, in addition the cumbersome permit system. He also proposed that the development of the renewable energy industry could be promoted by increasing the prices of renewable energy and relying on market pricing mechanisms.

## 6. Conclusions

These studies highlight the transformative potential of RES systems in positioning the built environment as a key driver of global energy transition. Throughout this Special Issue, we have seen how technological innovations must be supported by effective policies to achieve widespread adoption of RES. The research presented demonstrates significant advances in several critical areas: the optimization of building components for RES integration, particularly in innovative PV technologies; the successful implementation of renewable systems in various building types, including historically sensitive structures; and the emergence of energy communities as a model for collective renewable energy generation and management at the urban scale.

Moving forward, several priorities emerge for future research and development. Economic feasibility remains a crucial consideration, particularly in developing regions and heritage contexts. System integration challenges persist, especially regarding the harmonious incorporation of renewable technologies into existing built environments. Public engagement and acceptance continue to be vital factors in successful implementation. These challenges underscore the necessity of maintaining a multidisciplinary approach that combines technical expertise with social, economic, and cultural considerations.

The findings presented in this Special Issue provide a strong foundation for advancing the integration of renewable energies in the built environment. They also emphasize the critical importance of collaboration across disciplines—from engineering and architecture to urban planning and heritage conservation—in developing sustainable solutions that meet both energy needs and broader societal goals.

**Author Contributions:** Conceptualization, E.L.; methodology, E.L.; formal analysis, E.L., T.C. and W.Z.; data curation, E.L., T.C. and W.Z.; writing—original draft preparation, E.L., T.C. and W.Z.; writing—review and editing, E.L.; supervision, E.L.; project administration, E.L. All authors have read and agreed to the published version of the manuscript.

**Conflicts of Interest:** The author declares no conflicts of interest.

## List of Contributions

1. Yu, Z.; Wang, Y.; Zhao, B.; Li, Z.; Hao, Q. Research on Carbon Emission Structure and Model in Low-Carbon Rural Areas: Bibliometric Analysis. *Sustainability* **2023**, *15*, 12353. https://doi.org/10.3390/su151612353.
2. Milošević, V.; Marchwiński, J.; Lucchi, E. Strain Analysis of Membrane Structures for Photovoltaic Integration in Built Environment. *Sustainability* **2025**, *17*, 1041. https://doi.org/10.3390/su17031041.
3. Zhang, Y.; Chen, T.; Gasparri, E.; Lucchi, E. A Modular Agrivoltaics Building Envelope Integrating Thin-Film Photovoltaics and Hydroponic Urban Farming Systems: A Circular Design Approach with the Multi-Objective Optimization of Energy, Light, Water and Structure. *Sustainability* **2025**, *17*, 666. https://doi.org/10.3390/su17020666.
4. Han, X.; Qu, P.; Wu, J.; Su, B.; Qiu, N.; Zhang, L. Research on the Spatial Pattern of Carbon Emissions and Differentiated Peak Paths at the County Level in Shandong Province, China. *Sustainability* **2023**, *15*, 13520. https://doi.org/10.3390/su151813520.

5. Xie, L.; Fan, L.; Zhang, D.; Liu, J. Passive Energy Conservation Strategies for Mitigating Energy Consumption and Reducing $CO_2$ Emissions in Traditional Dwellings of Peking Area, China. *Sustainability* **2023**, *15*, 16459. https://doi.org/10.3390/su152316459.
6. Ziozas, N.; Kitsopoulou, A.; Bellos, E.; Iliadis, P.; Gonidaki, D.; Angelakoglou, K.; Nikolopoulos, N.; Ricciuti, S.; Viesi, D. Energy Performance Analysis of the Renovation Process in an Italian Cultural Heritage Building. *Sustainability* **2024**, *16*, 2784. https://doi.org/10.3390/su16072784.
7. Romano, G.; Baiani, S.; Mancini, F. Integrating Technological Environmental Design and Energy Interventions in the Residential Building Stock: The Pilot Case of the Small Island Procida, *Sustainability* **2024**, *16*, 8071. https://doi.org/10.3390/su16188071.
8. Abouaiana, A.; Battisti, A. Towards Clean Energy Transition: An Exploratory Case Study from Rural Egypt. *Sustainability* **2025**, *17*, 1597. https://doi.org/10.3390/su17041597.
9. Ahern, C.; Oliver, R.; Norton, B. Harnessing Curtailed Wind-Generated Electricity via Electrical Water Heating Aggregation to Alleviate Energy Poverty: A Use Case in Ireland. *Sustainability* **2024**, *16*, 4470. https://doi.org/10.3390/su16114470.
10. Brunelli, L.; Borri, E.; Pisello, A.L.; Nicolini, A.; Mateu, C.; Cabeza, L.F. Thermal Energy Storage in Energy Communities: A Perspective Overview through a Bibliometric Analysis. *Sustainability* **2024**, *16*, 5895. https://doi.org/10.3390/su16145895.
11. Osiichuk, D. The Obstacles to the Growth of the Renewable Energy Industry in the European Union. *Sustainability* **2023**, *15*, 14612. https://doi.org/10.3390/su151914612.

## References

1. Markard, J.; Raven, R.; Truffer, B. Sustainability transitions: An emerging field of research and its prospects. *Res. Policy* **2012**, *41*, 955–967. [CrossRef]
2. United Nations. *The Sustainable Development Goals Report*; United Nations: New York, NY, USA, 2022.
3. United Nations. The 17 Goals. Available online: https://sdgs.un.org/goals (accessed on 14 June 2023).
4. Chen, T.; Tai, K.F.; Raharjo, G.P.; Heng, C.K.; Leow, S.W. A Novel Design Approach to Prefabricated BIPV Walls for Multi-Storey Buildings. *J. Build. Eng.* **2023**, *63*, 105469. [CrossRef]
5. Zhang, W.; Wong, N.H.; Zhang, Y.; Chen, Y.; Tong, S.; Zheng, Z.; Chen, J. Evaluation of the photovoltaic potential in built environment using spatial data captured by unmanned aerial vehicles. *Energy Sci. Eng.* **2019**, *7*, 2011–2025. [CrossRef]
6. Zhang, W.; Zhang, Y.; Li, Z.; Zheng, Z.; Zhang, R.; Chen, J. A rapid evaluation method of existing building applied photovoltaic (BAPV) potential. *Energy Build.* **2017**, *135*, 39–49. [CrossRef]
7. Lau, S.Y.; Chen, T.; Zhang, J.; Xue, X.; Lau, S.K.; Khoo, Y.S. A New Approach for the Project Process: Prefabricated Building Technology Integrated with Photovoltaics Based on the BIM System. *IOP Conf. Ser. Earth Environ. Sci.* **2019**, *294*, 012050. [CrossRef]
8. Chen, T.; Sun, H.; Tai, K.F.; Heng, C.K. Analysis of the barriers to implementing building integrated photovoltaics in Singapore using an interpretive structural modelling approach. *J. Clean. Prod.* **2022**, *365*, 132652. [CrossRef]
9. Lucchi, E. *Solar Energy Technologies in Cultural Heritage*; Elsevier: Amsterdam, The Netherlands, 2025; p. 458. ISBN 9780443239892/eBook ISBN 9780443239885. [CrossRef]
10. Wang, X.; Wang, G.; Chen, T.; Zeng, Z.; Heng, C.K. Low-Carbon City and its Future Research Trends: A Bibliometric Analysis and Systematic Review. *Sustain. Cities Soc.* **2023**, *90*, 104381. [CrossRef]
11. Lucchi, E. Integration between photovoltaic systems and cultural heritage: A socio-technical comparison of international policies, design criteria, applications, and innovation developments. *Energy Policy* **2022**, *171*, 113303. [CrossRef]
12. Zhang, Y.; Cao, Y.; Chen, T.; Lucchi, E. Optimized Community-Level Distributed Photovoltaic Generation (DPVG): Aesthetic, Technical, Economic, and Environmental Assessment of Building Integrated Photovoltaic (BIPV) Systems. *J. Build. Eng.* **2025**, *103*, 112085. [CrossRef]
13. Xiao, Z.; Tian, Z.; Chen, T.; Ouyang, C.; Zhou, Y.; Heng, C.K.; Lucchi, E. Sustainable Adaptation of Heritage Buildings in Tropical Rainforest Climates: The Innovative Practice of Tanjong Pagar Railway Station in Singapore. *Energy Build.* **2025**, *335*, 115560. [CrossRef]

*sustainability*

MDPI

*Article*

# Towards Clean Energy Transition: An Exploratory Case Study from Rural Egypt

**Ahmed Abouaiana [1,*] and Alessandra Battisti [2]**

[1] Department of Architecture, Faculty of Engineering, Sinai University, Kantara Branch, Ismailia 41636, Egypt
[2] Department of Planning, Design, and Technology of Architecture, Sapienza, University of Rome, Via Flaminia 72, 00196 Rome, Italy; alessandra.battisti@uniroma1.it
* Correspondence: abouaiana.ahmed@gmail.com or ahmed.abouaiana@su.edu.eg

**Abstract:** Rural areas are ideal for renewable energy facilities, supporting sustainable development and energy transition. Egypt aims to reduce greenhouse gas emissions in the electricity sector by 37% and energy consumption by 17% by 2030. Rural Egypt, hosting two-thirds of the population and building stock, consumes one-third of the total electricity. Thus, this paper provides an exploratory study to diagnose and benchmark the energy-use intensity of rural buildings and quantify the correlation between residential electricity consumption, built environment elements, and socio-economic factors, in addition to promoting techno-economic assessments of renewable energy from photovoltaic panels in rural Egypt, supporting national policies amid rapid rural development. The study utilized different analytical and field methods and statistical analyses. A typical agriculture-based rural village in the Delta region, northern Egypt, was selected; the built environment, building types, and socio-economic factors were examined. The results revealed a significant correlation between lifestyle, built-up area, household size, and floor numbers with residential buildings' electricity consumption. The average annual electricity use intensity was benchmarked at 2.5–92.3 kWh/m$^2$ for six non-residential building typologies and at 22 kWh/m$^2$ and 6.67 kWh/dwelling for residential buildings. Under current regulations, rooftop solar panels can generate electricity significantly, but are not profitable. Eventually, insights for policymakers to inform energy transition policies and national initiatives for rural regeneration were provided. The research focused on a local context, but the methodology can be applied to rural settlements in similar contexts.

**Keywords:** benchmark; diagnose of built environment; energy-use intensity; regeneration of agriculture-based rural villages; renewable energy; research in field; socio-economic aspects; statistical analysis

Academic Editor: Rajesh Kumar Jyothi

Received: 27 December 2024
Revised: 22 January 2025
Accepted: 24 January 2025
Published: 14 February 2025

**Citation:** Abouaiana, A.; Battisti, A. Towards Clean Energy Transition: An Exploratory Case Study from Rural Egypt. *Sustainability* **2025**, *17*, 1597. https://doi.org/10.3390/su17041597



## 1. Introduction

### 1.1. Research Background

The effects of climate change in rural areas will be complex and difficult to predict. They will impact infrastructure, cause loss of life, and change agriculture and ecosystems [1]. The global energy scene is experiencing a rapid transformation, driven by the imperative to mitigate climate change and secure renewable energy supplies [2] in response to the raised global mitigation targets in the Paris Climate Accord, the shift to clean energy has gained attention for improving energy efficiency and contributing to climate change mitigation [3]. In 2022, The greenhouse (GHG) emissions were 53.85 billion tons; by sector, energy use in buildings represents 17.5% (residential buildings are 10.9%) of energy-related emissions from electricity generation [4].

Renewable energy (RE) is a vital component of the clean energy transition; the latter refers to the shift from fossil fuels to cleaner energy sources, which increases the use of cleaner alternatives and improves energy efficiency as it has been explicitly recognized as a promising means of advancing rural development as an essential aspect of revitalizing rural areas in the clean-energy transition implementation [5]. RE benefits include improving air quality, reducing household chore time, allowing for more study time, and improving safety and comfort. They also boost productivity, support local businesses [6], and positively influence local communities' financial well-being and employment [7].

In this light, efforts to mitigate climate change have increased globally, focusing on energy in rural areas. Thus, in recent years, many key strategies have been recognized for advancing sustainable development goals through employing renewable energies, such as agrivoltaics and renewable energy cooperatives [8]. Each country set targets to increase renewable energy shares in the energy mix. The global target is to reach 42% renewable energy sources for electricity generation; 68 countries will reach 100%. Europe set an increased goal of increasing the share of RE in final energy consumption from 32% to 42.5% by 2030, aiming to accelerate decarbonization and reinforce energy security [9]. However, rural policies face challenges in coordinating different policies and fitting their applications to target micro-scale rural areas because rural areas still account for almost half the world's population and have embedded historical significance that varies in various contexts, either in intangible aspects, like lifestyle and economic activities, or the built environments.

Therefore, combining efforts from a bottom-up perspective with top-down perspectives is inevitable to boost sustainable rural development practices [10], and bottom-up initiatives are apt to be more manifest in the energy transition [11]. It is crucial to take a bottom-up approach driven by scholars and researchers to empower rural communities to take charge of the energy transition, including promoting local initiatives, supporting community projects, and promoting decentralized energy systems, e.g., rural renewable energy communities.

Similarly, it is essential to benchmark the consumption pattern of different building typologies and explain the influences between the socio-economic aspects and rural built environment (as the building sector is one of the most predominant factors) on energy consumption, to obtain reliable data that can aid decision-making in managing an efficient strategic energy transition [12]. The debate on investigating the correlation has occurred for decades. [13], and it is still gaining fast-growing interest in academic research, which endeavors to reach an overall conclusion in the field [14]. The following section discusses these aspects.

#### 1.1.1. RE: Towards a Clean-Energy Transition

From a top-down perspective, for instance, Europe was at the forefront of investments in renewable energy communities for three decades, in many forms, like cooperatives, until the release of the EU Clean Energy Package in 2016, where the European Commission defined energy communities as collective actions where citizens take ownership of their vitality consumption. Energy communities can take three forms: individual self-consumption, which involves a renewable energy production plant where energy is self-consumed, promoting efficiency and sustainability; collective self-consumption, which involves multiple consumers in a building using a RE source; and energy communities, which include members collectively producing energy to meet consumption needs with renewable sources. RE communities are formed by self-consumption that focuses on maximizing the role of renewables among self-consumers; in other words, they enable individual users to meet self-consumption and to sell energy that exceeds the generated energy, in order to be prosumers. In this light, the European Commission initiated the Rural Energy Community

Advisory Hub, to accelerate the development of sustainable energy community projects in European rural areas [15], to drive them to enhance economic activity and create jobs to reduce energy poverty and accelerate the energy transition in the EU [15]. In this context, She and Pezzulo [16] summarized the academic research in the past 6 years by reviewing 86 articles to explore the transition from localized renewable energy to advanced, integrated systems; they found the RE systems and technologies crucial, and they summarized the implementations that accelerate the energy transition, like smart grids and energy storage systems, biogas, and PV [16].

Using a bottom-up approach, Furmankiewicz et al. [17] investigated how bottom-up rural development initiatives fulfill the goals of climate and energy policies in rural Poland, and found a lack of awareness of struggling policy implementations. They mentioned that attention should be paid to implementing a clean-energy transition to provide a significant prospect of increasing RE capacity, driving innovation, and filling the *"information gap"* of societies. In Germany, Akizu et al. [18] discussed how bottom-up initiatives can complement government-led efforts to achieve sustainability goals, including reducing primary energy consumption, lowering residential electricity and heat usage, and increasing renewable energy generation for self-sufficiency. Similarly, in rural areas in Latin America and Africa, a hybrid energy system to meet electrical energy was proposed [19,20] and rooftop photovoltaics for decarbonization and a clean energy transition in Europe, India, China, and America was provided [21]. In Saudi Arabia, Mustafa et al. [22] simulated the application of a decentralized hybrid energy system consisting of solar photovoltaics (PV) on a target-case building to support national green energy and economy.

In Egypt, Salah et al. [23] reviewed Egypt's energy landscape, including rural areas, examining its potential for sustainable energy, comparing it to global averages, and addressing challenges in developing the renewable energy sector. The hybrid RE and PV in rural areas are countable sources for remote areas because of high solar irradiation in Egypt, as presented in the References [24–26]. The authors developed a pilot zero-energy building using a stand-alone PV system in a social building in a rural village (the same target case as in this study) to pave the way to further implications and support the momentum of the clean-energy transition [27]. The First Author investigated national efforts in renewable energy-aligned sustainable rural development. In addition to investigating bottom-up practices for employing different RE sources at buildings and farms in rural Egypt as accelerators for energy transition and a decarbonized future, the study found solar energy applications predominated as RE sources in residential buildings; however, farms rely mainly on solar, then wind, and then biomass energy [28]. Kassem [29] emphasized the fact that integrating PV and biogas is the most effective RE solution to afford energy prices for rural dwellings in southern Egypt.

#### 1.1.2. Diagnosing the Built Environment and Building Consumption for Energy Transition

Diagnosing the correlation between the built environment and building energy consumption is essential for several domains. Firstly, it can support policies and inform practice. For example, Torabi Moghadam et al. [30] relied on existing census and district heating energy consumption data. They used GIS as a methodology to characterize the impact of building morphology on energy planning to understand better residential building energy usage and the implications of energy performance policies. Juan et al. [31] conducted field investigations and interviews, followed by analyzing the residential conditions and energy use classifications using both bottom-up and top-down approaches, as they found this way is efficient for diagnosing the rural buildings and regional level in China; their findings supported climate-related-building policies that, in turn, accelerate the energy transition [31]. Sena et al. [32] mentioned that efforts on this issue in developing countries (like Egypt) still need to be improved, and that more investigation is needed. So, to fill this gap, they inspected building characteristics and socio-demographic factors (e.g., education level and employment status) by blending field and questionnaire surveys with statistical methods.

Secondly, it supports sustainable rural development practices; researchers such as Wang et al. [33,34] accentuated the fact that the land cover and road network, the spatial pattern of the surrounding environment, and building physics characteristics can significantly influence building energy consumption and GHG emissions, using simulation and analytical methods. Osorio et al. [35] stated that operational energy consumption patterns could control energy consumption during rapid urban growth and tailoring the implementation of mitigation strategies, as they estimated energy use for building operations using energy metrics to explore the link between energy consumption and chosen built-environment characteristics. The correlation can also promote effective energy conservation measures and techniques, as concluded by Woo and Cho, who diagnosed the built environment's impact on the energy use of Korean buildings [36], but it can also decrease the connected GHG emissions and help to rationalize residential energy use and consequently reduce energy expenditures, as presented by Antonopoulos et al., who used a statistical model to explore the correlation between building features and demographics with building energy consumption in the USA [37].

Thirdly, it can help in the early planning stage for energy performance, as in Wang et al.'s approach [38], which assessed and quantified the influences of urban morphology on the energy performance of building stocks, blended field surveys to examine the status quo of the built environment and building characteristics, and then stimulated and modeled the energy demand. Likewise, Zhu et al. [39] recently characterized the impact of the urban morphological parameters on decisions to improve energy efficiency and outdoor thermal comforts via field methods and simulation tools, to quantify the effects of the built-environment morphology of dormitory blocks on energy consumption and renewable energy production [40].

Despite these broad benefits, numerous studies presented associated challenges. To give an example, Tam et al. [41] conducted a chronological review for six decades to diagnose energy-related occupant behavior and impacts on domestic consumption and its implications; their results found a substantial deviation between actual and predicted energy-consumption patterns in the European contexts; the measured energy consumption is larger by one-half than the predicted ones, in compliance with Heinonen and Junnila's study [42], which conducted holistic on-site and analytical methods to examine different housing types and energy use in urban and rural Finnish areas [42]. Another study by Geng et al. [43] presented a simplified method to assess building energy performance (diagnoses of buildings) and collect information on energy bills and simulations in rural China to diagnose energy consumption; they reported that the actual measurements indicated less energy consumption values than the simulations by 12%. On the other hand, household energy consumption varies significantly by lifestyle. Lifestyle is influenced by family composition, age, housing type, appliance ownership and usage, and habits during leisure time. Social activities significantly affect energy consumption, making it hard to identify the most impactful ones, as discussed by Nishida et al., in a model for annual power use from a lifestyle questionnaire for estimating household seasonal energy use [44].

#### 1.1.3. Diagnosing the Correlation for Energy Transition in Egypt

In the Egyptian context, the issue has been discussed by scholars; a secondary study by Alsaadani, entitled "*A statistical review of a decade of residential energy research in Egypt*" [45], provided a systematic literature review to discuss building or urban-scale elements' correlation with domestic electricity consumption in Egypt between 2010 and 2020. Among

124 eligible articles focusing on the local realm, she found the research tendencies concentrated on assessing the architectural and material parameters of the building scale with buildings' energy consumption, and few articles considered urban pattern factors. She recommended focusing on the impact of understudied areas, such as building systems, on whole-building performance to support future research and broaden the scope of Egyptian residential energy research. It is conspicuous that the contextualizing was not explicit (either urban or rural).

Nafeaa et al. [46] evaluated the influence of residential building morphology in urban areas in Egypt, utilizing simulation tools to quantify domestic consumption, either within their experiment or, as they reported, another study in a similar context that indicated a similar rate, as they highlighted the consumption affected by many factors like building systems, appliances, and behavior aspects. While Attia et al. [47] model the correlation between the building typology and domestic electricity consumption in different climatic zones, they do not explicitly state whether the correlation is urban or rural.

In the rural areas in the Delta region (similar to the target case's context in this study), rural commons, Hegazy et al. [48] examined the impact of the urban form in rural areas in this Delta region in Egypt, simulating different building typology scenarios (attached, semi-detached, and detached) with various floors and different orientations; they found that attached typologies consumed the lowest average, while detached type consumed the highest average as the building envelope was exposed to thermal load. Also, the more floors there were, the more consumption; however, they did not determine the current consumption pattern of specific case studies (e.g., electricity bill collection).

Salama and Hassan [49] examined the statistical correlation between socio-economic and demographic aspects and energy-saving behavior in rural settlements in Ismalia Governorate, the Canal Suez region, and found a positive correlation between age, education level, and family member education. They spotted a negative correlation between income level, room number, family size, and electricity-saving behavior; on the contrary, Abdelhay [50] studied the impact of electricity accessibility on socio-economic factors and urban construction in a rural settlement in Fayoum Governorate, South Egypt. Finally, Elsabbagh et al. [51] mentioned that changes in economic activity affect consumption patterns, as discussed in a case study in the Al-Sharkya Governorate in the Delta region.

### 1.2. State of the Art

By taking a look at the primary and secondary studies that globally and in Egypt discussed and developed scenarios to quantify the correlation between the built environment and energy consumption building, which go hand in hand with promoting harvesting renewable energy towards clean-energy transition in Egypt, we found, as far as we know, that no study has been conducted from a bottom-up perspective to collect data on electricity consumption patterns, quantifying the correlation using statistical methods, and focusing on rural contexts to examine the correlation in rural villages in Egypt. This is the first, even though some studies investigate different approaches in individual case studies in rural Egypt, simultaneously to proposing rooftop PV on-grid systems towards clean-energy transition. Beyond these bounds, with respect to the confirmed facts and arguments mentioned in the previous studies, the academic progress of the issue is slow in developing countries, as per the claims of Sena et al. [32] and Quan and Li [14], in addition to Alsaadani's research [45] in the Egyptian context and other insights regarding no clear benchmark for rural buildings, as per remarks from the studies of Attia et al. [47] and Nafeaa et al. [46]; their results significantly varied by about 80% for the same building typology in the same country. We would say that the bottom-up approach, coterminously exploring various perspectives, methodologies, and case studies, is inevitable in reducing the associated uncertainties, as we have presented. As well as the exploratory studies, this approach can also be set as the pillar of a beneficial research foundation that aims to yield a first inroad into an area that currently needs to be investigated and to "*call for a few afterthoughts*" [52], which is what our study provides.

### 1.3. Objective

This study promotes a pilot exploratory experiment to support the clean-energy transition in Egypt by determining the electricity consumption patterns of different building types and describing and observing the statistical correlation between residential electricity consumption and the rural built environment, focusing on urban characteristics, architectural typologies, socio-economic activities, and dwellers' lifestyles. Then, it carries out a techno-economic assessment of rural renewable-energy communities within the current regulatory frameworks, to be integrated with the arguments and discussions presented for the rural Egyptian context in Reference [28].

### 1.4. Hypothesis

The study's hypothesis in this exploratory study proposes that the current approach can be a game-changer tool in the Egyptian context, in line with the current development scene to meet the national goals to improve the electricity efficiency from buildings in rural areas and meet national energy and GHG emissions targets, as presented in (Section 3.1.1). This is in addition to the regional context, with the intention of being a regional hub for energy transition [53], and despite the associated uncertainties related to geopolitical tensions that increase year by year, where energy is at the core of this equation. Moreover, it will pave the way for scholars to develop different approaches, build reliable data drives for retrofitting practices, and guide energy-efficient buildings in early design stages to increase awareness of the issue, as confirmed by Tam et al. [41] and Furmankiewicz et al. [17].

### 1.5. Research Questions

The objective can be achieved by addressing each research question (RQ), as follows:

- RQ1. What is the current energy-use intensity (EUI) of rural buildings in Egypt, and what factors of built-environment elements and socio-economics influence the domestic energy consumption, mostly?
- RQ2. What is the techno-economic assessment of installing PV panels to promote renewable rural energy communities in Egypt within the current regulatory framework?

## 2. Methodology

### Study Structure and Methodology

Qualitative analysis and quantitative field data were combined to address the objective and to tackle research questions, namely:

Part 1—Theoretical and Analytical Methods

The aim is to define the context of the case study, which takes place by highlighting the main characteristics of the case study, energy profile, and buildings in a rural Egypt (Section 3.1.1) macro-context: the Delta region in northern Egypt (Section 3.1.2), diagnosing the target case study, Lasaifar Albalad village (Section 3.1.3), determining the built environment's parameters, and providing criteria as input for the questionnaire employed.

Part 2—Field Method and Statistical Analysis

The main aim is to benchmark the EUI, explore the influences on electricity consumption patterns, and investigate the correlation between these factors and electricity consumption. Hence, different tools have been utilized, namely on-site investigations, interviews, and a questionnaire, to carry out the following:

- Diagnose the current consumption patterns of residential and non-residential buildings (Sections 3.2.1 and 3.2.2).
- Designate the relatedness between dependents and independent variables resulting from the questionnaire, using two statistical models: the Pearson Correlation Coefficient to characterize the correlation between numerical building characteristics and electricity consumption, and the nominal one-way Analysis of Variance (ANOVA) to investigate electricity consumption, socio-economic activities, and urban settings (Section 3.2.3).

Part 3—Techno-economic assessment of installing PV panels

Defining the requirements to apply PV panels according to the regulatory framework and quantifying their economic and technical impacts (Section 3.2.4). Figure 1 visualizes the research design and the methodology.

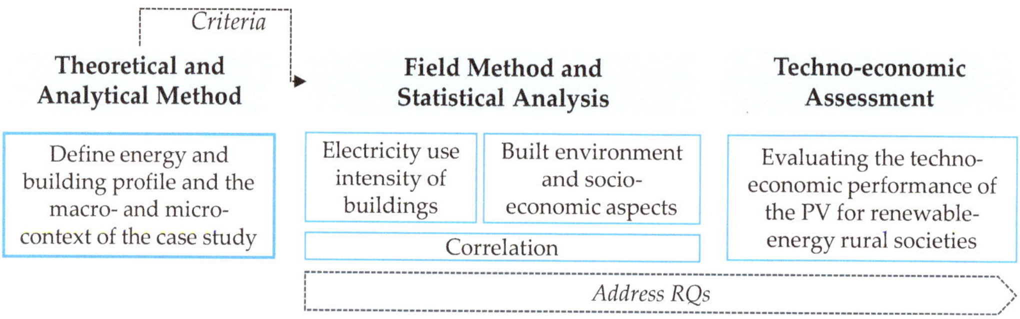


**Figure 1.** A visual summary of research design and methodology.

## 3. Results

### *3.1. Defining the Case Study*

#### 3.1.1. Energy Profile and Building in Rural Egypt, at a Glance

In Egypt, the GHG emissions account for 319.69 metric tons of carbon dioxide equivalent (Mt $CO_2$eq) in 2022, representing 0.7% of global emissions [54]; despite the low emission share compared to the worldwide average, Egypt is vulnerable due to climate change impacts, and mainly, northern rural areas are among the most fragile regions due to sea-level rise [55]. According to the Egypt's second Updated Nationally Determined Contribution report in June 2023 [56] and the Voluntary National Review in 2021 [57], Egypt set vital targets for 2030 compared to the business-as-usual scenarios, to reduce GHG by about one-third by 80–100 Mt $CO_2$eq, where the energy is a key enabler, in line with the sustainable development goal (SDG) 13—Climate Action, and to increase the share of RE in energy production to 30–32.5% (which will account for 42% by 2035), and in line with SDG 7—Clean Energy [58]. Concurrently, the second National Energy Efficiency Action Plan (NEEAP) 2018–2022 aims to reduce Egypt's energy consumption by 18%, by 2030 [59].

Along the same lines, energy policies in Egypt are directed to addressing energy transition as well as climate action, especially in the aftermath of hosting the 2022 United Nations Climate Change Conference—Conference of the Parties—COP 27, and with these actions are reflected to the sustainable development of rural areas, represented in two mainstreams: firstly, by doubling the agriculture lands and inhabited areas, and secondly, by revitalizing and regenerating the existing rural built environments [28], such as the national initiative Decent Life (the Arabic expression is Haya Karima), which focuses on several aspects such as accelerating poverty eradication, developing the quality of housing, and providing adequate building, and infrastructure, interacting with seven of the sustainable development goals (SDGs) [58].

The building sector is responsible for 5%, and the electricity and heat sector is responsible for 32% of the total GHG emissions [54]. The total final energy consumption is 2,582,513 terajoules in 2022 from different energy sources; behind the transportation and industry sectors, the residential consumption came in third place, at 20%, that is, electricity representing one-half of energy sources among oil products, natural gas, and biofuel and waste, as visualized in Figure 2a. The distributed electricity in 2020 accounted for 122,443 GWh (30% for rural areas), divided into eight sectors; the uppermost consumer sector is the household with 61,542 GWh (36% in rural areas), and the total share of RE in electricity generation is 12%, as depicted in Figure 2b.

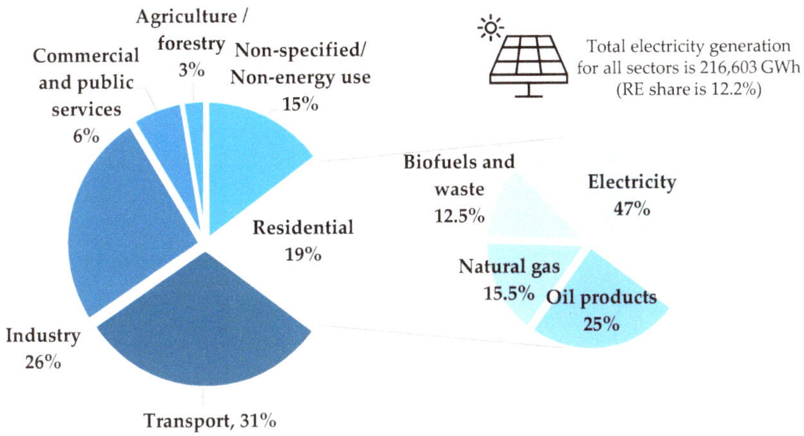


(a)

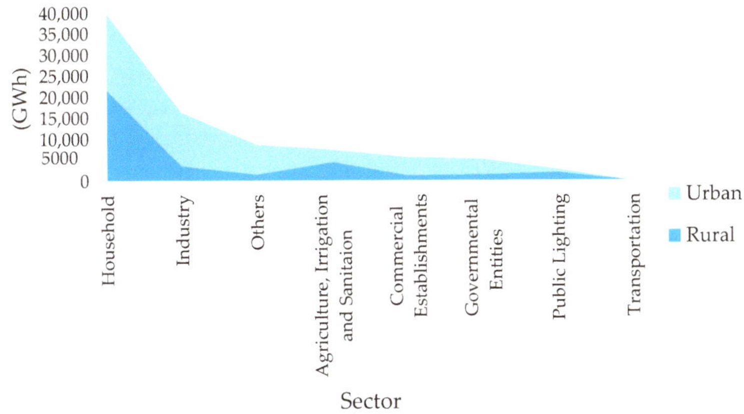


(b)

**Figure 2.** Energy and electricity profile of Egypt: (**a**) total final energy consumption by energy source, total electricity generation, and renewable share of electricity generation, in the reference year 2022, developed by the Authors, based on [60]; (**b**) distributed energy according to uses by rural–urban, in the reference year 2020, developed by the Authors, based on Reference [61].

According to the *Population, Housing, and Establishments Census*, Egypt has 16,185,063 buildings; the residential buildings are 13,467,333, representing 83% of the total; rural areas host 70% of the residential buildings [62]. These numbers are increasing yearly, due to rapid development rates and the establishment of new urban communities and, consequently, national housing projects; for example, 1.7 million public housing units and 0.45 million social housing units were constructed between 2017 and 2023 [63,64]. In the same vein, the rural settlement patterns in Egypt have been altered due to various aspects, resulting in anthropological activities and contemporary built-environment typologies leading to the consumption of further energy and the generation of more GHG emissions. This increases the uncertainties associated with the future of consumption patterns, which this study can explore, contributing towards achieving the national energy and climate targets, supporting electricity sector and the national rural-regeneration processes.

#### 3.1.2. Macro-Context: Delta Region, Egypt

Egypt's regions are grouped into seven physical planning regions, and every region comprises three hierarchies: governorates, districts, and villages/satellite villages. The Delta, one of these regions, comprises five governorates, 106 districts and cities, and 1404 villages, and is one of the most important agricultural and fisheries source in Egypt; the agricultural land covers 77% of the total area, and is characterized by several aspects, predominately rural areas. It hosts around one-fourth of the population in Egypt (more than 22 million) within a total area of 12,357 km$^2$, representing 14% of the total inhabited area and 1.2% of the total; the population density exceeds 1773 km$^2$/inhabitants. The Delta's rural area also has about 2.5 million buildings, representing 78% of the Delta region's buildings, 15% of the total national building stock, and 11% of the total land-use area [62].

The non-residential buildings in the region comprise public universities and many private higher-education buildings, in addition to numerous schools. There are about 1004, including all types, such as the Islamic religious sciences "Azhar", public, commercial, more than 300 healthcare buildings, and more than 500 cultural buildings, besides the social and recreational buildings such as youth centers, stadiums, and services, like post offices [65,66]. Residential buildings predominate in the region, and represent 95% of the total. By floor number, the average predominant typologies are the one-story and two-story buildings, representing the highest number, with 70% (915,788 and 918,040), then the three-story, with 21%, then the four- and five-floor buildings, with 7% and 2%, respectively [62].

The rural areas in the Delta region consume one-third of the distributed electricity (the primary energy source for domestic use); residential lighting uses the highest amount of energy, with 55% [67], similar to the national average. Regarding household electricity usage, 89% of Egyptian families in rural areas own refrigerators, 33% use electric water heaters (along with 15% who have gas water heaters), 19% possess automatic washing machines, 5% have microwaves and ovens, and 4% depend on air conditioners. The urban fabric of all Egyptian villages consists of three major patterns: the linear pattern, which aligns with agricultural land boundaries; the scattered urban pattern of the expansion areas; and the compacted pattern in the center of the villages, which is a typical pattern in the Delta region [68].

#### 3.1.3. Macro-Context: Lasiafar Albalad Village

Lasaifr Albalad is located between 31.18° N and 30.72° E, with an elevation map of 6 m. It is a typical rural agriculture-based village, representing the region's villages, following the Konaieset Alsaradoosy local unit in Desouq District, Kafr El-Shiekh governorate. According to our survey, the population was nearly 14,000 in 2021, and in 2006, the population was 8589 inhabitants [69], which increased by about 40% in 15 years. The inhabited area inside the urban boundary is about 0.35 $km^2$. The settlement encompasses the urban fabric typical of villages in the Delta region (scattered, compacted, and some linear). The local road network includes four width gradients, mostly from 8, 6, 4, and 2 m; we notice that that the width of some roads varies, as per the unclear borders. It has few public and many private in-between spaces (e.g., backyards and gardens).

The village features around 700 buildings, with about 4% non-residential and 96% residential. The non-residential structures include six religious buildings (mosques), four preparatory and primary (general and Azhar) schools, and governmental buildings such as the mayor's house and the agriculture association. Additionally, there are public service facilities like the post office and health unit, along with a youth sports center. In addition, some storage and commercial buildings are scattered along the water body.

These numbers have been verified in two ways; firstly, the First Author had an official layout from the agricultural association for the reference year 2006. He also used Open Street Maps to count building numbers; an example is available in the Reference [70], and he conducted a spatial analysis of the village to strengthen the collection of data. Finally, he provided the first open-source master layout and land use of the village in 2022 in AutoCAD format, version 2010, available as a reference [71].

The inhabitants' lifestyles can be grouped into four groups. First, the youth meet in the youth center, another private playground, coffee shops, or a walk. Second, older people and adults have similar lifestyles. The differences are fewer sports activities and more family visits. Third, women have a more enclosed life, characterized by home visits and shopping in the temporary public market on a weekly basis. Fourthly, playing outside is one of the preferable activities for children. The settlement is a main point for primary education schools in the surrounding area. Contrariwise, the older students move to the Desouk district for secondary schools and other centers for higher education, like Kafr Elshiekh University. The primary means of transportation are small buses that lead to nearby districts. The small three-wheel car is the leading means of transportation, especially inside the settlement. Figure 3 depicts examples of the built environment of the Lasaifar Albalad rural settlement.

### 3.2. *Field Study Results*

The field study was conducted between April and May 2021 in two stages. The first stage was for interviews and personal meetings (Section 3.2.1); the selection criteria aimed to include volunteer influencers like the manager of the agricultural association, to represent the local authority, the locals, an older citizen to act as a cultural expert who is fully aware of the village, and other locals who facilitated the conducting of the experiment, collecting monthly electricity bills and exploring their standpoints on the influences on electricity consumption. In the second stage, an online questionnaire was developed (Section 3.2.2); the aim was to diagnose the residential buildings (the predominate ones), namely, building characteristics, demographic data, socio-economic activities, and average electrical energy consumption in summer and winter residential-building characteristics, noting that the questionnaire was primarily developed based on the non-published Ph.D. thesis of the First Author, supervised by the Second Author [72].

#### 3.2.1. Phase 1: Interviews

The interviewed locals were asked open-ended questions to inspect the correlation between small-dimension agricultural economies and domestic electricity consumption, as follows:

What are the predominant agricultural activities? In this light, do you think there is a correlation between agricultural activities and monthly electricity bills? Finally, do you want to say anything related to this issue?

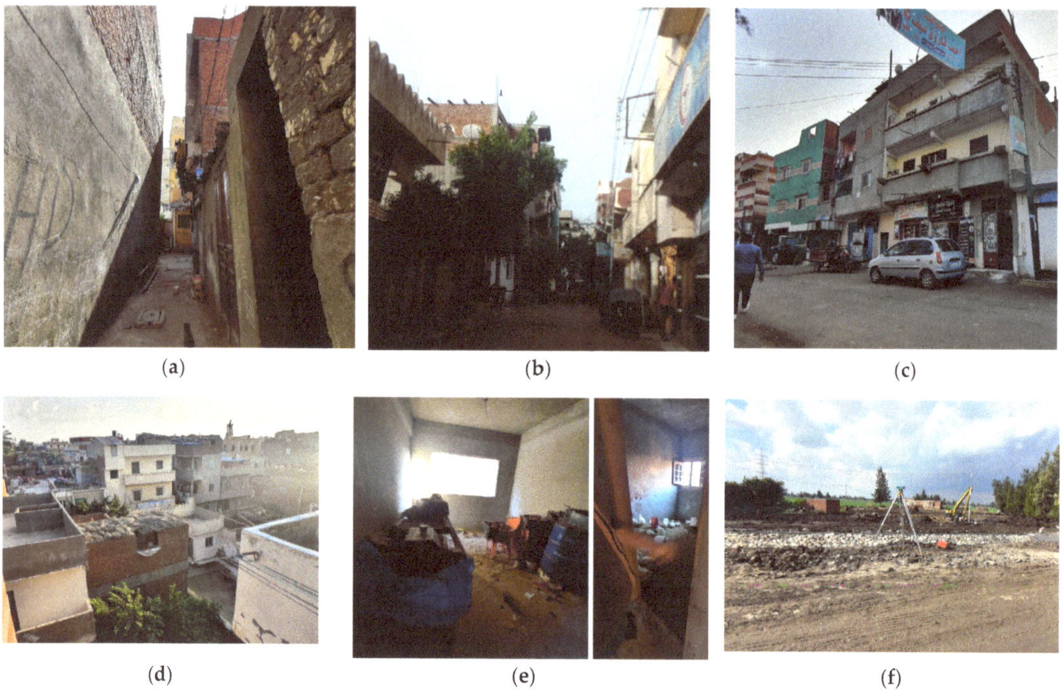

**Figure 3.** Examples of the built environment of the Lasaifar Albalab rural settlement (images' source and credit: Ahmed Abouaiana): (**a**) an inner road 2 m wide, made in the traditional organic fabric; (**b**) a road of 4 m width inside the traditional fabric, showing the mixed-used buildings (small economic activity on the ground floor); (**c**) a 6 m width road along with the water body, characterized by linear fabric, usually hosts the social activities like coffee shops; (**d**) in-between spaces of the residential buildings within the traditional fabric, with the upper roofs being vacant spaces; (**e**) micro-scale poultry farm inside a vacant dwelling; (**f**) the process of regenerating the infrastructure of the village within the Decent Life project.

A farmer described the predominant local farming cycles: "After harvesting the rice and wheat crops, they prepared them in the relative industry, namely, wheat grinding and rice bleaching in a factory in the village or nearby village/district for domestic use and to sell them". Also, the sugar beet crop is connected to the district's governmental sugar-production factories; nearly all the workers in this area are from the villages. Both indicated no impact of these activities on the monthly electricity consumption.

Another dweller reported that he relies on a micro-scale poultry farm inside the building in a vacant dwelling in an extended family house, Figure 3e. Also, he mentioned that some families are carrying out similar activity in light structures on the roof of the building, external yards, or vacant rooms, like those shown in Figure 3d. Poultry farming requires substantial lighting for eight months each year to enhance productivity (the annual cycle), doubling monthly electricity consumption. The larger poultry farming generally demands air conditioning (no farms exist inside the village's inhabited center). Figure 4 visualizes some of the non-residential buildings in the village.

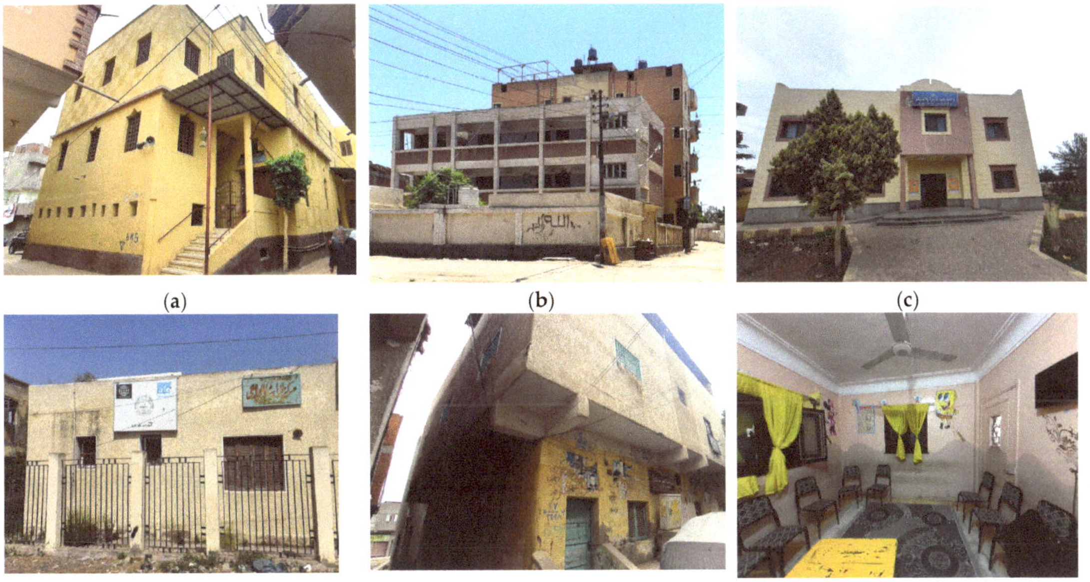

Figure 4. Non-residential building examples in the Lasaifar Albalad rural village (images' source and credit: Ahmed Abouaiana): (a) the grand mosque; (b) a school building; (c) the health unit; (d) the agriculture association; (e) the social development building; (f) a residential-building structure performs the social activities of an NGO.

Furthermore, the locals reported their monthly electricity bill expenditures, which are equivalent to an average of 200 kWh monthly, determined based on some collected electricity bills and the representatives' reports. Table 1 summarizes the investigated buildings' average consumption patterns and energy-use intensity.

Table 1. The characteristics and EUI of the non-residential buildings, developed by the Authors, after [27,72].

| Building Typology | Building Characteristics | | | Electricity Consumption | | | Description |
|---|---|---|---|---|---|---|---|
| | Footprint | Floor | Built-Up Area | Monthly | Annual | EUI | |
| | $m^2$ | Number | $m^2$ | kWh/ Month | kWh/Year | $kWh/m^2$/ Year | |
| The grand mosque | 300 | 1 | 300 | 220 | 2400 | 8 | The grand mosque is located in the village's core; its maximum capacity is nearly 250 worshippers, as shown in Figure 4a. |
| Preparatory school | 220 | 3 | 660 | 500 | 6000 | 9 | There are eight classrooms, each equipped with a computer and a smart display board, along with two computer labs, a science lab, a library, and executive rooms, as seen in Figure 4b. |
| Primary school | 210 | 3 | 630 | 250 | 3000 | 4.8 | It is positioned near the center of the village and comprises ten classes. |
| Health-unit building | 65 | 2 | 130 | 1000 | 12,000 | 92.3 | The center provides all basic healthcare services, like vaccinations, as depicted in Figure 4c. Consumption is higher than the norm because of the equipment used, like the medical fridges. |

Table 1. *Cont.*

| Building Typology | Building Characteristics | | | Electricity Consumption | | | Description |
|---|---|---|---|---|---|---|---|
| | Footprint | Floor | Built-Up Area | Monthly | Annual | EUI | |
| | $m^2$ | Number | $m^2$ | kWh/Month | kWh/Year | $kWh/m^2$/Year | |
| The agriculture association | 75 | 1 | 75 | 110 | 1320 | 17.6 | This government building assists farmers by offering finance and marketing processes while lowering production costs, see Figure 4d. |
| Social | 120 | 2 | 250 | 120 | 1440 | 6 | The social development facility houses an official unit for social affairs and a kindergarten. It operates officially for 35 h each week and is essential to serving the local community, see Figure 4e. |
| Social (NGO) [27] | 150 | 1 | 150 | 100 | 1200 | 8 | The building belongs to Albakyat Alsalihat, a non-governmental organization (NGO), see Figure 4f. It accommodates a residential building providing social activities for children, orphans, and widows. |
| Independent shops | 16–25 | 1 | 16–25 | 80–120 | 1200 | 2.5 | It encompasses various types of businesses, including restaurants, vehicle services, mechanics' workshops, and bakeries. These establishments are primarily located along the main road to serve commercial needs, operating for about 70 h weekly. |

#### 3.2.2. Phase 2: Questionnaire and Data Collection

Questionnaire Design

A close-ended structured questionnaire has been designed to investigate three mainstream factors: building characteristics, urban settings (built environment), and socio-economic factors, based on the criteria presented in Sections 3.1.2 and 3.1.3, to tackle the purpose of this study (Section 1.3). In addition, an optional open-ended question was designed to enable locals to express opinions about environmental issues like improving energy and buildings, or any other issue. The rationale behind this is to engage locals by collecting their preferences and insights, if any. The three groups and their sub-factors are summarized as the following:

- Building characteristics: floor area, floor number, building form, orientation, building type (residential or mixed-use and, if mixed-use, what kind), and monthly average in summer and winter for the dwelling and for the entire building.
- Socio-economic activities: household size in dwellings, household size in the buildings, if any, household breadwinner's work type, working as a farmer or not, mobility to work, breadwinner's work location, lifestyle, and finally, wife's work status.
- Urban characteristics: location by urban fabric and location by road network

Additionally, the scientific terms in the questionnaire have been simplified and written in Egyptian dialectic, and each question has a rationale behind it. These questions are summarized as the following:

Diagnosing building location by the urban fabric

- Is the building surrounded by numerous buildings, like those near the grand mosque? This indicates that the compacted (traditional) fabrics, like the mosque in the center of the village, are characterized by high building density, e.g., Figures 3d and 4a.
- Is the building by the main road? This indicates that the linear urban fabric and the main road in the village, which is usually typical for most Egyptian villages, are aligned with the water body, as seen in Figure 4c.

- Is it near the village's boundaries, like those near the healthcare buildings or cemeteries? This signifies the scattered urban fabric, e.g., Figure 4b.

  Diagnosing building location by road network

- What is the road width in front of your building: 2, 4, or 6 meters? The replies to this question indicated that the buildings located on 2 m roads are entirely located on the compacted urban fabric, e.g., Figure 3a; likewise, the 4 m roads complied with the position of the scattered, e.g., Figure 3b, and the 6 m road with the linear, e.g., Figure 3c,f.

  Diagnosing building type (all residential or mixed-use)

- Is there any activity on the ground floor, such as a shop, pharmacy, studio, or dairy factory? If yes (like the one shown in Figure 3c), indicate the activity type.

  Building Form

- Is the building regular (square form), elongated and regular (rectangular form), or is the house unusually elongated and irregular (with projections and recesses in the facade)? This answer relies on the landlord's perception and description.

  Thermal Comfort

- The respondents were asked whether they felt thermal comfort. Hence, the question was whether your building is "Bahary", meaning orientation towards the north seaside, and meaning good conditions, or "Kibli", meaning the building is oriented towards the south, expressing uncomfortable conditions.

It is worth mentioning that, regarding the query about the mean electricity consumption in winter and summer, the respondents were asked to report the consumption of the other units in the building; this is because most buildings typically house extended families or relatives, which are common characteristics of Egyptian rural settlements. Google Forms created the questionnaire; it clearly indicates that the proposal aims to improve energy efficiency in the village, and will be used in academic research, and no personal data, such as telephone number and name, are requested. Also, it is clear that the participants are village residents and that each family should answer only once. The questionnaires were distributed to three Facebook groups in the village in May 2021, one of which is as in Reference [73]. The original version of the questionnare is available online, at Reference [74].

Questionnaire Sample and Results

A total of 37 responses were collected. After excluding the invalid inputs, two respondents were not residents of the village, and three did not report energy consumption; 32 eligible replies remained. Analyzing the data collected (Supplementary Materials S1 and S2), the sample represents 83 families, with a total of 458 individuals. Figure 5a illustrates the attributes of buildings and urban settings. Figure 5b illustrates the socio-economic activity attributes of the households.

To detect the positioning of buildings, 12 buildings (38% of the total) are situated within the organic fabric, evenly spread across various road widths. Then, the linear fabric comprises 13 buildings (41% of the total), most facing the 6 m road. Finally, the scattered fabric bordered by agricultural areas contains seven buildings positioned along the 4 m and 6 m roads.

The respondents were asked to describe their feelings toward thermal comfort, and 59% of the respondents stated good thermal comfort, "Bahary"; the residents of the buildings located on linear and scattered fabric expressed good thermal comfort; on the contrary, the buildings' residents situated on the traditional fabric expressed non-comfortability. That indicated such a correlation between the urban fabric and thermal comfort.

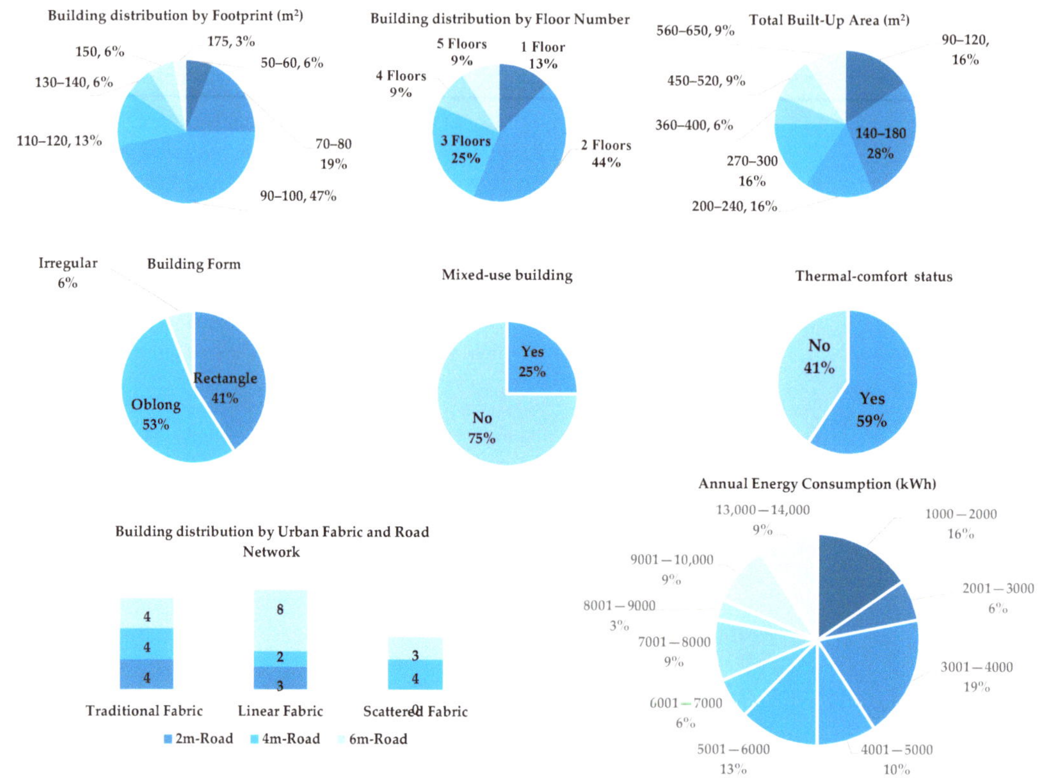


(**a**)

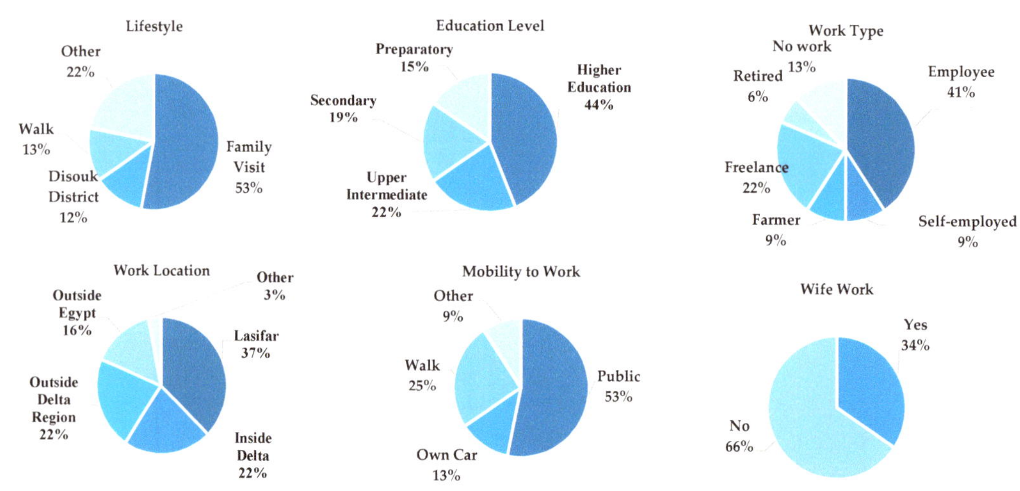


(**b**)

**Figure 5.** Sample characteristics of (**a**) buildings and urban settings; (**b**) socio-economic activities, developed by the Authors, after [72].

Concerning monthly energy usage, the household's consumption varies from 50 to 495 kWh in summer and from 50 to 560 kWh in winter, with an average of about

200 kWh, supporting the consumptions reported by the local people within the interviews (Section 3.2.1). The study classified the buildings by electricity consumption; the average annual energy consumption varies from 1140 to 14,964 kWh, with a mean of 5678 kWh. The consumption can be grouped as the following: the 3001 to 4000 kWh range accounts for 19%, followed closely by the 1000 to 2000 kWh range, at 16%. The less common range of 8001 to 9000 kWh is attributed to a single building, while the remaining categories are distributed fairly evenly. The findings indicate that the minimum annual energy intensity for rural dwellings is 7.9 kWh/m$^2$/year, and the maximum is 43.3 kWh/m$^2$/year, with an average of 22 kWh/m$^2$/year.

By work type, 41% are employees, 22% of respondents are freelancers (working depending on the task), like construction laborers, 13% are not working (unemployed and retired), and 9% are self-employed or own a business. Regarding work location, 37% work in the village, and 22% work either inside or outside the region. Finally, 16% of the population work outside Egypt (in the Arabic Gulf states). Regarding mobility to work, the majority, one-half, rely on public transportation (the so-called microbus), 23% walk to work, and 13% use their cars. The data indicates that workers in Lasifar go to work on foot and by public transportation.

Two-thirds of wives prefer to stay at home. The prevailing pattern of the respondents' recreational activities were family visits, 53%, which came behind other activities, including spending time in coffee shops and the youth center. Small families implement these activities. Walking and going to Disouk City, 12% for each, were in third place.

Notwithstanding, the responses to the optional question were irrelevant to electricity consumption and were not significant, as only three respondents (9%) answered and provided three recommendations: firstly, the essential need for eco-friendly waste management to preserve the environment; secondly, because of the insufficient garbage removal process, it was recommended that there should be a collaborative approach by the local people to make sure the garbage was collected sufficiently; thirdly, it is necessary to cover the sewage drains located in the center of the village, which have a negative environmental impact.

#### 3.2.3. Statistical Analysis

Statistical analysis is crucial for organizing and processing data scientifically and mathematically [75], and explains the relations between variables with four main measurement scales: ordinal, nominal (for qualitative data), ratio, and interval (for quantitative data) [76]. These can be summarized as the following:

Qualitative data type

- Ordinal scale measurements are data that have some orders and ranks.
- Nominal scale measurements refer to names of tags that do not have numerical values for categorizing such an object. In our study, they are represented by socio-economic variables (e.g., lifestyle, economic activity, and mobility to work) and built-environment variables (e.g., building position by road width or urban tissue and building form).

Quantitative data types

- Ratio (continuous) scale measurements with interval data qualities, like the natural zero point; in this study, they are represented by electricity consumption and building characteristics (e.g., footprint, floor number, household size, and built-up type).
- Interval scale measurements: the measured data on a scale showing the distance from one to another.

Therefore, according to the data types and variables, two statistical models were exploited to explain the relations between the dependent and independent investigated variables. Firstly, the Pearson Correlation Coefficient, revealed by Equation (1), is a commonly used technique for assessing the linear strength and direction of correlation between two continuous variables: $x$ represents the independent variable value, and $y$ represents the dependent variable value (electricity consumption). The $r$ value is the correlation coefficient that lies between ($-1$ to $+1$), where ($-1$) designates a perfect weak correlation, while (+1) means a strong linear correlation, and (0) indicates no existing relation [77,78]. Many open-source resource calculators can facilitate the conducting of Pearson correlation formulas, such as Reference [79].

$$r = \frac{n\sum xy - (\sum x)(\sum y)}{\sqrt{n\sum x^2 - (\sum x)^2}\sqrt{n\sum y^2 - (\sum y)^2}} \tag{1}$$

Secondly, the one-way ANOVA was exploited to observe the statistical correlation between the data and to enable the interpretation of one ratio variable (electricity consumption) with one nominal variable (e.g., urban fabric). It is among the most commonly utilized statistical methods for hypothesis examination [80], as it allows the descriptor analysis that poses significant differences between groups of one nominal independent variable ($x$), the so-called explanatory variable, in correlation with the ratio response (dependent) variable ($y$) [76,81]. In other words, the one-way ANOVA assesses the average variation within a group against the variation derived from group means. It also requires a minimum sample size of 30 [82] (the number of study responses is 32). The answers to each question are ranked by number, to set numerical values of the variable; for example, in the question on Lifestyle, the answers are ordered as (1) Family Visits, (2) Going to Desouk, (3) Walking, and (4) Other, and likewise, for each group of variables, where each group under investigated variables, is coded and ordered separately, as shown further on between Tables 3–8. The steps taken in the one-way ANOVA analysis have been described, simplified, and summarized in Appendix A.

The analysis took place using IBM SPSS Statistics software to predict, plan, and analyze data to confirm assumptions and accurate conclusions to resolve problems in business and research [83]. The IBM SPSS Statistics software (version 30.0.0) was developed by the International Business Machines Corporation (IBM); this tool facilitates steering simple and complex statistical and mathematical models, which proved accurate in steering multiple simple and complex statistical and mathematical analyses valid in academic research [84,85], and is commonly used to diagnose the correlation between EUI and buildings, similar to those in our study, such as Reference [86]. We confirm its effectiveness during the analysis.

The data utilized in performing the statistical analysis in this study are extracted from SPSS and made available in the Supplementary Materials (S1). The same file is converted to an Excel format and available at Supplementary Materials (S2).

Numerical Data: Pearson Correlation Coefficient

The analysis of the Pearson Correlation Coefficient between annual electricity usage and building characteristics revealed a very strong positive relationship between the number of floors and built-up area, with correlation values of +0.887 and +0.836, respectively. Additionally, the number of households and total buildings demonstrated strong correlations, of +0.604 and +0.6843. Conversely, the footprint of buildings exhibited a weak statistical correlation with electricity consumption, showing only a weak positive correlation of +0.34, with a probability value ($p$-value) equal to 0.055 (more than 0.05), which shows no statistical significance (more elaboration on the $p$-value is presented in Appendix A). Table 2 shows the correlation coefficient between residential building characteristics and electricity consumption.

**Table 2.** Correlation coefficient between numerical data and annual building electricity consumption, developed by the Authors, after [72].

| Aspect | r Value | *p* -Value | Statistical Significance |
|---|---|---|---|
| Footprint | +0.342 | 0.055 | No |
| Built-Up Area | +0.836 | 0.000 | Yes |
| Total Members | +0.604 | 0.000 | Yes |
| Household Number | +0.6843 | 0.000016 | Yes |
| Floor Number | +0.887 | 0.000 | Yes |

Socio-Economic Activities

For investigating the correlation between lifestyle and electricity consumption, the results seen in Table 3, which indicated that the *f*-Value was 4.432, at a probability value (*p*-value) with a significance level equivalent to 0.011 (less than 0.05), indicating statistical significance variations among variables. Thus, post hoc testing to enable multiple comparisons is conducted, where the table shows the increase in the mean consumption for the group who spent their recreational activities going to the Disouk district, with an annual mean of 8397 kWh, and in family visits, 6654.35 kWh out of the total sample's mean (5678.25 kWh). Therefore, post hoc testing is requested to decide which particular groups differed.

Table 4 shows the variance between lifestyle group variables, where the building residents perform family visits as a predominant social activity (Group 1), compared to Group 4, where the residents tend to carry out other activities, like spending time in the youth center and coffee shops, has a significant mean difference, of 4614.355 kWh, at a *p*-value of 0.018. Likewise, the difference between Group 2, where the residents prefer to spend recreational time in Disuk City, compared to Group 4, was 6357 kWh, at a *p*-value of 0.020. In comparison, there was no statistical significance between the other groups.

**Table 3.** Variation analysis between lifestyle and annual electricity consumption, developed by the Authors, after [72].

| Group Code | Group Description | Mean (kWh) | Standard Deviation (SD) | *f*-Value | *p*-Value | Statistically Significant |
|---|---|---|---|---|---|---|
| 1 | Family Visits | 6654.35 | 3577.10 | | | |
| 2 | Going to the Disouk | 8397 | 4390.266 | 4.432 | 0.011 | Yes |
| 3 | Walking | 5178 | 2834.736 | | | |
| 4 | Other | 2040 | 1074.430 | | | |

**Table 4.** Multiple comparisons between lifestyle variables and annual electricity consumption, developed by the Authors, after [72].

| Lifestyle Groups with Electricity Consumption | Mean Differences Between Groups (kWh) | *p*-Value | Statistically Significant |
|---|---|---|---|
| (2) Going to the Disouk, (4) Other | 6357 | 0.020 | Yes |
| (1) Family Visits, (4) Other | 4614.355 | 0.018 | |

Table 5 shows the variation analysis between building form and the annual electricity consumption; the *p*-value is 0.783 for the educational level and electricity consumption variables, which means there is no observed effect between them.

**Table 5.** Variation analysis relating building form to yearly electricity consumption, developed by the Authors, after [72].

| Group Code | Group Description | Mean (kWh) | SD | *f*-Value | *p*-Value | Statistically Significant |
|---|---|---|---|---|---|---|
| 1 | University Education | 5030.57 | 2448.254 | | | |
| 2 | Above Intermediate | 6516 | 3734.526 | 0.358 | 0.783 | No |
| 3 | Secondary | 6540 | 5021.819 | | | |
| 4 | Preparatory | 5284 | 5685.529 | | | |

Table 6 shows the variation analysis between work factor variables (work status, work location, mobility mean to work, wives' work status) and electricity consumption.

- The buildings where the people work as farmers have the highest annual electricity consumption mean, of 10,084 kWh, while the lowest mean in the buildings where their residents are not working is 3030 kWh. However, the statistical analysis results indicated that the *p*-value equals 0.235 (more than 0.05), which means there is no statistical significance between work type and electricity-consumption variables. In other words, no observed effect and no multiple comparisons are required.
- By work location, the buildings of the people working inside the village have the highest mean, of 7230 kWh, and the lowest occurred in the buildings where their residents work abroad, in Egypt, of 3096 kWh. However, the statistical analysis revealed that the *p*-value equals 0.087; therefore, there is no observed effect between the work location and electricity consumption.
- With respect to mobility means to work, the lowest electricity-consumption average happened in the buildings where residents rely on public transportation, at 5214.35 kWh, and the highest was for the buildings where their residents tend to walk to work, while the other means have a similar average. However, the statistical analysis revealed that the *p*-value equals 0.911; therefore, there is no effect between mobility to work and electricity consumption.
- Based on the wives' work status, the buildings where the wives work consume an average annual of 6133 kWh more energy than those where the wives do not work, which is 5440 kWh. However, the statistical analysis revealed that the *p*-value equals 0.626; therefore, there is no observed effect between the wife's work status and energy consumption.

**Table 6.** Variation analysis between work characteristics and annual electricity consumption, developed by the Authors, after [72].

| Group Code | Group Description | Mean (kWh) | SD | *f*-Value | *p*-Value | Statistically Significant |
|---|---|---|---|---|---|---|
| | | Work Type | | | | |
| 1 | Employee | 5592.92 | 2533.883 | | | |
| 2 | Self-employed | 6800 | 1705.872 | | | |
| 3 | Farmer | 10,084 | 7706.227 | 1.465 | 0.235 | No |
| 4 | Freelance | 5412 | 4318.061 | | | |
| 5 | Retired | 4170 | 466.690 | | | |
| 6 | No Work | 3030 | 2833.161 | | | |
| | | Work Location | | | | |
| 1 | Lasaifar Albalad | 7230 | 4827.620 | | | |
| 2 | Kafr Elshiekh Gov. | 7145.14 | 2480.721 | | | |
| 3 | Outside Delta | 3658.29 | 1808.054 | 2.277 | 0.087 | No |
| 4 | Outside Egypt | 3096 | 1648.417 | | | |
| 5 | No Work | 3840 | 0 | | | |
| | | Mobility to Work | | | | |
| 1 | Public Transportation | 5214.35 | 3101.115 | | | |
| 2 | Private Car | 6132 | 3639.398 | 0.177 | 0.911 | No |
| 3 | Walk | 6316.50 | 5490.798 | | | |
| 4 | Other | 6000 | 3186.785 | | | |
| | | Wife's Work Status | | | | |
| 1 | Working Wife | 6133.09 | 3074.036 | | | |
| 2 | Housewife | 5440 | 4092.959 | 0.242 | 0.626 | No |

Urban Fabric and Building Characteristics

Table 7 shows the variation analysis between the variables of urban fabric (urban fabric and road network) and building characteristics (building form and floor numbers) and electricity consumption

- With respect to building form, the results revealed that irregular buildings consumed the highest annual electricity consumption, a mean of 11,352 kWh, nearly double that of each different form. However, the statistical analysis results indicated that the *p*-value equals 0.078; hence, there is no observed effect between building form and electricity consumption.
- With respect to floor number, the results confirmed a positive link between the floor number and energy consumption; the five floors had the highest consumption of 14,072, while the one-floor buildings consumed an annual average of 1560 kWh. Although it makes sense that the higher the floor number, the more consumption, likewise, for the footprint variable, the results revealed that the *p*-value is less than 0.001 (a very high significance level). This means there is a statistical significance for between-group variables, so post hoc testing to enable multiple comparisons is conducted, to determine the mean differences. Table 8 shows the variance between the floor-number group variables, confirming this strong positive correlation.
- With respect to urban fabric, the buildings on the scattered fabric consume the highest annual average, of 6378.86 kWh, while the buildings at the linear fabric consume the lowest yearly average, of 4572.92 kWh. However, the statistical analysis results indicated that the *p*-value equals 0.626; hence, there is no observed effect between building location by urban fabric and electricity consumption.
- With respect to building position on the road network, buildings on roads with a width of 6 m consume the highest annual average, of 6870.40 kWh, followed by 4 m and 2 m width roads; this indicates a positive correlation between the width of the road and electricity consumption. However, the results showed that the *p*-value equals 2.218, meaning that no statistical effect was observed between location by urban fabric and electricity consumption.

Finally, Figure 6 depicts a visual summary of the mean electricity consumption of the factors examined in the questionnaire.

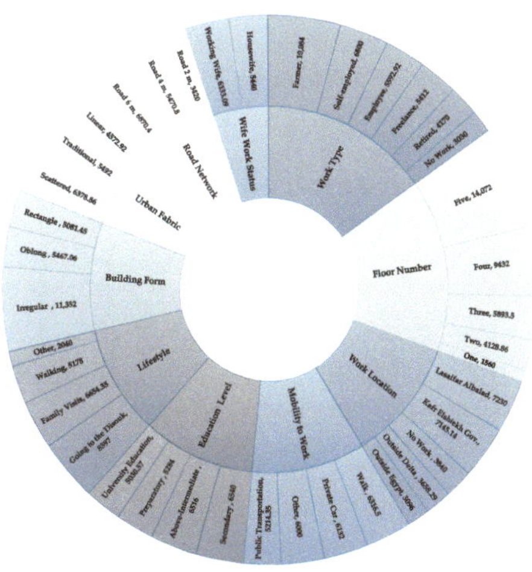


**Figure 6.** A visual summary of the mean electricity consumption of the buildings' sample with the examined factors in kWh, developed by the Authors.

**Table 7.** Variation analysis between building and urban characteristic variables and annual electricity consumption, developed by the Authors, after [72].

| Group Code | Group Description | Mean | (SD) | *f*-Value | *p*-Value | Statistically Significant |
|---|---|---|---|---|---|---|
| | | | Building Form | | | |
| 1 | Rectangular Form | 5081.45 | 3253.906 | | | |
| 2 | Oblong Form | 5467.06 | 3788.301 | 2.785 | 0.078 | No |
| 3 | Irregular Form | 11,352 | 2562.555 | | | |
| | | | Floor Number | | | |
| 1 | One Floor | 1560 | 5321.165 | | | |
| 2 | Two Floors | 4128.86 | 1983.013 | | | |
| 3 | Three Floors | 5893.50 | 1799.388 | 30.937 | <0.001 | Yes |
| 4 | Four Floors | 9432 | 806.146 | | | |
| 5 | Five Floors | 14,072 | 900.107 | | | |
| | | | Urban Fabric | | | |
| 1 | Traditional Fabric | 5492 | 3659.470 | | | |
| 2 | Linear Fabric | 4572.92 | 3823.805 | 0.242 | 0.626 | No |
| 3 | Scattered Fabric | 6378.86 | 4200.079 | | | |
| | | | Road Network | | | |
| 1 | Road 2 m | 3420 | 1535.187 | | | |
| 2 | Road 4 m | 5470.80 | 3089.142 | 2.218 | 0.127 | No |
| 3 | Road 6 m | 6870.40 | 4436.363 | | | |

**Table 8.** Multiple comparisons between floor-number variables and annual electricity consumption (kWh).

| Floor Number with Electricity Consumption | Mean Differences Between Groups | *p*-Value | Statistically Significant |
|---|---|---|---|
| (5) Five Floors: (1) One Floor | 12,512 | >0.001 | |
| (5) Five Floors: (2) Two Floors | 9943.143 | >0.001 | |
| (5) Five Floors: (2) Three Floors | 8178.500 | >0.001 | Yes |
| (5) Five Floors: (2) Four Floors | 4640 | 0.002 | |

### 3.2.4. Techno-Economic Assessment of Installing PV

In the current regulatory framework, the net-metering scheme, first adopted in 2013, allows for an energy purchasing agreement between three parties, the Egyptian Electric Utility, and the Consumer Protection Regulatory Agency, responsible for implementing policy decisions and administering licenses, and the consumer. This enables an energy-purchasing agreement to be made between a qualified third-party company (operating and managing the plant) and a client who benefits from the purchased energy and from making a contract with the electricity company.

In a nutshell, the Electricity Distribution Company has to be contacted for supervision and to check if the roof is suitable for PV. Then, the necessary permit is obtained from the local authority. The qualified third party (accredited by the Ministry of Electricity) is responsible for reviewing and approving plans to ensure compliance with the PV plant installation regulations. Eventually, the third party will notify the distribution company to conduct a final inspection and approve the installation. An approved certificate for the implementing company is issued by the New and Renewable Energy Authority, and the contract between the implementing company and the project owner is launched to implement the project within 6 months of approval; all required data and procedures are available on the official solar energy platform [87], which it also provides estimators for PV size capacity and all fees [88].

Technically, designing and optimizing PV systems for Egyptian rural communities is promising, as Egypt lies in the Sunbelt region, which ensures very high monthly mean sunshine hours [88]. The studied area (North of Egypt's Nile Delta) has about 180 kWh/m$^2$ monthly, and a 5 kW solar-power generation unit is suitable to be placed above an average house [24]; as well as this, it is the minimum PV system size accepted under a net-metering scheme. The technicalities of designing and installing PV systems for residential buildings in rural Egypt are discussed in depth in Reference [26], and for all buildings (residential and non-residential) in Reference [89], and have a lifetime of 25 years. Six scenarios have been developed to evaluate the technical performance based on the electricity-consumption tariff, employing a PV unit with 5 KWp, assuming that each segment's group consumes the maximum load.

To determine the financial aspects, the First Author contacted some qualified PV local suppliers available on the official website [90], to collect financial quotations to update the pricing and explore their insights on the issue. Three local suppliers have been engaged; a manager of one requested to contribute anonymously, and the other two qualified companies provided the prices of USD 3655 and 3750, including all equipment (PV cells, inverter, wires, fixation steel structure, and transportation to the village), including the two-way meter from the electricity company, in the same context. Gabr et al. [91] estimated the cost to be USD 3125 (in 2020), which is in accordance with the PV Hub's estimated price of USD 3045 [92]. The selling price of the kWh generated from RE is about USD 0.016. The residential electricity tariff is divided into seven segments, as shown in Table 9. Public buildings also belong to the residential tariff.

Two economic models were utilized to investigate the profitability of PV systems under six scenarios. First, the simple payback period (SPP) analysis, Equation (2), is the projected years required to regain the initial investment that occurs when the cumulative net cash flow equals zero [93]. The net present value (NPV), Equation (3), represents the present value of cash inflows minus cash outflows, determined by the expected net cash flow (CF) in a particular time ($t$). This is related to the project at a given discount rate, which is 27%, according to Central Bank of Egypt, which provides accurate profitability, considering different aspects like inflation and so on. Profitability is determined when the value becomes positive in a particular year, and vice versa; the project becomes unprofitable if the values become negative [94]. Table 10 shows the economic assessment of the PV units.

$$SPP = \frac{Initial\ Investement}{Net\ cash\ flows\ per\ period} \tag{2}$$

$$\sum\nolimits_{t=0}^{n} \frac{CFt}{(1+r)^t} \tag{3}$$

**Table 9.** The residential electricity tariff for the fiscal year 2024/2025, the equivalent expenditure of the maximum consumption per segment, and the selling price of electricity generated (to be read in conjunction with Table 10).

| Segment | 1st | 2nd | 3rd | 4th | 5th | 6th | 7th | Selling price of kWh/RE |
|---|---|---|---|---|---|---|---|---|
| Consumption range (kWh) * | 0–50 | 50–100 | 101–200 | 201–350 | 351–650 | 651–1000 | >1000 | |
| Price (EGP) * | 0.68 | 0.78 | 0.95 | 1.55 | 1.95 | 2.10 | 2.23 | 0.016 |
| Price (USD) ** | 0.014 | 0.016 | 0.019 | 0.031 | 0.040 | 0.043 | 0.045 | |
| Maximum expenditure according to consumption | | | | | | | | |
| Monthly (kWh) | 50 | 100 | 200 | 350 | 650 | 1000 | N/A | |
| Annually (kWh) | 600 | 1200 | 2400 | 4200 | 7800 | 12,000 | | |
| Monthly (USD) | 0.7 | 1.6 | 3.8 | 10.85 | 26 | 43 | | |
| Annually (USD) | 8.4 | 19.2 | 45.6 | 130.2 | 312 | 516 | | |

* According to the new Electricity Tariff on 9/2024 [95]. ** The exchange rate of 1 USD is about 49.26 EGP, according to the Central Bank of Egypt in November 2024 [96].

**Table 10.** The economic assessment of assumed scenarios for residential and non-residential buildings (to be read in conjunction with Table 9).

| Segment Group | Annual Consumption (kWh) | Annual Expenditure in Electricity (USD) | PV Unit Generation (kWh) | Surplus in Electricity Generation (kWh) | Selling Price (0.016 USD/kWh) | Economic Benefit (Saving + Generation) (USD) | Initial Cost (USD) | Economic Profitability | |
|---|---|---|---|---|---|---|---|---|---|
| | | | | | | | | SPP | NPV at 25 Years (USD) |
| First | 600 | 8.4 | | 8500 | 136 | 144.4 | | 24.9 | −3066 |
| Second | 1200 | 19.2 | | 7900 | 126.4 | 145.6 | | 24.7 | −3062 |
| Third | 2400 | 45.6 | 9100 | 6700 | 107.2 | 152.8 | 3600 | 23.6 | −3035 |
| Fourth | 4200 | 130.2 | | 4900 | 78.4 | 208.6 | | 17.3 | −2829 |
| Fifth | 7800 | 312 | | 1300 | 20.8 | 332.8 | | 10.8 | −2370 |
| Sixth | 12,000 | 516 | | −2900 | −46.4 | 469.6 | | 7.7 | −1865 |

The results indicated that the installed PV system led to positive energy consumption, with lower consumption and higher generation for PV units. Groups 1 and 2 achieved significant RE production, and the third and fourth achieved 279% and 117%, respectively. In the fifth group, the generated electricity represented 16% of the total, and the sixth indicated that the proposed system generated 75% of the annual consumption. The required size to achieve zero energy is 7 KWp. The SPP analysis revealed that in all scenarios within the lifetime of PV system (25-year period), the higher the RE generation, the longer the period. However, the NPV indicated that not all systems are profitable.

## 4. Discussion

In 2022, Egypt's GHG emissions totaled 319.69 mt $CO_2$eq. The electricity and building sectors are responsible for one-third of these emissions, showing how much energy-efficiency intervention in both sectors significantly contributes to reducing emissions and supporting the state's energy and climate policies related to building. This is in addition to reaching the targets and climate-action commitment by 2030, especially in rural areas that host three-fourths of the building sector, about 65% of the population in Egypt, and consume one-third of the distributed electricity. This study aimed to characterize the electricity consumption trends in rural buildings, and found its association with the built environment and socio-economic patterns by fostering a novel approach not conducted before in the local context. The Lasiafr Albalad rural settlement has been selected to represent typical agriculture-based settlements and rural–urban tissue, representing 339 main rural settlements in the Delta region that are predominately dedicated to agricultural activities and land areas.

The study provided a holistic exploratory approach, combining different methods: the analytical method identified the main characteristics of the ultimate context, using national census and official reports, which better understood the built environment in the Delta region, including tangible and intangible aspects, like urban setting, building types, socio-economic factors, and energy profile. Then, the field method, including an in situ survey, one-to-one interview, and questionnaire, diagnosed the intimate context of the village, including locals' lifestyle and building energy consumption, in addition to spatial analysis, leading to the production of the first online, open-source editable map. Then, the statistical analysis examined and quantified the relationship between all these factors, utilizing the Pearson Correlation Coefficient for numerical variables and one-way ANOVA to examine one qualitative variable with one numerical one. After that, technical and economic assessments were conducted to explore the rooftop PV under the net-metering scheme. This thoroughly and effectively addressed the research questions presented in the sections below.

### 4.1. Addressing RQ1

An intensive on-site investigation lasted two months; the urban characteristics and architectural typologies were diagnosed within this. A sample of the local community, who volunteered and were enthusiastic about contributing to improving the village, was informed about the intervention to explore socio-economic activities and understand the possible benefits of the research. They shared their thoughts about electricity consumption, and the interviews revealed that the small-dimension economic activity dramatically affects electricity consumption, requiring lighting to provide a suitable environment for raising birds and poultry. These findings are contrary to another study conducted by the First Author [97], to diagnose the correlation between small-dimension fishing activities in a rural coastal settlement in the Delta region, which indicated no relationship between small-dimension fishery economies and residential electricity consumption, as well as the fact that changing economic activities lead to different electricity consumption patterns, as has been discussed by Elsabbagh et al. [51], and Abdelhay [50].

The electricity-use intensity of the non-residential buildings is between 2.5 and 18 kWh; the health-unit building alone has the highest consumption. It was noticed that the lighting demand is not affected, due to daily operating hours being in the morning, unlike the dwellings, in which the lighting represents about one-half of the consumption, as per the official census. This also showed the tendencies of household consumption patterns; these tendencies have confirmed the finding of a bottom-up intervention conducted by the First Author to examine the local social acceptance of proposed retrofitting solutions [69]; likewise, our study contributes another knowledge dimension to this issue.

The data collected from the questionnaire showed that the sample's data were homogeneous and distributed normally. Based on the analysis, the annual electricity use intensity of the dwellings ranges between 7.9 and 43.3 kWh/m$^2$, with a mean of 21 kWh/m$^2$, where the population sample revealed that the yearly residential electricity is 534 kWh/capita, with a daily average of 6.67 kWh/dwelling.

These results complied with the study of Salama and Hassan, which reported that the monthly expenditure for a household is 232 EGP [49]. We converted this expenditure, which belongs to the fourth segment (according to 2020 tariff prices), into about a monthly 282 kWh and daily 8.5 kWh, which is above the average indicated in rural areas in different regions in Egypt. These are 5.075 kWh/day/dwelling [26] and from 1 to 2.2 kWh/day, with an average of 1.6 kWh/day [98] in the Asyut region, 3 kWh/day in the Greater Cairo region [24], and 2.936 kWh/day in the South Upper region [88].

For similar contexts, our results came below the average in rural Morocco, which has 11.26 kWh [99], and slightly above the studies of Ghafoor et al. [100] in Pakistan and those of Almutairi et al. [101] in Iran, who indicated 5.9 and 5.625 kWh/day/dwelling, respectively; they were significantly above the averages indicated by Mahmoud and Ibrik in Palestine, at 1.75 kWh [102], and those of Sukarno et al. in Indonesia, at 1 kWh/day/dwelling (the annual consumption is 300 kWh) [103]. Figure 7 depicts the energy-consumption benchmark of rural houses in Egypt and similar contexts.

These presented values can give a significant benchmark to the Egyptian rural units, considering that some of these contexts are remote villages with inadequate electrification, unlike the Delta's high-density and 100%-electrified villages. This indicates a correlation between economic activities and energy consumption, as found in our study and by Salama and Hassan in similar agriculture-based settlements in the Suez Canal region.

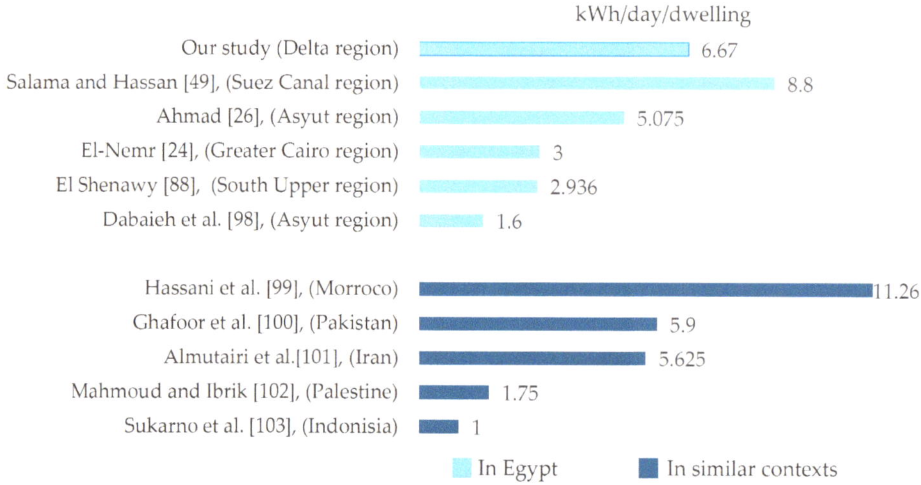


**Figure 7.** Benchmarking the daily energy consumption of rural dwellings in Egypt and similar contexts, developed by the Authors after [24,26,49,88,98–103].

The EUI in rural areas in Egypt remains too low compared to the urban contexts, at nearly less than one-fifth, compared to Nafeaa et al.'s findings [46], which indicated an EUI of 110 kWh/m$^2$. The average we found is within the range stated by Attia et al. [47]. However, again, neither study explicitly states whether the context is rural or urban. These variations also occur globally; for example, the EUI of residential, rural buildings in China is higher than that of urban areas by one-fourth [104]. Therefore, this study explored a case study to fill this gap and create momentum to classify and benchmark rural electricity consumption.

Another insight was collected from the study of Mustafa et al. [22] in Saudi Arabia, who benchmark the EUI of rural dwellings in the southwest region at 134 kWh/m$^2$, with a range of 125.6–182.4 kWh/m$^2$ for other regions. Thus, we advocate for scholars to fully define the consumption patterns of their case studies and data collection to support academic efforts in diagnosing rural building energy consumption, and which, in turn, fills significant gaps, as discussed within this article.

The study investigated the correlation between residential electricity consumption in numerous groups of variables, including five for numerical demographic and building features, presented in Table 2. There were also 10 nominal groups of variables, with a total of 41 sub-variables, as visualized in Figure 6, representing the socio-economic aspects and built environment. The diagnosed correlation can help pave the way to understanding the broader factors influencing energy behaviors and transition acceptability [105]. The study revealed an observed statistical correlation between social activities, namely home visits and going to nearby districts, with electricity consumption.

Meanwhile, a robust positive correlation has been confirmed between electricity consumption, household size, occupancy rate, floor numbers, and total built-up area. For the latter, our results are, notably, contrary to the findings of Hegazy et al. [48], whose results indicated a negative correlation for rural buildings in the same region; the higher built-up area has the lowest consumption. This variation might have occurred because their study was limited to simulation predictions with fixed operational activities for all scenarios.

Also, our findings supported Salama and Hassan's findings that the more members, the more the electricity consumption. Finally, some insights were observed, like a positive relationship between the buildings in which the residents expressed good thermal conditions and those located on linear and scattered fabric; vice versa, for the people who described the worst thermal conditions, most buildings were situated in traditional urban fabrics. This requires deep investigation at a regional scale to draw, in a better way, the tendencies of the most influencing aspects of domestic electricity consumption.

### 4.2. Addressing RQ2

PV is a game-changer in the clean-energy transition for different buildings: residential (the majority) and public buildings. Egypt's conducive climate, with high monthly sunshine hours, supports the installation of 5 kW solar units, suitable for average homes. The minimum PV system size aligns with net-metering regulations. The financial aspects were assessed using quotations from local suppliers; two economic models, the SPP and NPV, were used to evaluate the profitability of PV systems. The results revealed that the SPP could be feasible for the sixth-segment (651–1000 kWh) dwellings in year 11 and the seventh-segment (more than 1000 kWh) in year 8, as the SPP does not consider the discount rates and inflation. However, the NPV confirmed that PV rooftops on residential buildings are not economically-feasible, and the NPV of the projected life cycle is USD −1865 in the optimistic scenario, namely for the highest segment consumption, above 1000 kWh monthly.

Our finding supports Tsuchiya et al.'s conclusion that investments do not adequately cover current operational expenses costs of PV for rural villages in Tanzania [106], and are similar to those in Aref et al.'s study [107], which indicated that the SPP obtained in year 7.2 and the net present cost was, throughout the 25-year prediction period, under a discount rate of up to 15% (the year thet the profitability achieved was not explicit). However, their study differs from ours, as they conducted off-grid solutions for all the dwellings in a small, remote, rural Pakistani village, populated by 800.

Nafeh [108] opposes our finding that off-grid and on-grid PV is advantageous and ideal for long-term investments in rural, remote areas in Egypt. This is similar to Çamur et al. [109], who indicated that the prices of the energy generated from on-grid rooftop PV in Lebanese dwellings are competitive. Despite that, we would emphasize that the economic assessment of both studies estimated the discount rate (10% and zero) in Egypt and Lebanon, respectively.

In Poland (one of the economies most reliant on non-RE energy in Europe) Rej-Witt and Dębska simulated installing PV on a rural dwelling with the same power size as that used in our study (5 kW); they found the payback period at 4.3 years, because they conducted the economic assessment based on 20% of the total investment amount, assuming that the relevant financing scheme of the European Commission can co-finance the remaining 80% [110]. This shows the essential role of top-down support.

The long period for a return on investment raises crucial questions: who should initiate it? We would say that the government can lead the integration of PV solutions into national rural-development projects. Also, the private sector can invest in carbon offset and carbon credits in the field of carbon transformations, and work in this way to benefit from selling the carbon credits. The latter, fortunately, became a fact: in August 2024, the first voluntary market of carbon credit was established to promote green energy, which might encourage the private sector in investing in installing PV.

Although the potential benefits from PV toward clean-energy transition, especially in rural areas in Egypt and in similar climatic conditions, as well as the consumption patterns, was benchmarked, and its correlation with built environments and socio-economic factors diagnosed, many risks should be considered while conducting similar approaches. For example, Abdulhady and Metwally [111] classified the risks into the appropriate designs to adapt to the effects of climate conditions on PV system performance; dust accumulation significantly affects PV output. Measurements show a 10% reduction from rain and about 17% due to storms and dust. The impact is more significant in desert areas and less severe in regions with frequent rainfall (like the Delta region), and the shading is the most influential parameter on the PV module efficiency. Degradation manufacturers of solar PV systems typically guarantee a performance lifespan of 25 years. They assure that panels will produce at least 90% of their rated capacity in the first 10 years and around 80% in the following 10 to 15 years. This requires developing operational and technical manuals and building capacities for regular maintenance, besides designing PV systems to fit each climatic zone. On the other hand, we recommend that the qualified companies provide risk mitigation plans, case by case.

A recent policy paper was released by the IDSC Public Policy Forum, the Information and Decision Support Center—the Egyptian Cabinet has stated that Egypt struggles to attract foreign direct investment, and investors face difficulties due to unstable exchange rates and the significant disparity between the EGP official and black market rates, which inflate costs for components and raw materials. This is especially the case, in that regulatory barriers and intense market competition also hinder investment. The cost of solar panels and import and customs fees have risen significantly, impacting the overall affordability of home solar systems [112].

In this realm, a representative of a local PV supplier (who did not authorize us to mention his company's name in the study) emphasized the fact that the issue is too challenging, based on three concerns: the readability of infrastructure in rural Egypt, building construction conditions, and the very high initial cost versus the extended return on the investment period. Consequently, this barrier should be considered in terms of the holistic economic situation. The discount rate is too high, meaning that the return on initial investment takes a long time, requiring that scholars in the financial domain review this issue. For example, if the discount rate dropped to 10%, as it was for many years before the current economic conditions, a higher profitability of greater than 1000 kWh/month would be achieved in the 16th year, which is also high.

Not only are the economic aspects a barrier, but we also advocate that considering the social aspects, namely, raising awareness of the local community, is crucial. An apparent example, fortunately, is that the authors conducted a pilot experiment to install a roof-top PV unit, off-grid in 2022, in a social building; the building was zero-energy consumption for nearly two years, the project financed by the private sector through a participatory approach [27]. Unfortunately, while writing this article, the landlady informed us that someone had broken the PV panel, as seen in Figure 8.

**Figure 8.** A broken PV panel reflects the importance of raising awareness of the local community. (Image source and credit: Ms. Dalia Hamouda, landlady of the building).

## 5. Conclusions and Implications

To conclude, this study has investigated a typical built-environment pattern in rural settlements in the Delta region of Egypt. The pilot experiment succeeded in quantifying domestic electricity consumption and finding novel statistical observations and correlations between the built environment and domestic electricity consumption. It provided a better understanding that can pave the way to achieve the ambitious goal of retrofitting the rurally built environment in Egypt, as an initial step from a bottom-up perspective, which supports the growing research tendencies at local levels and global ones [14]. These results led to a better understanding of domestic energy consumption, as a readiness step to improve energy efficiency, as emphasized in a similar context [113]. In addition to the technical efficiency of buildings and related activities, it is vital to fulfill global sustainability goals [114] and to inform practices like retrofitting and energy planning.

The novelty of this study is characterized by its holistic nature, and it defined the current status quo of electricity consumption and its role in associated GHG emissions in rural Egypt; it also diagnosed the built-environment patterns of rural settlements, as the EUI benchmarked, in an exploratory study. As far as we know, this is the first time this has been carried out in Egypt, engaging stakeholders (local community and the private sector), as it fostered different data-collecting and processing methodologies, and benefits energy transition [115].

The low electricity consumption, indicated from the results, could be a solid potential for reaching the goal of zero-energy buildings and positive clean-energy production. This can act as a tool for revitalizing rural commons and serving urban areas, especially by generating clean energy from rooftop PV panels, under the net-metering scheme. This proved the significant surplus in the generated clean energy, as the study promoted rooftop PV on-grid electricity in 100%-electrified rural settlements, unlike the prevailing literature in Egypt, which focused on off-grid solutions for remote areas. Moreover, the study filled an essential gap, by determining the most-influencing factors and correlations to accelerate the clean-energy transition, besides setting a benchmark consumption to support tailoring energy-efficiency strategies and improve energy efficiency, in compliance with the arguments presented in Sections 1.1.2 and 1.1.3. The analysis was performed for rooftop PV with an installed capacity of 5 KW under the regulatory framework of the net-metering scheme. It estimated six electricity-generation scenarios for residential and non-residential buildings, following the maximum consumption of each consumption segment.

Manifestly, financial aspects burden the local people; in other words, the local people may prefer to invest in banks with a high interest rate (exceeding 20%) rather than in PV, apart from the already low monthly expenditure on electricity bills. Accordingly, the economists' inputs are beneficial in addressing this issue; likewise, anthropologists and all relevant domains regarding clean-energy transition prepare the local people (for example, the possibility of breaking the PV panel) and consider the local people's preferences. The role of scholars from the architecture and urban-planning disciplines who work in energy efficiency and retrofitting rural buildings is crucial to integrating innovative RE solutions for such interventions. In fact, we would emphasize the fact that fostering transdisciplinary approaches is inescapable for the clean-energy transition.

What is more, the private sector can make a significant contribution by providing business plans to help accommodate RE solutions or energy-efficient technologies, such as long-term payment plans, and by grasping the opportunities of numerous banks providing green loans to apply for PV panels for residential buildings.

The Authors would accentuate that no roadmap exists without exploratory studies and narration. Also, the scholars' role as "knowledge brokers" and the bottom-up practices remain essential to support the momentum of creating a self-sufficient dogma that has become a demand not only for Egypt, but for the entire world, amid all the unprecedented extreme climate and geopolitical fluctuations. Into the bargain, this period, particularly, is a golden opportunity, because the state is already developing the national project Decent Life; the First Author intends to present this study to the decision-makers.

An apparent example that many scholars were advocating to localize the PV industry, is summarized by the First Author in reference [28]; fortunately, in 2024, the Prime Minister's Resolution No. 2732 has recently established a national council called the "National Council for Localizing the Technology of Manufacturing Electronic Chips and Solar Cells". The council aims to review legislation regulations, follow up on the implementation of governmental commitments, and identify investment opportunities and the necessary solutions to overcome the obstacles to investment in this field. This reflects the tendencies toward clean-energy transition and liberalizing and decentralizing the electricity sector.

### 5.1. Policy Implications

Diagnosing electricity consumption patterns and their correlation with socioeconomic activities can provide valuable insights for policymakers. This information can inform the development of effective energy policies that promote sustainable energy use, reduce energy poverty, and support economic growth, supporting the argument of Gouveia et al., which addressed energy policy gaps associated with non-residential buildings within the Mediterranean climate [116], and policy- and design-oriented notions of energy planning [117], as well as rural areas. These are essential for thriving energy transitions, and vice versa; they can boost sustainable rural development [118]. Our findings support filling the gap between predicted and actual consumption patterns, similar to Tam et al.'s approach.

Although the landscape in Egypt is transiting towards a clean energy future and a liberalized electricity sector, it is quite evident that, under the current net-metering schemes, the minimum size required is 5 kW, which is not feasible at all for individuals. Therefore, the policies can be tailored to enable the use of energy cooperatives in agricultural practices (e.g., farms and structures) or buildings. Maybe the agriculture association could lead this role, particularly as it already plays a significant role in agriculture-based communities. Also, affording different RE sources, like biomass, can be beneficial, through updating the NEEAP and relevant policies with a specific target, such as raising awareness, training, incentives, facilitating retrofitting buildings and farm structures, and so on, or as implemented with solar energy (the most used RE source), in addition to collaboration with the private sector, like incentives. Actually, top-down development and government support are inevitable to afford clean energy, as discussed by Rej-Witt and Dębska [110], and these factors can increase the mix, reaching 42% RE sources.

This intervention can provide guidelines for developing infrastructure and rural built environments linked with national projects, such as Decent Life, or within planning new agricultural communities, a core strategy of the Egyptian rural development policies to expand the inhabited areas. Also, policymakers can design targeted energy-efficiency programs that address the needs of different socio-economic groups. Also, to tailor buildings, the energy-efficiency codes that provide minimum energy-efficacy requirements for new and existing buildings, according to the national macro-climatic classification [119], might be insufficient, due to the variety of micro-climates as crucial influencers of energy consumption [120].

Our study quantifies the domestic electricity-consumption patterns, making rural commons qualified to be energy communities, and can act as energy baskets for serving urban areas. Eventually, understanding the factors driving electricity consumption can help policymakers develop effective strategies to reduce GHG emissions associated with energy use and to downscale these practices into local administrative-unit levels, which have to incubate more bottom-up initiatives and engage more stakeholders who, in turn, can support the overall policies, and will meet the targets of NEEAP and climate action in 2030, and the RE share in 2035, and beyond.

### 5.2. Limitation and Future Direction

Despite the study's holistic nature, the authors declare it has limitations related to the context, as different target cases can lead to different results. Also, the examined non-residential buildings vary notably in terms of orientation, size, village population, occupancy, and operational activities; consequently, they might lead to a different EUI. Furthermore, examining different factors and widening the population sample and case studies are encouraged, to support the reliability of the results. On the other hand, using more advanced statistical models like two-way ANOVA and multi-criteria decision analysis can discover more indicators of more rural settlement patterns, like the remote ones [121].

Also, an important indicator was observed from the open-ended question in the questionnaire; despite it being based on a few respondents' standpoints, it can give an indicator that the locals may prefer to mitigate the negative environmental impacts of the waste and drainage before improve energy efficiency, which refers to the importance of widening the in situ investigations by scholars from environmental science domains to be integrated with our study. More elaboration to examine the thermal comfort factor, hand-in-hand with defining the equipment and appliances in further studies, would be beneficial to classifying domestic consumption. The study focused on solar energy; hence, considering different RE sources like biomass and wind energy, can be helpful.

Finally, the research discussed a specific local context; the methodology can be replicated in all settlement patterns in the Egyptian governorates, which account for 4741 villages and 30,888 satellite ones. It could also provide indicators for similar contexts in the built environment within the Mediterranean climate, either for the authors or interested scholars, accelerating the solving of a real-world problem and implementing an on-the-ground practice for leading energy development practices to mitigate climate change and the high environmental challenges in rural settlements.

**Supplementary Materials:** The following supporting information can be downloaded at https://www.mdpi.com/article/10.3390/su17041597/s1, S1: Collected Sample in SPSS (.SAV) format, and S2 the collected data in Excel Format.

**Author Contributions:** Conceptualization, A.A.; methodology, A.A. and A.B.; statistical analysis, A.A.; validation, A.A. and A.B.; formal analysis, A.A. and A.B.; investigation, A.A.; resources, A.A.; data curation, A.A.; writing—original draft preparation, A.A.; writing—review and editing, A.A. and A.B.; visualization, A.A.; supervision, A.B. All authors have read and agreed to the published version of the manuscript.

**Funding:** This research received no external funding.

**Institutional Review Board Statement:** Experimental protocols followed the internationally recognized ethical standards for examinations and were verified from the Research Ethical Committee, Sinai University, Cairo, Egypt with Memorandum No. (SU.REC.2024 (29)). The date of approval is 24 December 2024.

**Informed Consent Statement:** Informed consent was obtained from all subjects involved in the study, by the First Author.

**Data Availability Statement:** The original data presented in the study are openly available in the manuscript and the Supplementary Materials.

**Acknowledgments:** The First Author acknowledges the technical support from Doraid Magdy Abosalem, Managing Director, Solaracil Egypt Renewable Energy, https://solaracil.com/ (accessed on 21 January 2025), and Karim Ashour, Engineer, Clenergy, https://www.clenergy-mena.com (accessed on 21 January 2025) in providing designing the PV system, and for technical and financial offers related to supply and applying PV to the Lasiafr Albalad village, besides sharing their insights on their experiences in the previous application, within the net metering scheme.

**Conflicts of Interest:** The authors declare no conflicts of interest.

## Abbreviations

The following abbreviations are used in this manuscript:

| | |
|---|---|
| ANOVA | Analysis of Variance (statistical formula) |
| EUI | Energy Use Intensity (energy efficiency indicator |
| GHG | Greenhouse gas |
| Mt $CO_2$eq | Metric ton of carbon dioxide equivalent (measurement unit) |
| NEEAP | National Energy Efficiency Action Plan |
| NPV | Net present value (economic assessment indicator) |
| PV | Photovoltaic |
| R value | Correlation coefficient (statistical indicator) |
| RE | Renewable energy |
| RQ | Research Question |
| SD | Standard deviation (statistical measure) |
| SDG | Sustainable Development Goal |
| SPP | Simple payback period (economic assessment indicator) |
| *p* value | Probability value (statistical measure) |

## Appendix A

In a nutshell, we simplified the steps, especially for those readers that do not have a statistical or mathematical background. In the following few paragraphs, Figure A1 illustrates the steps for using ANOVA, while Tables A1 and A2 recap the steps to start conducting the ANOVA.

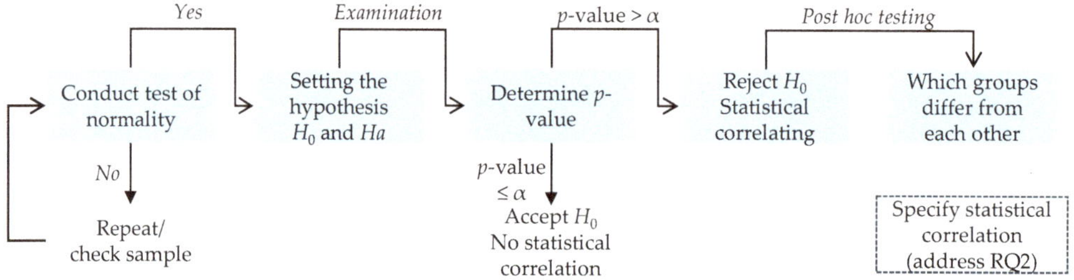


**Figure A1.** A flow chart to highlight the main steps while conducting the ANOVA and to show how it can answer the RQ2.

Before beginning, the values of the dependent variable should be normally distributed and randomly sampled, which can be determined by the normality test (an essential step to accepting conducting statistical analysis). This test can be offered by standard models like the Shapiro–Wilk test, as it is the most proper method for small sample sizes, with less than 50 samples; once the model shows a significant level of more than 0.05, it implies variables' homogeny [122], and, vice versa, if it is more than 0.05, it means less homogeneity, resulting in less reliability of findings. It is essential to elaborate on the significance level. It is referred to by the ($\alpha$) sign, which assumes the probability of error with a percentage, usually 5%, in the study's results. In other words, if the study is repeated 100 times, the same results will occur 95 times.

After that, two hypotheses will be set, the null hypothesis ($H_0$) and the alternative hypothesis ($H_a$), at a significance $\alpha$ = 0.05. In our study, both hypotheses are $H_0$: no statistical correlation between residential electricity consumption and the examined factors. $H_a$ indicates there is a statistical correlation.

The $p$-value (an absolute value) should be determined to explore the hypothesis. Meanwhile, the $p$-value is determined by the $f$-value. Some factors that determine the $f$-Value are shown in Table A1. To interpret the results, if the $p$-value > $\alpha$, do not reject $H_0$ (no statistical significance occurred). If the $p$-value $\leq \alpha$, reject $H_0$ in favor of $H_a$ (the study is statistically significant). Once the study has statistically significant differences between variables (for each group), the ANOVA presumes that at least one group means is significantly dissimilar. Still, it does not indicate which variable is the main one. Hence, to determine which groups differ, it is acceptable to continue with the post hoc testing (multiple comparisons) to identify which groups differ, precisely. [123]. The ANOVA is conducted according to the ANOVA table, Table A1. Table A2 shows the abbreviations used, summarizes the steps taken in the analysis, and simplifies them.

**Table A1.** ANOVA Table (should be read with Table A2).

| Source | *DF* | *SS* | *MS* | *F* | *p* |
|---|---|---|---|---|---|
| Between-Group (factor) | $m-1$ | *SS* (Between) | *MST* | *MST*/*MSE* | absolute value |
| Within-Group (Error) | $n-m$ | *SS* (Error) | *MSE* | | |
| | $n-1$ | *SS* (Total) | | | |

**Table A2.** Description of the components of the ANOVA analysis.

| Acrynom | Description | Equation | Definition and Rationale |
|---|---|---|---|
| $H_0$ | Null hypothesis | (A1) | Indicates no difference among group means (no statistical correlation observed). |
| $H_a$ | Alternative hypothesis | (A2) | Indicates statistically significant result occurs (one group varies significantly, with respect to all means of the dependent variable). |
| $\alpha$ | Significance level (alpha) | N/A (absolute value) | Assumes the probability of error with a percentage, usually 5%, in the study's results; in simple words, if the study is repeated 100 times, the same results will occur 95 times. |
| $f$-value | Probability value | (A3) | In the ratio of the between-group and within-group variance, a higher $f$-value indicates the higher difference between sample means, relative to the variation within the samples. The $f$-value is essential to determine the $p$-value (see below, in this table). |
| *SST* | Sum of Squares (Total) | (A4) | The total variation that can be assigned to different factors (the essential factor to determine the $f$-value) is determined by the sum of squares between groups. |
| *SSB* | Sum of Squares (Between) | N/A (input to Equation (A4)) | The sum of squares between the group means and the grand mean. |
| *SSE* | Sum of Squares (Error) | N/A (input to Equation (A4)) | The sum of squares between the data and the group means. |
| *MS* | Mean Square | N/A | To evaluate the variance between sample means with respect to the overall sample mean; in other words, it is the variance between groups (input in ANOVA table, Table A1). |
| *MSB* | Mean Square due to Treatment | (A5) | Estimating sample mean variance from the overall mean (input in ANOVA table, Table A1). |
| *MSE* | Mean Square due to Error | (A6) | Estimating the sample variances for the population variance (input in ANOVA table, Table A1). |
| $p$-value | Probability value | Absolute value | It examines the hypothesis (the probability assuming no difference between variables—the null hypothesis). Where if $p$-value $\leq \alpha$, reject $H_0$ in favor of $H_a$, while if $p$-value > $\alpha$, do not reject $H_0$. |
| *DF* | Degree of freedom | N/A (input to ANOVA table) | To determine the dependent variables which can be used in the analysis. If n total variable data are collected, the $DF = n-1$. If there are $m$ groups for comparison, the $DF = m-1$ is associated with the interest factor, and finally, if there are $n$ total variable data collected and $m$ groups for comparison, the $DF = n-m$. |

$$Ho = \mu_1 = \cdots = \mu_n \tag{A1}$$

$$Ha = \mu_1 \neq \cdots \neq \mu_n \tag{A2}$$

$$F\ value = \frac{MSB}{MSE} \tag{A3}$$

$$SST = SSB + SSE \tag{A4}$$

$$MSB = \frac{SSB}{(m-1)} \tag{A5}$$

$$MSE = \frac{SSE}{(n-m)} \tag{A6}$$

## References


1. Dasgupta, P.; Morton, J.; Dodman, D.; Karapinar, B.; Meza, F.; Rivera-Ferre, M.G.; Toure Sarr, A.; Vincent, K.E. Rural Areas. In *Climate Change 2014: Impacts, Adaptation, and Vulnerability. Part A: Global and Sectoral Aspects. Contribution of Working Group II to the Fifth Assessment Report of the Intergovernmental Panel on Climate Change*; Field, C.B., Barros, V.R., Barros, D.J., Dokken, K.J., Mach, M.D., Mastrandrea, T.E., Bilir, M., Chatterjee, K.L., Ebi, Y.O., Estrada, R.C., et al., Eds.; Cambridge University Press: Cambridge, UK; New York, NY, USA, 2014; pp. 613–657.
2. IEA. *World Energy Outlook 2023*; IEA: Paris, France, 2023.
3. Li, M.; Jin, T.; Liu, S.; Zhou, S. The Cost of Clean Energy Transition in Rural China: Evidence Based on Marginal Treatment Effects. *Energy Econ.* **2021**, *97*, 105167. [CrossRef]
4. Ritchie, H.; Rosado, P.; Roser, M. $CO_2$ and Greenhouse Gas Emissions. Available online: https://ourworldindata.org/co2-and-greenhouse-gas-emissions (accessed on 28 May 2024).
5. Clausen, L.T.; Rudolph, D. Renewable Energy for Sustainable Rural Development: Synergies and Mismatches. *Energy Policy* **2020**, *138*, 111289. [CrossRef]
6. Patnaik, A.; Kudal, P.; Dawar, S.; Dawar, P. Sustainable Rural Development through Renewable Energy Technologies: Identification of Potential Methods to Increase Rural Incomes. In *IOP Conference Series: Earth and Environmental Science, Jaipur, India, 24–25 June 2022*; IOP Publishing: Bristol, UK, 2022; Volume 1084, p. 012020.
7. Cebotari, S.; Cristea, M.; Moldovan, C.; Zubascu, F. Renewable Energy's Impact on Rural Development in Northwestern Romania. *Energy Sustain. Dev.* **2017**, *37*, 110–123. [CrossRef]
8. Kothari, S.; Strachan, N. *Cooperatives as Bottom-Up Energy System Actors for Increasing Renewables Adoption*; EnergyREV, University of Strathclyde Publishing: Glasgow, UK, 2023.
9. IEA. *Renewables 2023 Analysis and Forecast to 2028*; IEA: Paris, France, 2024.
10. Gao, Y. Top-down and Bottom-up Processes for Rural Development and the Role of Architects in Yunnan, China. *Buildings* **2016**, *6*, 47. [CrossRef]
11. Ghorbani, A.; Nascimento, L.; Filatova, T. Growing Community Energy Initiatives from the Bottom up: Simulating the Role of Behavioural Attitudes and Leadership in the Netherlands. *Energy Res. Soc. Sci.* **2020**, *70*, 101782. [CrossRef]
12. Fleiter, T.; Rehfeldt, M.; Herbst, A.; Elsland, R.; Klingler, A.-L.; Manz, P.; Eidelloth, S. A Methodology for Bottom-up Modelling of Energy Transitions in the Industry Sector: The FORECAST Model. *Energy Strategy Rev.* **2018**, *22*, 237–254. [CrossRef]
13. Davis, M. Rural Household Energy Consumption: The Effects of Access to Electricity—Evidence from South Africa. *Energy Policy* **1998**, *26*, 207–217. [CrossRef]
14. Quan, S.J.; Li, C. Urban Form and Building Energy Use: A Systematic Review of Measures, Mechanisms, and Methodologies. *Renew. Sustain. Energy Rev.* **2021**, *139*, 110662. [CrossRef]
15. Centre, E.C.J.R.; Perpiñá Castillo, C.; Hormigos Feliu, C.; Dorati, C.; Kakoulaki, G.; Peeters, L.; Quaranta, E.; Taylor, N.; Uihlein, A.; Auteri, D.; et al. *Renewable Energy Production and Potential in EU Rural Areas*; Publications Office of the European Union: Luxembourg, 2024.
16. Shi, Z.; Liang, F.; Pezzuolo, A. Renewable Energy Communities in Rural Areas: A Comprehensive Overview of Current Development, Challenges, and Emerging Trends. *J. Clean. Prod.* **2024**, *484*, 144336. [CrossRef]
17. Furmankiewicz, M.; Hewitt, R.J.; Kazak, J.K. Can Rural Stakeholders Drive the Low-Carbon Transition? Analysis of Climate-Related Activities Planned in Local Development Strategies in Poland. *Renew. Sustain. Energy Rev.* **2021**, *150*, 111419. [CrossRef]
18. Akizu, O.; Bueno, G.; Barcena, I.; Kurt, E.; Topaloğlu, N.; Lopez-Guede, J.M. Contributions of Bottom-up Energy Transitions in Germany: A Case Study Analysis. *Energies* **2018**, *11*, 849. [CrossRef]
19. Bernal, A.; Prin, M.; Ortiz, J. Bottom-Up Energy Transitions: Managing the Rise of Energy Communities in Latin America. Available online: https://www.iea.org/commentaries/bottom-up-energy-transitions-managing-the-rise-of-energy-communities-in-latin-america (accessed on 16 October 2024).

20. Ayodele, T.R. Sustainable Electricity Generation in Rural Communities Using Hybrid Energy System: The Case Study of Ojataye Village. *J. Renew. Energy Smart Grid Technol.* **2016**, *11*, 43–56.
21. Kapsalis, V.; Maduta, C.; Skandalos, N.; Bhuvad, S.S.; D'Agostino, D.; Yang, R.J.; Parker, D.; Karamanis, D. Bottom-up Energy Transition through Rooftop PV Upscaling: Remaining Issues and Emerging Upgrades towards NZEBs at Different Climatic Conditions. *Renew. Sustain. Energy Transit.* **2024**, *5*, 100083. [CrossRef]
22. Mustafa, J.; Almehmadi, F.A.; Alqaed, S.; Sharifpur, M. Building a Sustainable Energy Community: Design and Integrate Variable Renewable Energy Systems for Rural Communities. *Sustainability* **2022**, *14*, 13792. [CrossRef]
23. Salah, S.I.; Eltaweel, M.; Abeykoon, C. Towards a Sustainable Energy Future for Egypt: A Systematic Review of Renewable Energy Sources, Technologies, Challenges, and Recommendations. *Clean. Eng. Technol.* **2022**, *8*, 100497. [CrossRef]
24. Nemr, M.E.; Gebaly, A.E.; Ghazala, A. Optimal Sizing of Standalone PV-Wind Hybrid Energy System in Rural Area North Egypt. *J. Eng. Res.* **2021**, *5*, 46–56. [CrossRef]
25. Ibrahim, K.H.; Hassan, A.Y.; AbdElrazek, A.S.; Saleh, S.M. Economic Analysis of Stand-Alone PV-Battery System Based on New Power Assessment Configuration in Siwa Oasis–Egypt. *Alex. Eng. J.* **2023**, *62*, 181–191. [CrossRef]
26. Ahmad, G.E. Photovoltaic-Powered Rural Zone Family House in Egypt. *Renew. Energy* **2002**, *26*, 379–390. [CrossRef]
27. Abouaiana, A.; Battisti, A. Agile-Transdisciplinary Conceptual Framework for Retrofitting Mediterranean Built Environments. In *Mediterranean Architecture and the Green-Digital Transition, Selected Papers from the World Renewable Energy Congress Med Green Forum 2022, Florence, Italy, 20–22 July 2022*, 1st ed.; Springer: Cham, Switzerland, 2023.
28. Abouaiana, A. Rural Energy Communities as Pillar towards Low Carbon Future in Egypt: Beyond COP27. *Land* **2022**, *11*, 2237. [CrossRef]
29. Kassem, R.; Mahmoud, M.M.; Ibrahim, N.F.; Alkuhayli, A.; Khaled, U.; Beroual, A.; Saleeb, H. A Techno-Economic-Environmental Feasibility Study of Residential Solar Photovoltaic/Biomass Power Generation for Rural Electrification: A Real Case Study. *Sustainability* **2024**, *16*, 2036. [CrossRef]
30. Torabi Moghadam, S.; Toniolo, J.; Mutani, G.; Lombardi, P. A GIS-Statistical Approach for Assessing Built Environment Energy Use at Urban Scale. *Sustain. Cities Soc.* **2018**, *37*, 70–84. [CrossRef]
31. Juan, X.; Weijun, G. Analysis on Energy Consumption of Rural Building Based on Survey in Northern China. *Energy Sustain. Dev.* **2018**, *47*, 34–38. [CrossRef]
32. Sena, B.; Zaki, S.A.; Rijal, H.B.; Alfredo Ardila-Rey, J.; Yusoff, N.M.; Yakub, F.; Ridwan, M.K.; Muhammad-Sukki, F. Determinant Factors of Electricity Consumption for a Malaysian Household Based on a Field Survey. *Sustainability* **2021**, *13*, 818. [CrossRef]
33. Wang, P.; Yang, Y.; Ji, C.; Huang, L. Influence of Built Environment on Building Energy Consumption: A Case Study in Nanjing, China. *Environ. Dev. Sustain.* **2024**, *26*, 5199–5222. [CrossRef]
34. Wang, P.; Yang, Y.; Ji, C.; Huang, L. Positivity and Difference of Influence of Built Environment around Urban Park on Building Energy Consumption. *Sustain. Cities Soc.* **2023**, *89*, 104321. [CrossRef]
35. Osorio, B.; McCullen, N.; Walker, I.; Coley, D. Understanding the Relationship between Energy Consumption and Urban Form. *Athens J. Sci.* **2017**, *4*, 115–141. [CrossRef]
36. Woo, Y.-E.; Cho, G.-H. Impact of the Surrounding Built Environment on Energy Consumption in Mixed-Use Building. *Sustainability* **2018**, *10*, 832. [CrossRef]
37. Antonopoulos, C.; Trusty, A.; Shandas, V. The Role of Building Characteristics, Demographics, and Urban Heat Islands in Shaping Residential Energy Use. *City Environ. Interact.* **2019**, *3*, 100021. [CrossRef]
38. Wang, M.; Yu, H.; Yang, Y.; Jing, R.; Tang, Y.; Li, C. Assessing the Impacts of Urban Morphology Factors on the Energy Performance for Building Stocks Based on a Novel Automatic Generation Framework. *Sustain. Cities Soc.* **2022**, *87*, 104267. [CrossRef]
39. Zhu, S.; Ma, C.; Wu, Z.; Huang, Y.; Liu, X. Exploring the Impact of Urban Morphology on Building Energy Consumption and Outdoor Comfort: A Comparative Study in Hot-Humid Climates. *Buildings* **2024**, *14*, 1381. [CrossRef]
40. Xie, M.; Wang, M.; Zhong, H.; Li, X.; Li, B.; Mendis, T.; Xu, S. The Impact of Urban Morphology on the Building Energy Consumption and Solar Energy Generation Potential of University Dormitory Blocks. *Sustain. Cities Soc.* **2023**, *96*, 104644. [CrossRef]
41. Tam, V.W.Y.; Almeida, L.; Le, K. Energy-Related Occupant Behaviour and Its Implications in Energy Use: A Chronological Review. *Sustainability* **2018**, *10*, 2635. [CrossRef]
42. Heinonen, J.; Junnila, S. Residential Energy Consumption Patterns and the Overall Housing Energy Requirements of Urban and Rural Households in Finland. *Energy Build.* **2014**, *76*, 295–303. [CrossRef]
43. Geng, Y.; Ji, W.; Lin, B.; Hong, J.; Zhu, Y. Building Energy Performance Diagnosis Using Energy Bills and Weather Data. *Energy Build.* **2018**, *172*, 181–191. [CrossRef]
44. Nishida, K.; Takeda, A.; Iwata, S.; Kiho, M.; Nakayama, I. Household Energy Consumption Prediction by Feature Selection of Lifestyle Data. In Proceedings of the 2017 IEEE International Conference on Smart Grid Communications (SmartGridComm), Dresden, Germany, 23–27 October 2017; IEEE: Piscataway, NJ, USA, 2017; pp. 235–240.
45. Alsaadani, S. A Statistical Review of a Decade of Residential Energy Research in Egypt. *Energy Rep.* **2022**, *8*, 95–102. [CrossRef]

46. Nafeaa, S.; Mohamed, A.; Fatouha, M. Assessment of Energy Saving in Residential Buildings Using Energy Efficiency Measures under Cairo Climatic Conditions. *Eng. Res. J.* **2020**, *166*, 320–349. [CrossRef]
47. Attia, S.; Evrard, A.; Gratia, E. Development of Benchmark Models for the Egyptian Residential Buildings Sector. *Appl. Energy* **2012**, *94*, 270–284. [CrossRef]
48. Hegazy, I.R.; Moustafa, W.S.; Akl, M.H. Effect of Building Forms on Energy Consumption in Coastal Rural Communities in Egypt. *J. Urban Regen. Renew.* **2019**, *12*, 269–285. [CrossRef]
49. Salama, R.A.; Hassan, M.M. Implementation of the Rural Family's Household Electrical Energy Conservation Practices Ismailia Governorate. *Alex. Sci. Exch. J.* **2021**, *42*, 1909–1927. [CrossRef]
50. Abdelhay, A.A. The Economic and Social Impacts of Electricity Use in Fayoum. *J. Fac. Arts—Beni Suef Univ.* **2020**, *4*, 423–447. (In Arabic) [CrossRef]
51. Elsabbagh, T.A.A.; Elsayed, N.R.I.; Elshenawy, M.A.M.E. Spatial Analysis of Electricity Consumption Patterns in Industrial Activity in Sharkia Governorate—An Applied Study in Energy Geography Using Geographic Information Systems. *J. Humanit. Lit. Stud. Fac. Arts Kafrelsheikh Univ. Arab Repub. Egypt* **2023**, *28*, 1–69. (In Arabic) [CrossRef]
52. Swedberg, R. *Exploratory Research. The Production of Knowledge: Enhancing Progress in Social Science*; Cambridge University Press: Cambridge, UK, 2020; Volume 2, pp. 17–41.
53. European Commission. *EU-Egypt Joint Statement on Climate, Energy and Green Transition—STATEMENT/22/3703*; European Commission: Brussels, Belgium, 2022.
54. Simões, H.M.; Stanicek, B. *Egypt's Climate Change Policies State of Play Ahead of COP27*; European Parliament: Brussels, Belgium, 2022.
55. Masria, A.; Negm, A.; Iskander, M.; Saavedra, O. Coastal Zone Issues: A Case Study (Egypt). *Procedia Eng.* **2014**, *70*, 1102–1111. [CrossRef]
56. GoE. *Egypt's Second Updated Nationally Determined Contributions*; International Energy Agency: Paris, France, 2023.
57. MPED. *Egypt's 2021 Voluntary National Review*; The Ministry of Planning and Economic Development, Egypt: Cairo, Egypt, 2021.
58. UNDESA. Decent Life (Hayah Karima): Sustainable Rural Communities (Government) #SDGAction43597. Available online: https://sdgs.un.org/partnerships/decent-life-hayah-karima-sustainable-rural-communities#:~:text=%E2%80%9CDecent%20Life%E2%80%9D%20is%20an%20initiative,multidimensional%20poverty%20and%20unemployment%20rates (accessed on 27 September 2024).
59. MOEE. *Second: National Energy Efficiency Action Plan (NEEAP)—2018/2019–2021/2022*; ESMAP-RISE: Cairo, Egypt, 2016. (In Arabic)
60. IEA. Countries & Regions/Africa/Egypt—Energy System of Egypt. Available online: https://www.iea.org/countries/egypt (accessed on 21 January 2025).
61. CAPMAS. *Annual Bulletin of Electricity and Energy Statistics 2019/2020*; Ref. No. 71-22231-2020; Central Agency for Public Mobilization and Statistics: Cairo, Egypt, 2022. (In Arabic)
62. CAPMAS. *Egypt Population, Housing, and Establishments Census 2017*; Central Agency for Public Mobilization and Statistics: Cairo, Egypt, 2017. (In Arabic)
63. CAPMAS. Total Number of Housing Units Implemented by the Sector (Government—Public/Public Works) (In Arabic). Available online: https://www.capmas.gov.eg/Pages/IndicatorsPage.aspx?Ind_id=5803 (accessed on 27 September 2024).
64. CAPMAS. Total Number of Housing Units Implemented for the Social Housing Project (One Million Units) (In Arabic). Available online: https://www.capmas.gov.eg/Pages/IndicatorsPage.aspx?Ind_id=5807 (accessed on 27 September 2024).
65. GOPP. *The Strategic Plan for Urban Development for the Delta Region*; Ministry of Housing, Utilities and Urban Communities represented by the General Organization for Physical Planning: Cairo, Egypt, 2017. (In Arabic)
66. GOPP. *Development Strategy for the Governorates of Egypt, South Upper Egypt Region: Sohag Governorate—Qena Governorate—Aswan Governorates—Red Sea Governorate*; General Authority for Urban Planning, Ministry Of Housing, Utilities & Urban Development, Egypt: Cairo, Egypt, 2008. (In Arabic)
67. CAPMAS. *Annual Bulletin of Electricity and Energy Statistics 2017/2018. Issue: February-2021—Ref. No. 71-22231-2018*; Central Agency for Public Mobilization and Statistics: Cairo, Egypt, 2021.
68. Mahgoub, Y. The Transformation of Traditional Rural Settlements in Egypt. In *Traditional Environments in a New Millennium: Defining Principles and Professional Practices, Proceedings of the Second International Symposium in the Built Environment, Amasya, Turkey, 20–23 June 2001*; IZTECH Library: İzmir, Turkey, 2001.
69. Abouaiana, A. Retrofitting Rural Dwellings in Delta Region to Enhance Climate Change Mitigation in Egypt. *Environ. Clim. Technol.* **2021**, *25*, 136–150. [CrossRef]
70. Open Street Map. Changeset: 116508706 Building Update, Case Study Ahmed Abouaiana. 2022. Available online: https://www.openstreetmap.org/changeset/116508706#map=18/31.182792/30.719345 (accessed on 21 January 2025).
71. Abouaiana, A. *Lasaifar Albalad Settlement Map, Kafr El-Sheikh, Delta Region, Egypt*; Research Gate: Berlin, Germany, 2022.

72. Abouaiana, A. Agile Methodology as a Transdisciplinary Retrofitting Approach for Built Environment in Traditional Settlements in Mediterranean Region. Ph.D. Thesis, Sapienza University of Rome, Rome, Italy, 2022.
73. Youth of Lasifer Youth Center [In Arabic]. Available online: https://www.facebook.com/groups/226713854189879/posts/1640009669526950/ (accessed on 19 January 2025).
74. Abouaiana, A. Proposal to Improve Energy Efficiency in Lasiafar Village [In Arabic]. Available online: https://docs.google.com/forms/d/e/1FAIpQLSfppAiRC_n-j5l2WWsozb8TS_CE3ZWIQRFsyqtTJpUEZdwMZw/viewform?fbclid=IwY2xjawH1gDpleHRuA2FlbQIxMAABHQbc1q3HQa36gSBSQdknNCVgCYejBhvY48FCWX82V-lXh1-jpJOGH7LGPw_aem_ORjBzP96GGXh0gUU3HNvxA (accessed on 19 January 2025).
75. DePoy, E.; Gitlin, L.N. Introduction to Research E-Book: Understanding and Applying Multiple Strategies. In *Introduction to Research E-Book: Understanding and Applying Multiple Strategies*; Elsevier Health Sciences: Amsterdam, The Netherlands, 2019; ISBN 0323612482.
76. Boslaugh, S. *Statistics in a Nutshell: A Desktop Quick Reference*, 2nd ed.; O'Reilly Media, Inc.: Sebastopol, CA, USA, 2012; ISBN 1449361153.
77. Khamis, H. Measures of Association: How to Choose? *J. Diagn. Med. Sonogr.* **2008**, *24*, 155–162. [CrossRef]
78. Hauke, J.; Kossowski, T. Comparison of Values of Pearson's and Spearman's Correlation Coefficient on the Same Sets of Data. *Quaest. Geogr.* **2011**, *30*, 87–93. [CrossRef]
79. Social Science Statistics. Pearson Correlation Coefficient Calculator. Available online: https://www.socscistatistics.com/tests/pearson/ (accessed on 19 January 2025).
80. St, L.; Wold, S. Analysis of Variance (ANOVA). *Chemom. Intell. Lab. Syst.* **1989**, *6*, 259–272.
81. David, V. 3—*Structure as Groups of Objects/Variables*; Valérie David; Elsevier: Amsterdam, The Netherlands, 2017; pp. 35–55. ISBN 978-1-78548-239-7.
82. Ross, A.; Willson, V.L. One-Way Anova. In *Basic and Advanced Statistical Tests: Writing Results Sections and Creating Tables and Figures*; Ross, A., Willson, V.L., Eds.; SensePublishers: Rotterdam, The Netherlands, 2017; pp. 21–24. ISBN 978-94-6351-086-8.
83. IBM. IBM SPSS Software. Available online: https://www.ibm.com/spss (accessed on 19 January 2025).
84. Strunk, K.K.; Mwavita, M. *Design and Analysis in Educational Research: ANOVA Designs in SPSS®*; Routledge: London, UK, 2020; ISBN 0429432798.
85. Sen, S.; Yildirim, I. A Tutorial on How to Conduct Meta-Analysis with IBM SPSS Statistics. *Psych* **2022**, *4*, 640–667. [CrossRef]
86. Feng, W.; Chen, J.; Yang, Y.; Gao, W.; Zhao, Q.; Xing, H.; Yu, S. The Impact of Building Morphology on Energy Use Intensity of High-Rise Residential Clusters: A Case Study of Hangzhou, China. *Buildings* **2024**, *14*, 2245. [CrossRef]
87. PV-HUB. Egypt PV HUB Platform. Available online: https://pv-hub.org/ (accessed on 19 January 2025).
88. El Shenawy, E.T.; Hegazy, A.H.; Abdellatef, M. Design and Optimization of Stand-Alone PV System for Egyptian Rural Communities. *Int. J. Appl. Eng. Res.* **2017**, *12*, 10433–10446.
89. Fetyan, K.M.; Hady, R. Performance Evaluation of On-Grid PV Systems in Egypt. *Water Sci.* **2021**, *35*, 63–70. [CrossRef]
90. PV-HUB. Eligible Companies. Available online: https://pv-hub.org/?page_id=6927 (accessed on 19 January 2025).
91. Gabr, A.Z.; Helal, A.A.; Abbasy, N.H. Economic Evaluation of Rooftop Grid-connected Photovoltaic Systems for Residential Building in Egypt. *Int. Trans. Electr. Energy Syst.* **2020**, *30*, e12379. [CrossRef]
92. PV-HUB. Calculate the Estimated Cost of a Solar Cell System. Available online: https://pv-hub.org/?page_id=7318 (accessed on 19 January 2025).
93. Schweser, K. *SchweserNotes™ 2019 Level I CFA® Book 1: Ethical and Professional Standards and Quantitative Methods*; Kaplan, Inc.: Fort Lauderdale, FL, USA, 2018; ISBN 978-1-4754-7871-6.
94. CFA. Institute Capital Budgeting: Level I | 2021 Curriculum | CFA Program Level I | Corporate Finance. Available online: https://www.cfainstitute.org/en/membership/professional-development/refresher-readings/capital-budgeting-l1 (accessed on 18 June 2021).
95. EgyptERA. Electricity Tariff for the Fiscal Year 2024/2025* Starting from August Consumption. Available online: https://egyptera.org/ar/TarrifAug2024.aspx (accessed on 21 January 2025).
96. Central Bank of Egypt. Exchange Rates' Average Market Rate in EGP. 30 November 2024. Available online: https://www.cbe.org.eg/en/economic-research/statistics/exchange-rates (accessed on 30 November 2024).
97. Abouaiana, A.; Mendonça, P. Retrofitting Dwellings in Traditional Coastal Settlements in Egypt and Portugal Using Nature-Based Solutions and Conventional Thermal Insulation Materials: Technical and Economic Assessment. *J. Archit. Eng.* **2022**, *28*, 05022005. [CrossRef]
98. Dabaieh, M.; Makhlouf, N.N.; Hosny, O.M. Roof Top PV Retrofitting: A Rehabilitation Assessment towards Nearly Zero Energy Buildings in Remote off-Grid Vernacular Settlements in Egypt. *Sol. Energy* **2016**, *123*, 160–173. [CrossRef]
99. El Hassani, S.; Charai, M.; Moussaoui, M.A.; Mezrhab, A. Towards Rural Net-Zero Energy Buildings through Integration of Photovoltaic Systems within Bio-Based Earth Houses: Case Study in Eastern Morocco. *Sol. Energy* **2023**, *259*, 15–29. [CrossRef]

100. Ghafoor, A.; Munir, A. Design and Economics Analysis of an Off-Grid PV System for Household Electrification. *Renew. Sustain. Energy Rev.* **2015**, *42*, 496–502. [CrossRef]
101. Almutairi, K.; Hosseini Dehshiri, S.S.; Hosseini Dehshiri, S.J.; Mostafaeipour, A.; Issakhov, A.; Techato, K. Use of a Hybrid Wind–Solar–Diesel–Battery Energy System to Power Buildings in Remote Areas: A Case Study. *Sustainability* **2021**, *13*, 8764. [CrossRef]
102. Mahmoud, M.M.; Ibrik, I.H. Techno-Economic Feasibility of Energy Supply of Remote Villages in Palestine by PV-Systems, Diesel Generators and Electric Grid. *Renew. Sustain. Energy Rev.* **2006**, *10*, 128–138. [CrossRef]
103. Sukarno, I.; Matsumoto, H.; Susanti, L. Household Lifestyle Effect on Residential Electrical Energy Consumption in Indonesia: On-Site Measurement Methods. *Urban Clim.* **2017**, *20*, 20–32. [CrossRef]
104. He, B.; Yang, L.; Ye, M.; Mou, B.; Zhou, Y. Overview of Rural Building Energy Efficiency in China. *Energy Policy* **2014**, *69*, 385–396. [CrossRef]
105. Steg, L.; Perlaviciute, G.; Van der Werff, E. Understanding the Human Dimensions of a Sustainable Energy Transition. *Front. Psychol.* **2015**, *6*, 805. [CrossRef] [PubMed]
106. Tsuchiya, Y.; Swai, T.A.; Goto, F. Energy Payback Time Analysis and Return on Investment of Off-Grid Photovoltaic Systems in Rural Areas of Tanzania. *Sustain. Energy Technol. Assess.* **2020**, *42*, 100887. [CrossRef]
107. Arif, S.; Taweekun, J.; Ali, H.M.; Ahmed, A.; Bhutto, A.A. Building Resilient Communities: Techno-Economic Assessment of Standalone off-Grid PV Powered Net Zero Energy (NZE) Villages. *Heliyon* **2023**, *9*, e21426. [CrossRef]
108. Nafeh, A. Design and Economic Analysis of a Stand-Alone PV System to Electrify a Remote Area Household in Egypt. *Open Renew. Energy J.* **2009**, *2*, 33–37. [CrossRef]
109. Çamur, H.; Kassem, Y.; Alessi, E. A Techno-Economic Comparative Study of a Grid-Connected Residential Rooftop PV Panel: The Case Study of Nahr El-Bared, Lebanon. *Eng. Technol. Appl. Sci. Res.* **2021**, *11*, 6956–6964. [CrossRef]
110. Rej-Witt, M.; Dębska, L. The Use of a Photovoltaic System in a Single Family House in Poland–Case Study. In Proceedings of the E3S Web of Conferences, Meknes, Morocco, 9–10 December 2021; EDP Sciences: Les Ulis, France, 2022; Volume 336, p. 00010.
111. Abdulhady, H.; Metwally, H. Review of the Factors Affecting the Solar Energy Yield in Egypt. *Egypt. Int. J. Eng. Sci. Technol.* **2020**, *29*, 51–60. [CrossRef]
112. IDSC. *Public Policy Forum*; IDSC: Cairo, Egypt, 2024.
113. Rodrigues, F.; Cardeira, C.; Calado, J.M.F. The Daily and Hourly Energy Consumption and Load Forecasting Using Artificial Neural Network Method: A Case Study Using a Set of 93 Households in Portugal. *Energy Procedia* **2014**, *62*, 220–229. [CrossRef]
114. IEA. *Energy Efficiency 2020*; International Energy Agency: Paris, France, 2020.
115. Lode, M.L.; Felice, A.; Alonso, A.M.; De Silva, J.; Angulo, M.E.; Lowitzsch, J.; Coosemans, T.; Camargo, L.R. Energy Communities in Rural Areas: The Participatory Case Study of Vega de Valcarce, Spain. *Renew. Energy* **2023**, *216*, 119030. [CrossRef]
116. Gouveia, J.P.; Aelenei, L.; Aelenei, D.; Ourives, R.; Bessa, S. Improving the Energy Performance of Public Buildings in the Mediterranean Climate via a Decision Support Tool. *Energies* **2024**, *17*, 1105. [CrossRef]
117. Riva, F.; Tognollo, A.; Gardumi, F.; Colombo, E. Long-Term Energy Planning and Demand Forecast in Remote Areas of Developing Countries: Classification of Case Studies and Insights from a Modelling Perspective. *Energy Strategy Rev.* **2018**, *20*, 71–89. [CrossRef]
118. Naumann, M.; Rudolph, D. Conceptualizing Rural Energy Transitions: Energizing Rural Studies, Ruralizing Energy Research. *J. Rural. Stud.* **2020**, *73*, 97–104. [CrossRef]
119. HBRC. *Egyptian Code for Improving the Efficiency of Energy Use in Buildings, Part 1: Residential Buildings*; ECP306-2005; Ministry of Housing: Cairo, Egypt, 2008. (In Arabic)
120. Javanroodi, K.; Nik, V.M. Impacts of Microclimate Conditions on the Energy Performance of Buildings in Urban Areas. *Buildings* **2019**, *9*, 189. [CrossRef]
121. Xu, Z.; Wang, X.; Tang, S.; Chen, Y.; Yang, Y. Construction and Case Analysis of a Comprehensive Evaluation System for Rural Building Energy Consumption from an Energy–Building–Behavior Composite Perspective. *Sustainability* **2024**, *16*, 6959. [CrossRef]
122. Mishra, P.; Pandey, C.M.; Singh, U.; Gupta, A.; Sahu, C.; Keshri, A. Descriptive Statistics and Normality Tests for Statistical Data. *Ann. Card. Anaesth.* **2019**, *22*, 67–72.
123. Brown, A.M. A New Software for Carrying out One-Way ANOVA Post Hoc Tests. *Comput. Methods Programs Biomed.* **2005**, *79*, 89–95. [CrossRef] [PubMed]

sustainability

MDPI

Article

# Strain Analysis of Membrane Structures for Photovoltaic Integration in Built Environment

**Vuk Milošević [1], Janusz Marchwiński [2] and Elena Lucchi [3,4,*]**

1 Faculty of Civil Engineering and Architecture, University of Niš, 18000 Niš, Serbia; vukamer@yahoo.com
2 Faculty of Architecture, University of Technology and Arts in Applied Sciences in Warsaw, 00-792 Warsaw, Poland; j.marchwinski@wp.pl
3 Dipartimento di Ingegneria Civile e Architettura (DICAr), University of Pavia, 27100 Pavia, Italy
4 Dipartimento di Architettura, Ingegneria delle Costruzione e Ambiente Costruito (ABC), Politecnico di Milano, 20133 Milano, Italy
* Correspondence: elena.lucchi@polimi.it

**Abstract:** The integration of photovoltaic (PV) systems into tensioned membrane structures presents a significant advancement for sustainable applications in the built environment. However, a critical technical challenge remains in the substantial strains induced by external loads, which can compromise both PV efficiency and the structural integrity of the membrane. Current design methodologies prioritize stress, deflection, and ponding analysis of tensioned membranes. Strain behavior of whole structures, a key factor for ensuring long-term performance and compatibility of PV-integrated membranes, has been largely overlooked. This study addresses this gap by examining the whole membrane structure designed for PV integration, with the aim of optimizing the membrane to provide suitable conditions for efficient energy transfer while minimizing membrane strains. For this purpose, it provides a comprehensive strain analysis for full-scale hyperbolic paraboloid (hypar) membrane structures under various design parameters and external loads. Employing the Finite Element Method (FEM) via Sofistik software, the research examines the relationship between-load type, geometry, material properties, and patterning direction of membranes to understand their performance under operational conditions. The findings reveal that strain behavior in tensioned membrane structures is strictly influenced by these parameters. Wind loads generate significantly higher strain values compared to snow loads, with positive strains nearly doubling and negative strains tripling in some configurations. Larger structure sizes and increased curvature amplify strain magnitudes, particularly in parallel patterning, whereas diagonal patterning consistently reduces strain levels. High tensile-strength materials and optimized prestress further reduce strains, although edge type has minimal influence. By systematically analyzing these aspects, this study provides practical design guidelines for enhancing the structural and operational efficiency of PV-integrated tensioned membrane structures in the built environment.

**Keywords:** photovoltaics; tensioned membranes; strains; architectural membranes; flexible photovoltaic

Academic Editors: Marc A. Rosen and Yuanda Cheng

Received: 29 November 2024
Revised: 6 January 2025
Accepted: 24 January 2025
Published: 27 January 2025

**Citation:** Milošević, V.; Marchwiński, J.; Lucchi, E. Strain Analysis of Membrane Structures for Photovoltaic Integration in Built Environment. *Sustainability* **2025**, *17*, 1041. https://doi.org/10.3390/su17031041



## 1. Introduction

Creating a sustainable future for communities is a critical priority in addressing the global challenges of decarbonization and optimization of resources. One of the most promising strategies for achieving sustainability is the integration of renewable energy technologies (RET) into the built environment, an approach that aligns with the Sustainable

Development Goals (SDGs) of the United Nations, particularly SDG 7 (Affordable and Clean Energy) and SDG 11 (Sustainable Cities and Communities) [1]. Among RETs, photovoltaic (PV) systems have gained particular attention in building integration due to their advantages in sustainability, energy generation, flexibility, aesthetic appeal, lightness, and adaptability across diverse applications [2]. The evolution of PV technology progressed beyond conventional systems to include lightweight, flexible, and customizable solutions, opening new opportunities for integration with non-traditional surfaces [3]. Initially dominated by silicon-based solar cells (first generation), PV systems evolved to incorporate thin-film technologies (second generation), such as amorphous silicon (a-Si), cadmium telluride (CdTe), and copper indium gallium selenide (CIGS) [2]. These systems offered reduced material usage and enhanced adaptability to non-traditional surfaces. The third generation is based on a range of advanced technologies, such as dye-sensitized solar cells (DSSCs), organic photovoltaic (OPV) cells, quantum dot solar cells, and perovskite solar cells. They introduced innovation in flexibility, transparency, colors, patterns, and lightweight design. Among them, perovskite and OPV cells are particularly suited for applications in tensioned membrane structures due to their tunable properties and compatibility with flexible substrates [3].

The idea of applying PV to structures appeared in the previous century. Its development is closely related to the improvements in the PV technology. Rigid and heavy PV panels could only be attached to rigid and heavy buildings. Thus, the application of PV systems in the built environment has predominantly focused on heavyweight systems, typically categorized as Building Applied Photovoltaics (BAPV) and Building Integrated Photovoltaics (BIPV) [4]. While these approaches address the integration of solar energy into architecture [5,6], they primarily rely on rigid, heavyweight technologies, limiting their suitability for geometrically complex or delicate structures [7]. The development of flexible PV technologies has significantly expanded the possibilities for the integration of PV systems in architecture due to their ability to conform to non-linear geometries and their lightweight construction, which reduces the load on supporting structures [4,8]. Flexible and lightweight PV technologies have expanded possibilities for architectural integration, particularly in preserving the architectural value of retrofit projects through non-invasive and adaptive designs [5]. The attached PV was a large step forward, but the real integration was achieved when membrane-printed PV were created. Despite advancements in parametric modeling, prototype development, and performance evaluation of flexible PV membranes, their structural integration into tensioned membranes remains underexplored [2].

The critical gap lies in the lack of strain-focused analyses for double-curved, tensioned membranes. Unlike conventional structures, which are often produced through mass customization, PV tensioned membranes exhibit unique mechanical behavior that requires a fully customized approach tailored to the specific project [9]. Their properties result in significant strain variations, which can negatively impact the energy efficiency of even flexible PV systems. Research in this area has primarily focused on PV surface-mounted systems or small-scale samples, neglecting the holistic performance of fully integrated systems under real-world conditions [4]. Furthermore, studies often focus on the technological optimization of PV materials rather than adapting structural designs to accommodate existing PV systems. These aspects hold significant potential for the built environment, as fully integrated PV membranes offer a transformative opportunity by reducing material redundancy [10], improving structural efficiency, and creating multi-functional surfaces that generate energy while maintaining the lightweight and tensile properties characteristics of the membranes [11]. Previous studies have demonstrated the feasibility of integrating flexible PV materials [12], such as thin-film technologies, directly into membrane substrates, offering advantages in adaptability to curved and dynamic geometries. Research in this field has focused on the compatibility of PV designs with advanced parametric models to enhance the energy efficiency of BIPV tensile membrane structures [10]. Tools such as Grasshopper, ANSYS, EASY, and ABAQUS have been employed to analyze the structural and optical performance of these systems [9]. Similarly, prototypes of lightweight membranes incorporating OPV technologies have been developed [13], exploring fabrication, printing, and pattern designs [14]. A critical aspect of this research refers to the optimization of the shape of PV tensile membrane structures through advanced form-finding processes. Dynamic relaxation, finite element analysis [15], and force density techniques [16] have been applied to refine geometries for maximizing energy efficiency [17]. Additionally, dynamic structural analyses are used for evaluating the stability and adaptability of curved shapes in inflatable membrane structures. However, these systems are typically surface-mounted rather than fully integrated into the membrane's structure. Despite these advancements, technical challenges such as mechanical strain, thermal expansion, and material compatibility persist. This gap underscores the need for innovative approaches that move beyond conventional applications, enabling PV systems to become intrinsic to structural design [18]. A need for flexible and lightweight solutions for energy improvement has already been noticed [19]. Studies into the possible integration of PV and membranes have been published. Structural behavior of woven membrane with PV has been analyzed [20]. Membrane-embedded PV systems were examined in the context of structural and electrical performance [21]. Testing of PV-integrated membrane samples showed the relation between the strain and the electrical efficiency [22]. Physical models of ETFE membrane with PV [23] have been used for several studies regarding thermal [24], structural [25], electrical [26], and photothermal performance [27]. A study on the feasibility of flexible PV emergency shelter has been carried out [28]. Life cycle cost analysis [29] and carbon footprint assessment [30] of non-double-curved membrane-integrated PV have been conducted. There have still not been any studies in which the whole PV-integrated textile membrane structure is examined. From the published studies, it can be concluded that PV-membrane integration is possible on a smaller scale.

The integration of PV systems into double-curved tensioned membranes is currently constrained by differences in their mechanical properties. Unlike conventional structural designs, membrane structures are lightweight, span large areas, and function as both load-bearing and covering elements [31]. Their design relies on tension to resist external loads. These unique mechanical characteristics [32] result in geometric changes and large strains, which are incompatible with the design of most PV systems, even flexible ones, as such strains reduce energy conversion efficiency [33]. Flexible PV technologies can tolerate bending but not the large strains inherent to membrane structures [34]. These strains can severely impact their efficiency, creating a critical challenge for successful integration [31]. One of the possible ways to facilitate the integration that has not been scientifically explored so far is the strain analysis of the membrane structures, with the aim of limiting their strains.

This study addresses these gaps by introducing strain analysis into the design methodology for PV-integrated tensioned membranes. Unlike traditional approaches, which emphasize stress, deflection, and ponding, this research prioritizes the relationships between strain behavior and structural parameters such as geometry, material selection, prestress, and patterning. By analyzing how design decisions influence strain under external loads, this research develops actionable guidelines for optimizing the integration of PV systems into tensile membranes. The study limits its scope to textile membrane structures, excluding foil membranes, and prioritizes a methodology that focuses on the relationships and trends among structural parameters. By understanding how design decisions influence strain behavior under external loads, the study aims to provide scientifically grounded recommendations that enable sustainable PV-membrane integration. This integration offers mutual benefits: PV systems benefit from large, adaptable application areas, while membranes access renewable energy for on-site use or grid transfer. Beyond the systems themselves, this research supports broader environmental goals (e.g., SDG 7 "*Affordable and Clean Energy*", SDG 11 "*Sustainable Cities and Communities*", SDG 9 "*Industry, Innovation, and Infrastructure*", and SDG 13 "*Climate Action*"), contributing to global decarbonization and climate neutrality initiatives. By fostering the use of renewable energy and advancing decarbonization efforts in innovative architectural contexts, this research supports the practical implementation of climate neutrality design solutions for the built environment.

## 2. Materials and Methods

The research investigates strain behavior in tensioned membrane structures intended for integration with PV through numerical simulations using Finite Element Method (FEM) software SOFiSTiK 2023 to model tensioned membrane structures and analyze their mechanical properties under various load conditions. In the software, a geometrically non-linear analysis has been carried out. The membrane material has been modeled as linear elastic and orthotropic, according to [35]. Prestress can be defined along the two principal directions of the material and is applied during the form-finding process. Supports are rigid and fixed, except for edge cables. The connection between the cable and the membrane does not allow for the membrane to slip. Quadrilateral membrane finite elements are used for the membrane, and their size is approximately 0.10 × 0.10 m. The elements are formulated to handle in-plane tensile forces, which dominate in tensioned membrane structures, and are compatible with geometrically nonlinear analysis, capturing large deformations accurately. Temperature effects, material creep and aging, and dynamic effects are not considered. Connections between the pieces of the membrane are not modeled; therefore, the model consists of one piece of membrane. This is common practice in membrane structures research and analysis, as cutting patterns are defined at a later stage of development. The study uses a hyperbolic paraboloid (hypar) shape membrane as the base model, which is commonly used in tensile membrane structures. By exploring six key design parameters (size, curvature, material properties, prestress intensity, patterning, and edge types) [36] and their influence on structural strains, the research aims to provide actionable recommendations for reducing strain and ensuring compatibility with strain-sensitive PV systems. A range of values for each parameter is tested to identify trends in the relationships between these parameters and the resulting strains under load. This approach ensures that the limitations of current PV technologies, particularly regarding strains, are respected in the design of PV-integrated membrane structures. Additionally, the findings remain relevant as PV technologies advance and their mechanical properties improve.

Based on this methodology, the study is structured in the following sections:

- Numerical simulation that explains the use of FEM software (in SOFiSTiK 2023) for modeling membrane structures and their loads (Section 2.1).
- Model setup that provides details about the initial model (hypar-shaped membrane, dimensions, material properties) and explains the choice of parameters and applied loads (Section 2.2).
- Variable parameters that describe the six parameters analyzed and their relevance in influencing strain under load (Section 2.3).
- Simulation of multiple scenarios that analyzes a series of 88 different models, each with varying values for the six parameters (Section 2.4).

### 2.1. Numerical Simulation

As a first step in the analysis, an initial numerical model of the membrane structure is created. It is used to test the mesh convergence and as a base point for defining the variable parameters. The initial model is created with a hypar geometry, based on [37], with a square base measuring 6.00 × 6.00 m and a height of 2.00 m. The material properties include a 0.001 m thick membrane with an elastic modulus of 600.00 kN/m, a shear modulus of 30.00 kN/m in both principal directions, and a Poisson's ratio of 0.4. The warp direction runs between the two elevated corners. The membrane is discretized into finite elements of approximately 0.10 × 0.10 m, and the mesh has been verified for convergence under applied loads. The membrane's edges are reinforced with cables of 0.012 m diameter, prestressed to 30.00 kN, and made from steel with an elastic modulus of 205.00 kN/mm$^2$. The analysis was performed using the Direct Parallel Sparse Solver (PARDISO) in SOFiSTiK 2023. The nonlinear solver in SOFiSTiK was configured specifically for the analysis of tensioned membrane structures, accounting for their inherent geometric nonlinearity through a third-order theory. The solver was set with a maximum iteration limit of 90 iterations, ensuring it could handle the iterative nature of solving nonlinear equilibrium equations. To achieve a high level of accuracy, a tolerance of 0.001 was specified, requiring the residuals to fall below 0.1% of their initial values. These settings provided a robust framework for capturing the equilibrium states of the system under applied loading and large deformations. In practice, the solver consistently achieved convergence well within the defined iteration limits, demonstrating its reliability and accuracy in resolving the nonlinear behavior of the structure. The initial model is visualized in Figure 1.

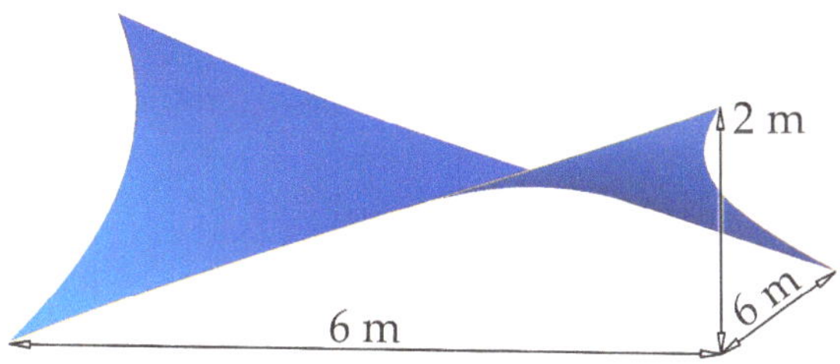


**Figure 1.** Visualization of the initial model (source: Authors' elaboration).

### 2.2. Model Setup

In addition to self-weight and constant prestress, tensioned membrane structures are subjected to external loads. Snow and wind loads are the two most common external loads acting on these structures. While recent research has highlighted the importance of concentrated loads on membranes [38], these are not considered here, as concentrated loads (typically representing a person standing on the membrane) are not relevant for PV-integrated systems.

The snow load is defined with an intensity of 0.60 kN/m$^2$, applied uniformly across the membrane in a vertical downward direction. The wind load is modeled with an intensity of 1.00 kN/m$^2$, acting perpendicular to the membrane surface with uplift action. Both loads are treated as static and are applied independently. The definitions of these loads align with parameters used in previous research [39], without a specific location set for the structure.

### 2.3. Variable Parameters

Starting from the defined initial model, a set of parameters was identified for analysis. The values of these parameters were varied to investigate their influence on membrane strains under load. Six parameters have been proven to affect the behavior of tensile membrane structures and are thus expected to also influence strains. These parameters are:

- Size of the structure (Section 2.3.1).
- Curvature of the structure (Section 2.3.2).

- Membrane material properties (Section 2.3.3).
- Prestress intensity (Section 2.3.4).
- Membrane patterning direction (Section 2.3.5).
- Membrane edges (Section 2.3.6).

The first step in designing any membrane structure is usually choosing the general shape of the membrane. In this study, hypar membranes are analyzed, although other shapes are commonly used, e.g., cone membrane shapes. Once the shape is selected, several other parameters need to be specified for the membrane to be defined. All the important parameters have been set as the six variables in this study, with only a few defined with constant values. All these parameters are determined by the designer of the tensile membrane structure. Designers have the flexibility to adjust these parameters to meet structural, aesthetic, and functional requirements. This flexibility also provides an opportunity to fine-tune them to reduce membrane strains under load, making integration with strain-sensitive PV systems feasible. The precise relationship between these parameters and resulting strains remains insufficiently understood, forming the focus of this research.

#### 2.3.1. Size of the Structure

The original size of the membrane was 6.00 × 6.00 × 2.00 m. In order to analyze the influence of size on strains, different sizes had to be analyzed. While membranes can span large areas, it is not likely that individual membranes with integrated PV will be extremely large, due to the strain limitations of PV. Therefore, the span of 10 m has been selected as the maximal in this research. At the same time, membranes smaller than 2 m would not provide a large area for integration of PV. Therefore, a range of usable and realistic values has been selected. It is important to note that the curvature of the membrane must not change so that the size parameter can be independently examined. Five different sizes of the model were defined, two smaller and two larger than the initial model. These are:

- S1: 2.00 × 2.00 × 0.67 m.
- S2: 4.00 × 4.00 × 1.33 m.
- S3: 6.00 × 6.00 × 2.00 m.
- S4: 8.00 × 8.00 × 2.67 m.
- S5: 10.00 × 10.00 × 3.33 m.

In addition to selecting equal steps of increase, these sizes correspond to real-life applications of tensile membrane (Figure 2).

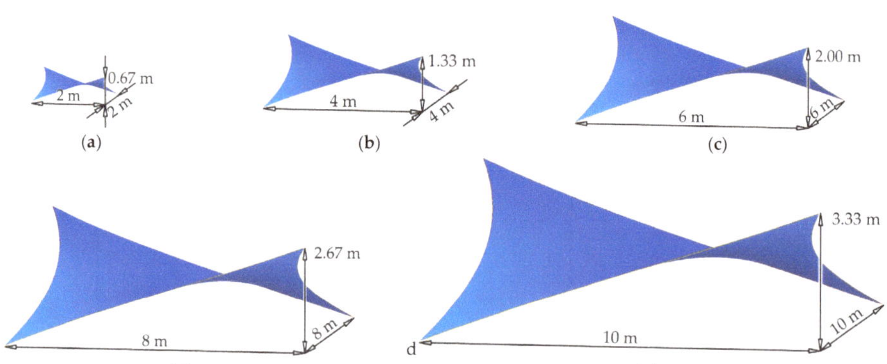


**Figure 2.** Analyzed sizes of the tensioned membrane structure: (**a**) S1 = 2.00 × 2.00 × 0.67 m; (**b**) S2 = 4.00 × 4.00 × 1.33 m; (**c**) S3 = 6.00 × 6.00 × 2.00 m; (**d**) S4 = 8.00 × 8.00 × 2.67 m; (**e**) S5 = 10.00 × 10.00 × 3.33 m (source: Authors' elaboration).

### 2.3.2. Curvature of the Structure

The geometry of the membrane is a big factor in defining how it will react to external loading. There are two simple ways to change the curvature: to increase the size of the base of the structure or to increase the height of the structure. In this research, the second option is selected, as this is what designers of the structure usually prefer. The height of the original structure is 2.00 m. For the upper limit, half of the length of the base side is selected. Flat membranes are not considered here, as they are prone to ponding, so an incremental step size of 0.50 m and a minimal height of 0.50 m was selected for the analysis. Analyzed heights of the structure are:

- H1: 0.50 m.
- H2: 1.00 m.
- H3: 1.50 m.
- H4: 2.00 m.
- H5: 2.50 m.
- H6: 3.00 m.

All the structures have a base of 6.00 × 6.00 m. Visual representation is given in Figure 3.

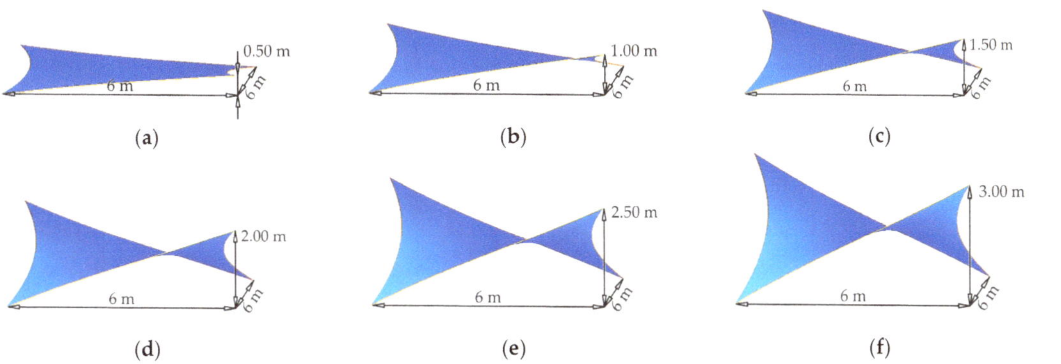


**Figure 3.** Analyzed heights of the tensioned membrane structure: (**a**) H1 = 0.50 m; (**b**) H2 = 1.00 m; (**c**) H3 = 1.50 m; (**d**) H4 = 2.00 m; (**e**) H5 = 2.50 m; (**f**) H6 = 3.00 m (source: Authors' elaboration).

### 2.3.3. Membrane Material Properties

The selection of membrane material for a specific structure depends on various factors, including expected lifespan, aesthetic considerations, cost, expected loads, and resistance to adverse effects. This aspect is one of the key outcomes of structural analysis. In numerical simulations, membrane materials are usually modeled as linear elastic and characterized by the following properties: elastic modulus in warp and weft direction, shear modulus, and Poisson's coefficient. Based on the guidelines provided in [37], six material configurations were taken to cover the whole range of membrane materials used in practice. According to the same reference, the rule states that the shear modulus can be defined in relation to the elastic modulus, as 1/20 of the value of the elastic modulus. The calculation value of Poisson's coefficient has a negligible impact on the simulation results and is therefore treated as a secondary consideration in this analysis. The following values for the membrane material are chosen for the analysis to represent a comprehensive range:

- M1: Elastic modulus 100.00 kN/m, shear modulus 5.00 kN/m, Poisson's coefficient 0.4.
- M2: Elastic modulus 300.00 kN/m, shear modulus 15.00 kN/m, Poisson's coefficient 0.4.
- M3: Elastic modulus 600.00 kN/m, shear modulus 30.00 kN/m, Poisson's coefficient 0.4.
- M4: Elastic modulus 1000.00 kN/m, shear modulus 50.00 kN/m, Poisson's coefficient 0.4.

- M5: Elastic modulus 2000.00 kN/m, shear modulus 100.00 kN/m, Poisson's coefficient 0.4.
- M6: Elastic modulus 5000.00 kN/m, shear modulus 250.00 kN/m, Poisson's coefficient 0.4.

The elastic modulus is considered to be the same in the warp and weft directions of the membrane in all cases.

#### 2.3.4. Prestress Intensity

Tensioned membrane structures rely on maintaining tension within the membrane. This is achieved by applying prestress. Prestress introduces a controlled level of tension to the membrane, ensuring it remains in tension even when subjected to external loads. Its intensity varies depending on design requirements and is determined through structural analysis. Existing recommendations define minimum values [40,41]. Prestress can differ between warp and weft directions of the membrane or remain equal, depending on the material's anisotropic or isotropic properties. In this analysis, they are assumed equal in both warp and weft directions, as the material used has identical properties in both directions. The initial model adopts a prestress value of 3.00 kN/m. To assess the effects of varying prestress levels, four additional values are selected for analysis. The recommended minimum prestress for tensioned membranes is 1.00 kN/m. For this study, an incremental step of 1.00 kN/m is selected. Therefore, the analyzed prestress values are:

- P1: 1.00 kN/m.
- P2: 2.00 kN/m.
- P3: 3.00 kN/m.
- P4: 4.00 kN/m.
- P5: 5.00 kN/m.

It is worth mentioning that the unproportional change in prestress between the warp and weft directions can alter the geometry of the membrane structure. In this study, such changes were intentionally avoided to exclude the influence of curvature on this phase of the analysis. By maintaining equal prestress in both directions, the analysis isolates the effects of prestress intensity without introducing geometric variations that could complicate the interpretation of results.

#### 2.3.5. Membrane Patterning Direction

The issue of patterning is very specific for woven membrane structures due to their orthotropic behavior. This behavior depends on the construction of the membrane, where the fibers are arranged in two perpendicular directions. Thus, the mechanical properties of the membrane vary depending on the orientation. During fabrication, the flat membrane material is cut into strips and joined to create the desired double-curved membrane surface. The material is cut along the warp fibers; therefore, the "patterning direction" effectively defines the orientation of the warp fibers in the membrane. In square-based hyperbolic paraboloid (hypar) membranes, the warp direction can be aligned with either one of the diagonals or one of the sides of the structure. When the membrane material has the same properties in both warp and weft directions, and the prestress intensities are equal, the number of patterning directions is reduced to two. This is because a 90° rotation of the membrane does not produce any structural difference. These two patterning directions are analyzed in this research and shown in Figure 4. They are effectively called:

- DP: diagonal patterning.
- PP: parallel patterning.

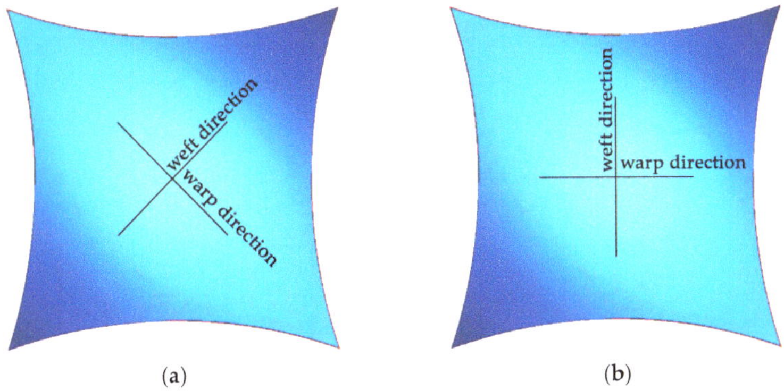


**Figure 4.** Analyzed patterning directions: (**a**) diagonal patterning DP; (**b**) parallel patterning PP (source: Authors' elaboration).

#### 2.3.6. Membrane Edges

One of the main advantages of tensioned membrane structures is their lightness, both in the structural and visual sense. To achieve this, prestressed cables are used as edge supports. However, there are cases where it is necessary to have straight edges of the structure, and this cannot be achieved by using cables. One of the possible examples is when a membrane needs to be attached to a neighboring structure or another membrane. As both edge types are possible and frequently used, it is decided to include both of them in the analysis. They are marked with:

- FE: flexible edges.
- RE: rigid edges.

The difference in models with two types of edges is given in Figure 5. Straight edges are modeled as rigid and fixed. Flexible cable edges are more complex as they affect the overall geometry of the structure. The prestress of the cables should be set in accordance with the prestress of the membrane and the size of the structure. The cable sag from the initial model was kept constant throughout the analysis. This effectively meant that the prestress intensity of edge cables had to be changed alongside the prestress of the membrane and the change in structure size.

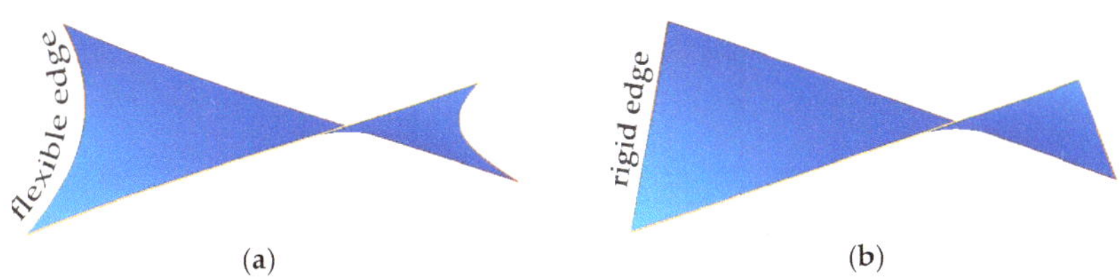


**Figure 5.** Analyzed edges of the membrane: (**a**) flexible edges FE; (**b**) rigid edges RE (source: Authors' elaboration).

### *2.4. Simulation of Multiple Scenarios*

In this research, six variable parameters are defined, each with specific values selected for analysis. The values of these parameters are varied independently, with each variation based on an initial reference model. The parameters and their respective variations are as follows:

- Structure size: 5 analyzed values (Section 2.3.1).
- Height of the structure: 6 analyzed values (Section 2.3.2).

- Membrane material properties: 6 analyzed values (Section 2.3.3).
- Prestress intensity: 5 analyzed values (Section 2.3.4).

These variations result in a total of 22 unique models.

Additionally, two analyzed parameters—type of edges and patterning direction—do not have numerical values but instead feature two distinct configurations each. The combinations of these configurations produce 4 variations (2 edge types × 2 patterning directions). Each of the 22 base models is combined with all 4 configurations of the categorical parameters, resulting in a total of 88 unique models. For each of these models, two different load cases are applied, resulting in a total of 176 analyzed load cases on 88 different models (88 models × 2 load cases). The results are presented in the next section.

## 3. Results

All numerical models of membrane structures have been loaded as planned, and resulting strains were recorded. Strains in two principal directions marked with S-I and S-II have been monitored for each load case. Both the layout of strain values and the extreme values have been analyzed. First, the results for the initial model are presented (Section 3.1). Then, results are given for the following parameters:

- Size of the structure (Section 3.2).
- Curvature of the structure (Section 3.3).
- Membrane material properties (Section 3.4).
- Prestress intensity (Section 3.5).
- Membrane patterning direction (Section 3.6).
- Membrane edges (Section 3.7).

Some graphical results are given in Appendix A.

### 3.1. *Initial Model*

The initial model served as a starting point for all other combinations. Therefore, its results are presented here in detail. Figure 6 shows the strains of the initial model, in both principal directions and under each applied load. The results are given for all points of the membrane. Extreme values can be read from the color scale on the left side of each graph. On the right and bottom side of each graph, a measurement scale is given. Membranes are presented in top view because this view provides the best visibility of results. The downside of this representation is that the height of the given membrane is not visible. In all graphs where the results are presented in this way, the elevated supports are located in the upper left and lower right corners.

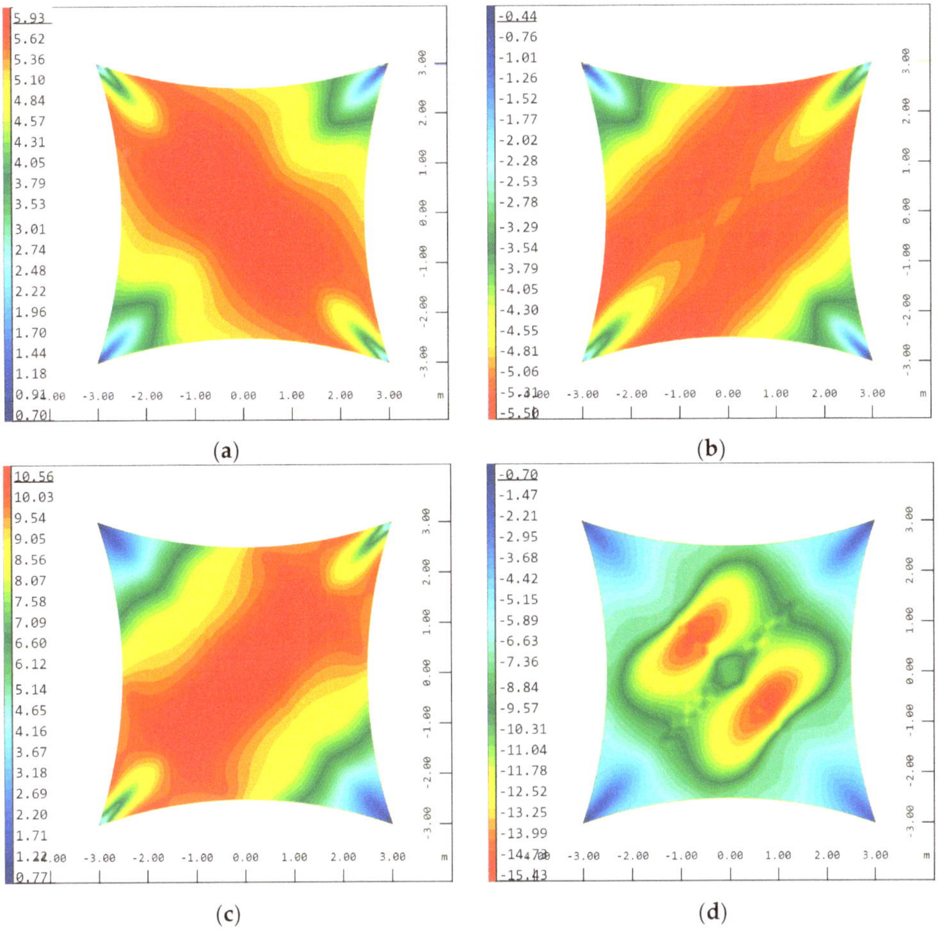


**Figure 6.** Strains (in m/m) of the initial model: (**a**) S-I strains under snow load; (**b**) S-II strains under snow load; (**c**) S-I strains under wind load; (**d**) S-II strains under wind load (source: Authors' elaboration).

### 3.2. Size of the Structure

The first analyzed parameter is the size of the membrane structure. The results for this parameter are given in Figure 7. Unlike the results for the initial model, here the values are not shown for each point of the membrane. Instead, extreme values of strain in S-I and S-II are given for both analyzed loads. In each graph, results for five different models are shown. The models differ in size, while other parameters are kept constant. In this way, the effect of changing size on resulting strains is isolated and can be analyzed independently. The horizontal axis shows the size of the model, while the vertical axis shows the value of strains. Four graphs are presented in Figure 7 because there are four combinations of values for patterning direction and type of edges.

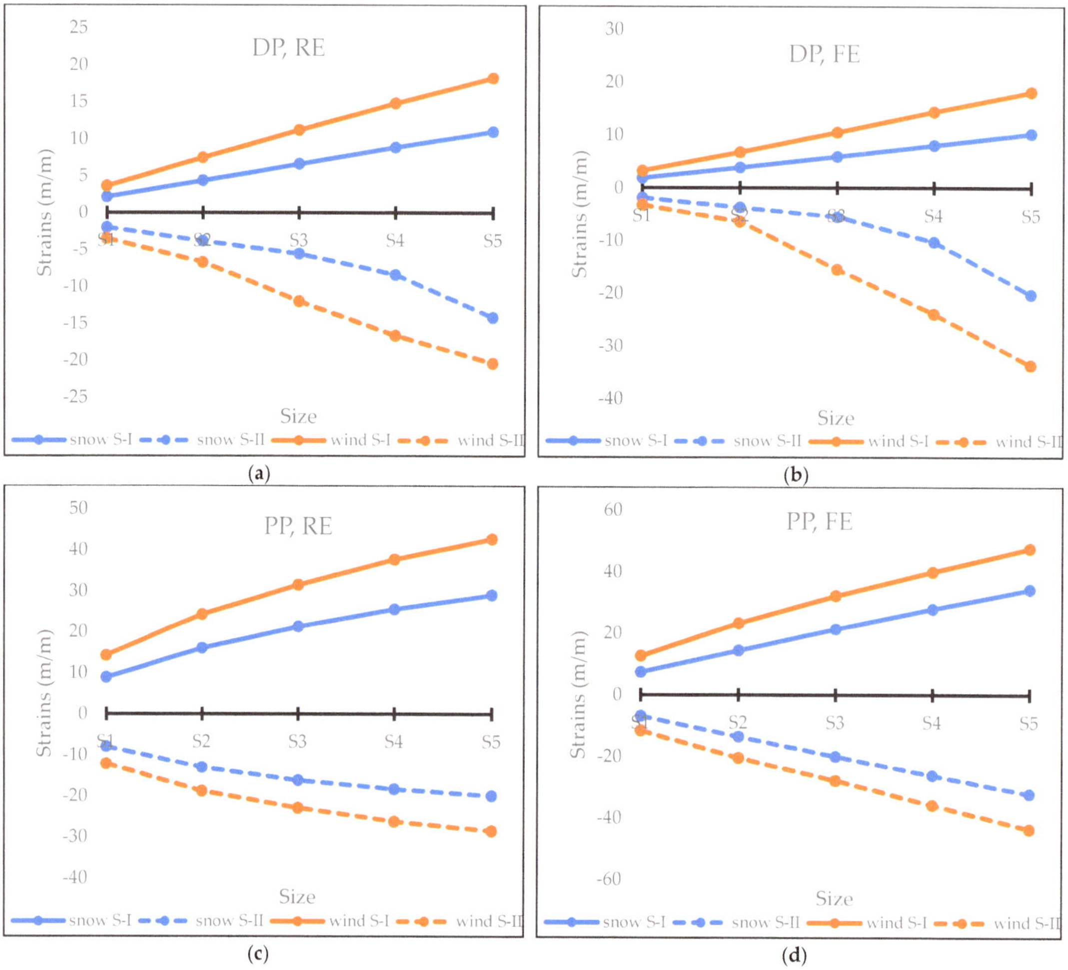


**Figure 7.** Strains dependence on the size of the model: (**a**) models with diagonal patterning and rigid edges; (**b**) models with diagonal patterning and flexible edges; (**c**) models with parallel patterning and rigid edges; (**d**) models with parallel patterning and flexible edges (source: Authors' elaboration).

### 3.3. *Curvature of the Structure*

The curvature of the membrane was the second analyzed parameter. The results are organized in a similar fashion as for the previous parameter. The main difference is that there are six analyzed values for the height of the structure, as opposed to five analyzed sizes. Here the heights are marked on the horizontal axis. The results for different curvatures of the model are given in Figure 8.

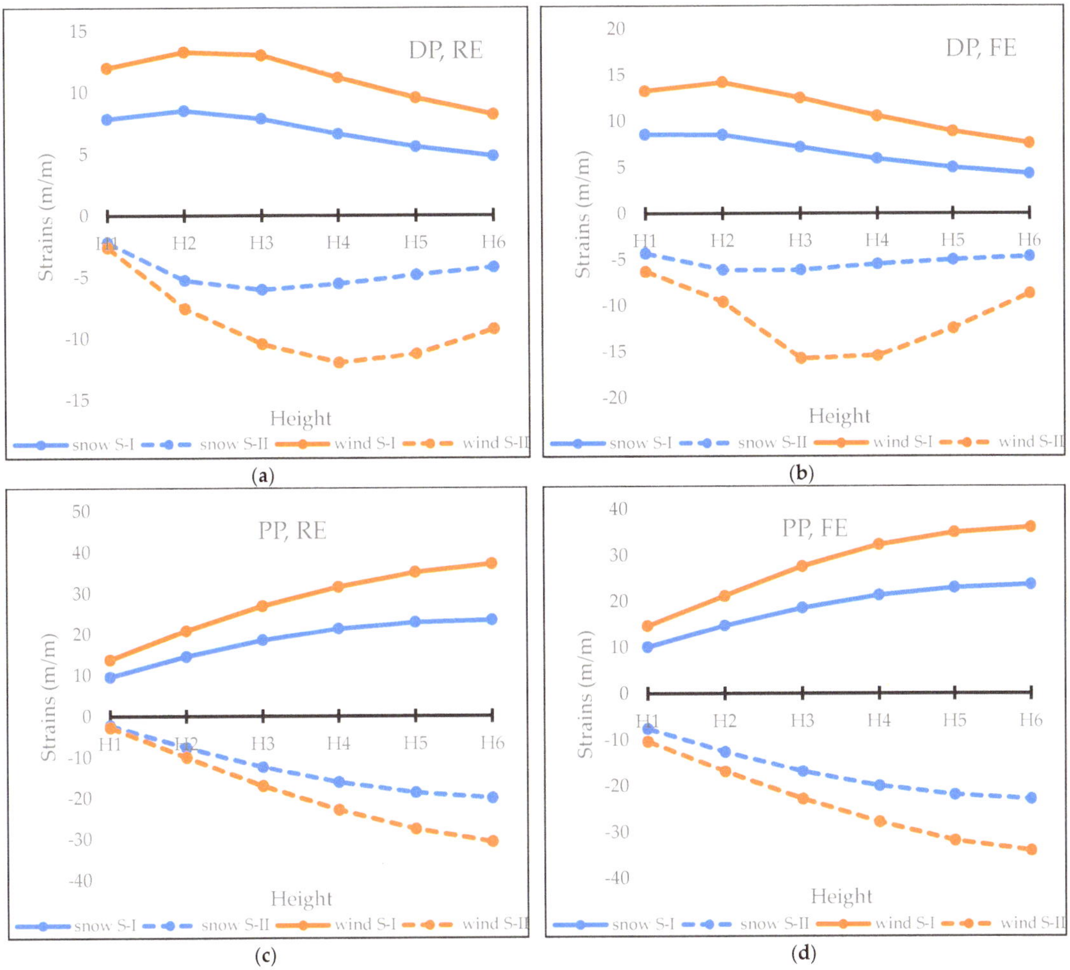


**Figure 8.** Strains dependence on the curvature of the model: (**a**) models with diagonal patterning and rigid edges; (**b**) models with diagonal patterning and flexible edges; (**c**) models with parallel patterning and rigid edges; (**d**) models with parallel patterning and flexible edges (source: Authors' elaboration).

### *3.4. Membrane Material Properties*

The membrane material was the next analyzed parameter. It also has six analyzed values, as the previous one. The results are presented using the same system. Extreme strain values under snow load are given in blue color, while the orange color is used for the results under wind load action. The obtained results are given as dots. Lines are used to connect the measured results in order to make the trends more visible. Lines in S-I are shown as continuous, and the lines in S-II as dashed. The results for different membrane materials are given in Figure 9.

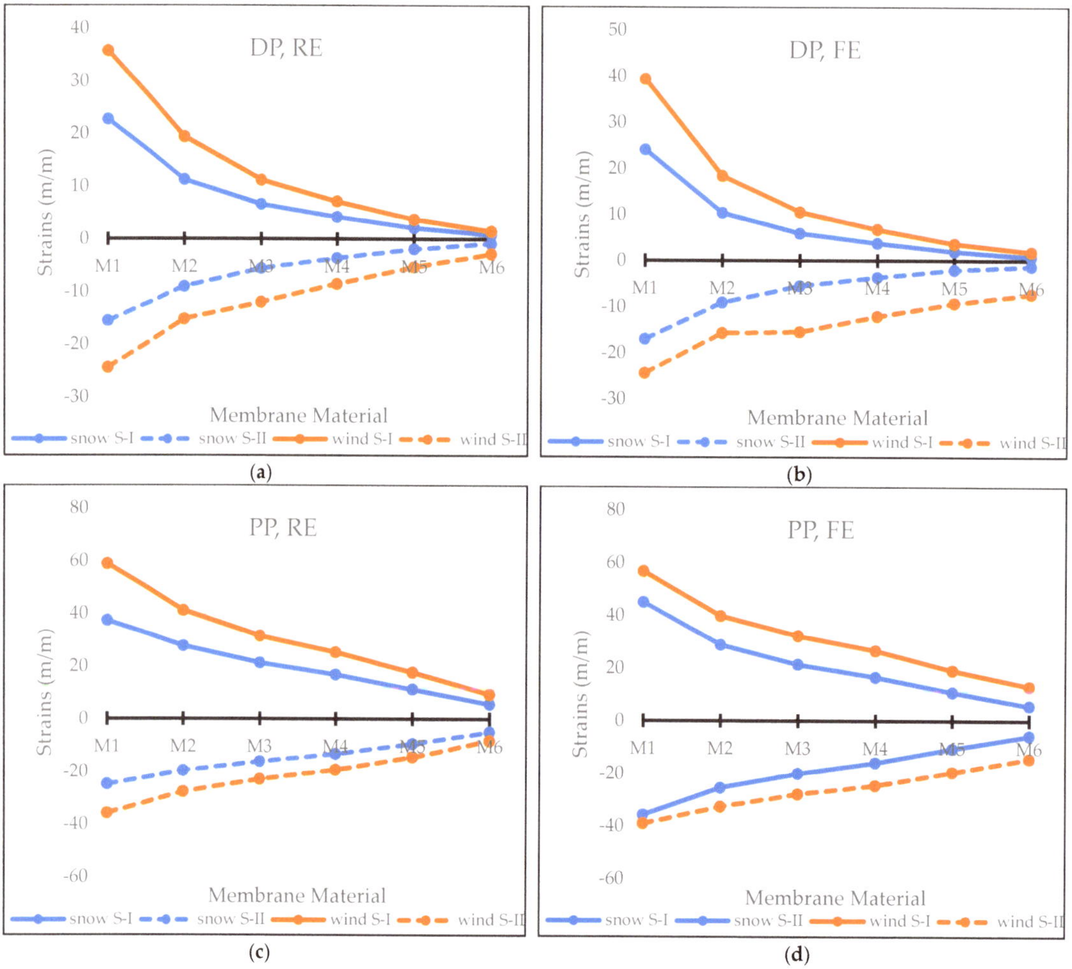


**Figure 9.** Strains dependence on the membrane material: (**a**) models with diagonal patterning and rigid edges; (**b**) models with diagonal patterning and flexible edges; (**c**) models with parallel patterning and rigid edges; (**d**) models with parallel patterning and flexible edges (source: Authors' elaboration).

### 3.5. Prestress Intensity

Prestress is the last analyzed parameter with numerical values. Therefore, the results of the analysis of this parameter are presented in the same way as the previous ones. There are five analyzed values, similar to the parameter of membrane size. The results are given in Figure 10. Once again, four graphs are given, as the results differ significantly in models with different patterning and edges. The analysis of the results is given in the discussion section of the paper.

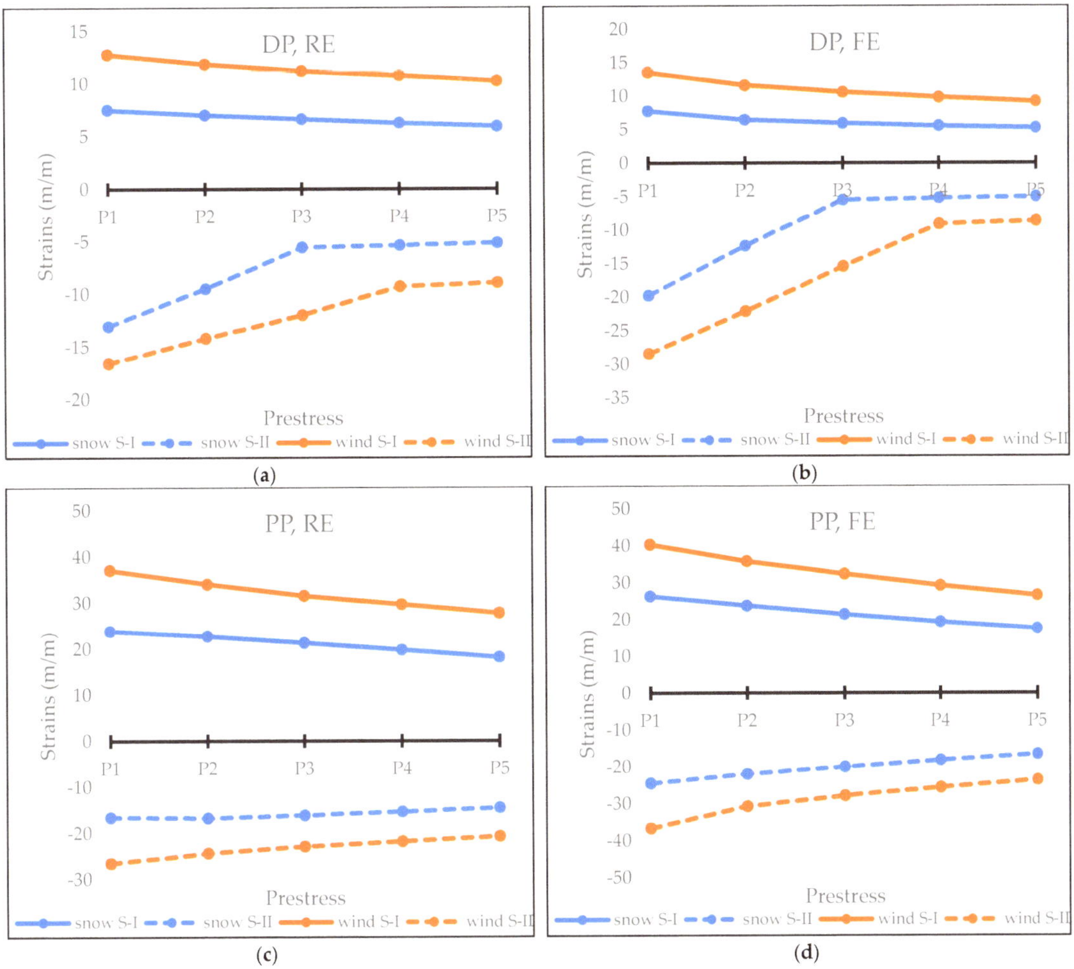


**Figure 10.** Strains dependence on the prestress: (**a**) models with diagonal patterning and rigid edges; (**b**) models with diagonal patterning and flexible edges; (**c**) models with parallel patterning and rigid edges; (**d**) models with parallel patterning and flexible edges (source: Authors' elaboration).

### *3.6. Membrane Patterning Direction*

The results of the parameter of patterning direction have already been presented with the results of all the previous parameters. Since it is obvious that this parameter influences the strains, it is decided that it is needed to show the results for different patterning in more detail. The results are shown for each point on the membrane. The diagonal and parallel patterning are shown one next to another. First, the results under snow load for both S-I and S-II are shown in Figure 11. Then, the results are shown for the wind load in Figure 12.

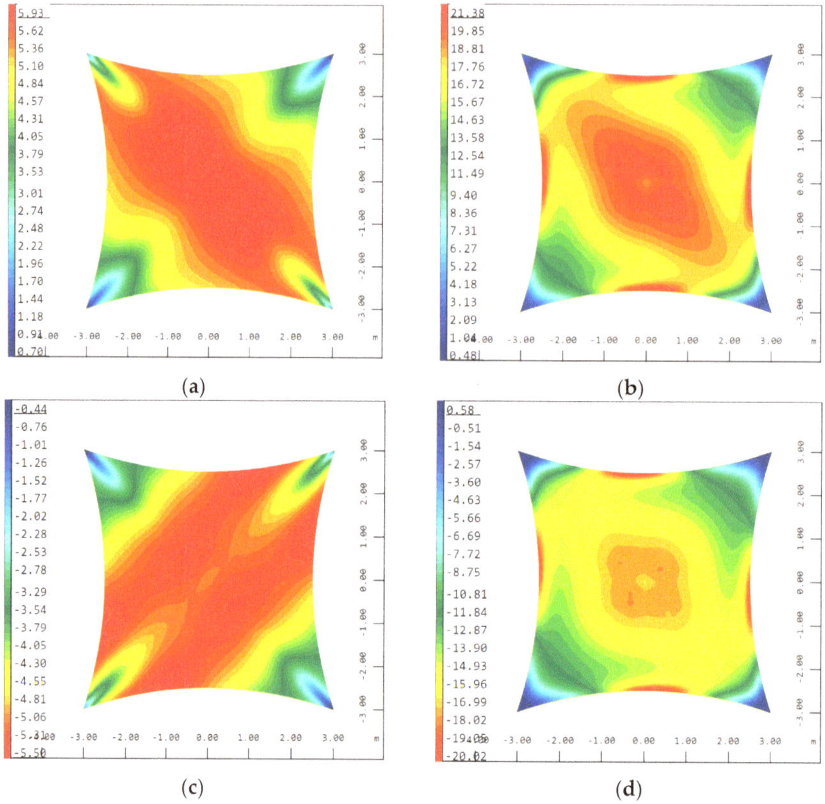


**Figure 11.** Strains (in m/m) of the models under snow load: (**a**) S-I strains under snow load of model with diagonal patterning; (**b**) S-I strains under snow load of model with parallel patterning; (**c**) S-II strains under snow load of model with diagonal patterning; (**d**) S-II strains under snow load of model with parallel patterning (source: Authors' elaboration).

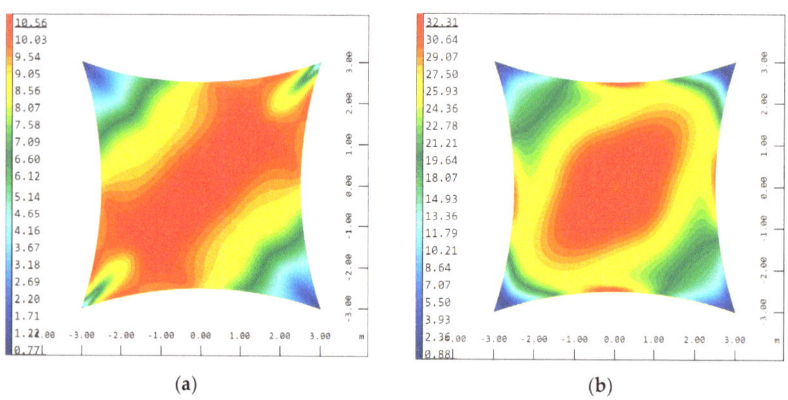


**Figure 12.** *Cont.*

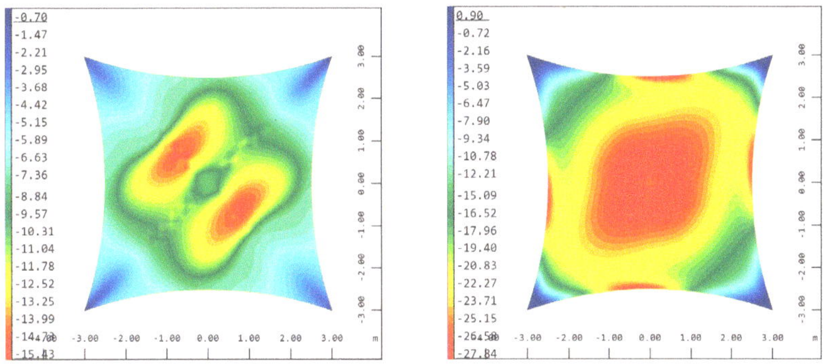

(c) (d)

**Figure 12.** Strains (in m/m) of the models under wind load: (**a**) S-I strains under snow load of model with diagonal patterning; (**b**) S-I strains under snow load of model with parallel patterning; (**c**) S-II strains under snow load of model with diagonal patterning; (**d**) S-II strains under snow load of model with parallel patterning (source: Authors' elaboration).

### 3.7. Membrane Edges

The results for the two analyzed edge types have also been already shown with the first five parameters. However, due to the different shapes the membrane takes, the results are presented here in more depth. Figures 13 and 14 show models with flexible and rigid edges parallelly. Figure 13 presents the result under snow load and Figure 14 under wind load. In each graph, the most extreme value is shown in red color, regardless of whether its value is positive or negative.

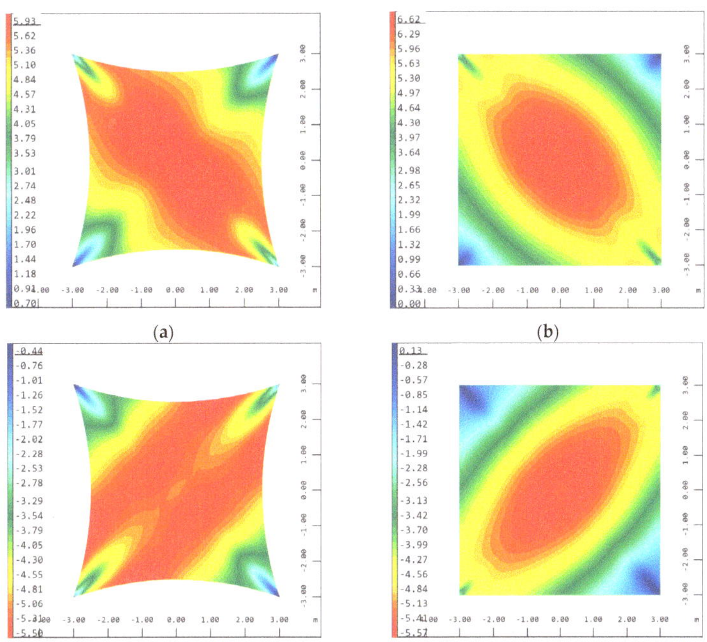

(a) (b) (c) (d)

**Figure 13.** Strains (in m/m) of the models under snow load: (**a**) S-I strains under snow load of model with flexible edges; (**b**) S-I strains under snow load of model with rigid edges; (**c**) S-II strains under snow load of model with flexible edges; (**d**) S-II strains under snow load of model with rigid edges (source: Authors' elaboration).

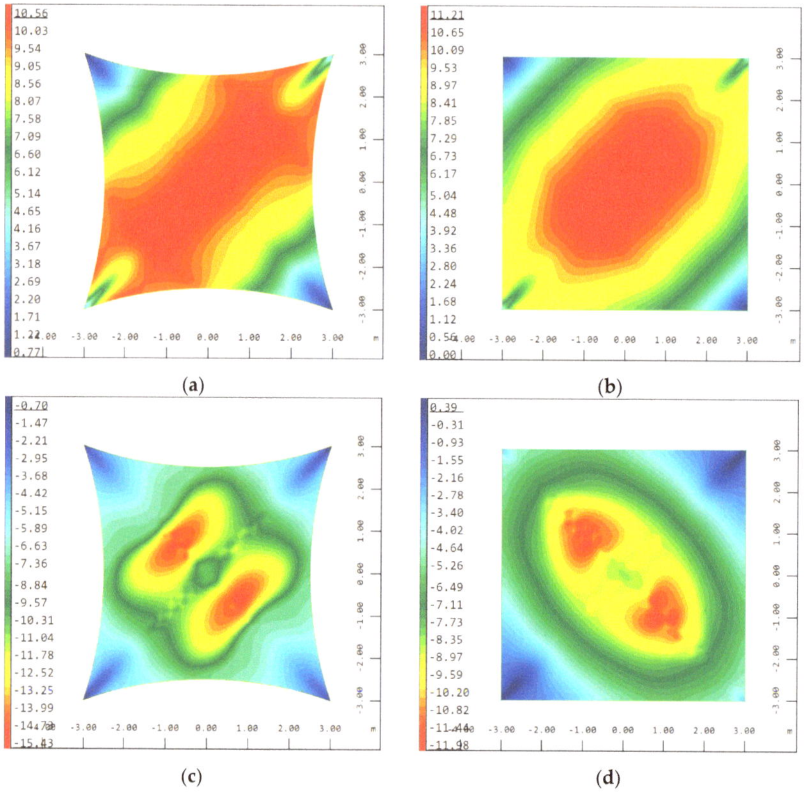


**Figure 14.** Strains (in m/m) of the models under wind load: (**a**) S-I strains under snow load of model with rigid edges; (**b**) S-I strains under snow load of model with rigid edges; (**c**) S-II strains under snow load of model with rigid edges; (**d**) S-II strains under snow load of model with rigid edges (source: Authors' elaboration).

## 4. Discussion

The results are discussed in this section in the same order as presented in the previous section. For certain aspects of the discussion, data provided in Appendix A are referenced.

First, the action of snow load is directly proportional to the vertical projection of the membrane. This relationship applies both in the numerical simulations and in real-world applications [42]. Specifically, changes in the vertical projection area, influenced by parameters such as size and edge type, result in proportional changes to the total snow load acting on the membrane. In contrast, wind load behavior differs as it is related to the surface area of the membrane. Consequently, wind load varies not only with parameters like size and edge type but also with the curvature of the structure.

Another important aspect is to note that all strains presented in this study represent the transition from the prestressed state to the loaded state of the membrane. The prestressing process itself induces strains that are not included in this analysis. If PV systems are installed after the membrane has been prestressed, these initial strains can be disregarded.

The mechanical behavior of woven membranes is extremely complex and cannot yet be fully replicated through numerical simulations. Therefore, the strain values should be interpreted as indicative guidelines rather than exact representations of real-world behavior. Although some of the resulting strains in this research are too high for practical application, they provide us with important information about the behavior of membranes.

The primary contribution of this research is in identifying the relations between the analyzed parameters and the resulting strains under load. The findings are not intended to replicate physical testing results but rather to establish a framework for understanding these interactions. The ultimate output of this research is a set of design guidelines for future PV-integrated membrane structures, derived from the discussion that follows, which explores the influence of various parameters on strain behavior under load.

The initial model shows slightly higher positive than negative strains under snow load. For positive strains, the maximum values are concentrated in the central part of the membrane, and high strains stretch towards the high supports (Figure 6). Strains in the S-I principal direction are the lowest close to lower supports. In the S-II direction, the area of large negative strains is oriented from one low support towards the other. Maximum values occur next to the edges, close to low corners of the membrane. The lowest strains in S-II are close to high corners. Under wind load, negative strains have larger absolute values than positive strains. High positive strains are located in the central part of the membrane and spread towards the low supports. Minimal strain values in S-I are next to the high corners, thus making this arrangement resemble strains in S-I under snow load, rotated by 90°. Areas of high negative strains are very concentrated and located close to the middle of the membrane, towards the high corners of the structure. The lowest negative strains are around all four supports. The behavior of strains can be explained physically: under snow load, the fibers connecting the high supports get elongated due to the downward load. Fibers parallel to the diagonal connecting the low supports have an upward sag, and their length decreases under snow load. In contrast, wind loads act upward, causing significant stretching of fibers previously under reduced tension. The peculiar strains in S-II are explained by the wind load acting perpendicular to the membrane, thus creating the largest deflections in the middle and at the half distance between the middle and the high supports. Extreme negative strains occur at the midpoint between these points, as this is the location where the membrane will shorten the most. Extreme values of strains under wind load are significantly larger compared to the snow load. Positive strains are almost twice as high, while negative strains are almost three times bigger.

The size of the area to be covered is usually given to the designer, but this is not necessarily equal to the size of the structure. The designer has the freedom to divide this area into several smaller structures, in case it is justified by some reason. The question posed in this research is whether the size of the membrane structure has any impact on strains under load, and if so, what is the relation between these two? The size of the model proved to have a large influence on the value of strains, based on the obtained results provided in Figure 7. The first observation that can be made is that in all analyzed cases, the increase in the size of the structure leads to an increase in strains. This is valid for both applied loads. The reason for this is that not only does the span increase, but also the total applied load, as explained earlier in this section. The next trend that can be noted is that strains in both principal directions are always more extreme under wind load. This is explained by the higher intensity of wind load action. In models with parallel patterning, positive strains have higher absolute values compared to negative strains, regardless of the edge type. Extreme strains in models with diagonal patterning are visibly lower than the strains of models with parallel patterning. This shows the significant influence of patterning direction on strains and will be checked on the rest of the obtained results. Figure A1 in Appendix A shows the layout of strains under snow load of models with different sizes. The size does not affect the strain's layout significantly. It can be observed that with the increase in size, the high strains, including both positive and negative strains, get more localized. This is more emphasized in S-II, where the position of negative strains does not change with the increase in size, but the extreme value gets higher than the rest of the strains, therefore the change of color on the biggest part of the largest models. Wind load results shown in Figure A2 show large positive strains along the diagonal connecting the two low supports. Maximal negative strains are located near the center of the membrane, symmetrically moved towards the corners.

Next analyzed was the curvature of the membrane, defined by different heights of the structure. In the case of this parameter, the behavior of the membrane differs considerably for models with different patterning directions. Models with parallel patterning react to increases in height with the increase in extreme positive and negative strains under both loads, as can be seen in Figure 8. This is not the case with models with diagonal patterning. At first, they show an increase in strains with the increase in height; however, further height increases lead to a reduction of strains. This behavior is not simultaneous in S-I and S-II strains. As it happens under both loads, it cannot be explained by differences in loads but rather by the combined effect of two factors. One is the increase in curvature, which should help the membrane resist external loads, and the other is the increase in the fiber length caused by the increase in height. The second factor is not as important in membranes with parallel patterning, since in this case the fibers are straight and not curved. Two observations made during the analysis of the first parameter are confirmed here: strains under wind load are more extreme than under snow load, and strains of models with diagonal patterning are lower than those with parallel patterning. The change of the edge type does not result in important changes in the extreme strain values. Figure A3 shows an interesting behavior of S-II strains under snow load. The position of the maximal negative strains moves with the increase in curvature, starting from the position next to the cables and close to the high corners, towards the center of the membrane, and then back next to the edge cables, but this time close to the low corners of the membrane. A similar movement of the position of maximal negative strains is observed under wind load, presented in Figure A4. In this case, the location of maximal negative strains moves from the edge close to the low corners towards the center of the membrane, and with an increase in curvature, it gets closer to the center.

While there are several factors influencing the choice of membrane material for any specific structure, it is the structural analysis that sets the minimum values for tensile strength. More resistant materials can be chosen for reasons other than structural. One of the possible reasons to select a material with higher mechanical properties may be the need to facilitate the integration of PV. This research aimed to answer whether this is justified, and the results showed clear trends. The change of membrane material results in a very significant change in maximal strains (Figure 9). With the increase in the membrane material mechanical values, the strains drop. This is especially noticeable in models with diagonal patterning. Among the graphs presented in Figures A5 and A6, especially interesting are the ones marked as (f), showing the S-I strains of a model with the highest values of membrane material. Here, the position of the maximal strains is unlike any other presented model, once again showing the uniqueness and importance of the parameter of membrane material. The results under snow and wind load are similar in layout, only rotated by 90°. The position of high strains moves with the increase in membrane material values from the area near high corners to the center of the membrane and then to the area near the low corners in S-I. The converse trajectory is present in the S-II direction. Once again, the change of edge type does not result in noteworthy changes of strains. Strain in models with diagonal patterning is here as well, lower compared to the same models with parallel patterning.

The increase in prestress intensity is a parameter usually defined by the structural needs. When structural analysis shows that the membrane will lose tension under expected external load, the pretension value is increased. The results obtained by this research show that the increase in prestress intensity can also help in reducing strains under load (Figure 10). The largest positive strains under both loads show almost linear behavior with changes in the prestress. Their value is reduced slightly with the increase in prestress. Negative strains in models with diagonal patterning show more rapid decrease up until a point where further increase in prestress has negligible effect on them. In models with parallel patterning, negative strains show almost linear decrease with the increase in prestress. Figure A7 shows that under snow load the position of high strains does not change significantly with the increase in prestress value. However, the areas of high strains expand in both the S-I and S-II directions, making a larger part of the membrane more strained, although the value of maximal strains decreases gradually. Similar behavior occurs under wind load (Figure A8). The influence of edge type shows a somewhat larger impact in models with low prestress values, especially in the values of negative strains. In other cases, the effects of changing the edge type are very low.

Patterning is a necessity in woven tensile membrane structures. The direction of patterning can have a subtle esthetical impact on the appearance of the structure. It also affects the structural properties, such as deflections under load. Sometimes the patterning orientation is defined in a certain way to ease the installation of the membrane. Results shown in Figures 11 and 12 provide results for the models that differ only in patterning direction. The presented results show significantly larger strains in models with parallel patterning. This is valid for both positive and negative strains. Under snow load strains, the parallel patterning is almost 4 times larger. Under wind load, positive strains are three times larger, and negative almost double with parallel patterning compared to the diagonal. Figures 11 and 12 show that the behavior of the membrane is different under the two patterning directions, and this is depicted by differences in strain layouts. When using parallel patterning, extreme strains under both loads are close to the middle of the edges and, under wind load, additionally in the central part of the membrane. In models with diagonal patterning, principal strains are mostly aligned with the warp and weft direction. Textile membranes have fibers in two orthogonal directions. These fibers are load-carrying, and due to such structure of the membrane, it is numerically modeled as orthotropic. The fibers provide higher strength in their directions, compared to the off-fiber directions. The largest strains occur when the load is applied diagonally to the fibers. This should not be confused with diagonal patterning of the membrane, where the fibers are aligned with the diagonals of the structure. Diagonal patterning ensures that fibers are in the direction connecting the corners. When corner supports are connected by a stronger direction of the membrane, it is reasonable to expect lower strains. In addition, when fibers are connecting the corners of the hypar membrane, they have curvature, unlike in parallel patterning. This amplifies the lower straining of the membrane under load compared to parallel patterning, which is in line with the obtained results.

Different membrane edges affect the geometry and structural properties. Among the affected structural properties are the strains under loading. However, the results have shown that, compared to other analyzed parameters, this one has a low impact on the strain values. If we consider the strains given in Figures 13 and 14, we can justify this conclusion. On the right side of the figures, the model with rigid edges is presented to compare it easily with the one with flexible edges. Only small differences in strain values can be noted when analyzing these results. In addition, the layout of the strains is very similar, except for the larger membrane area in models with rigid edges.

*Guidelines for Strain Analysis of Photovoltaics-Integrated Tensioned Membrane Structures*

Based on the findings of this study, the following design recommendations are classified according to different design stages for guiding the integration of PV systems into tensioned membrane structures:

- Strain management should be a critical focus during the conceptual and design stages to ensure compatibility between the PV system and the membrane structure. The following elements should be considered:
    1. Strain limits determination:
        a. Establish strain limit values specific to the PV system to be integrated, based on its tolerances for mechanical deformation and electrical efficiency degradation.
        b. Define acceptable strain thresholds by quantifying the impact of strain on the electrical output and lifespan of the PV system.
        c. Obtain limit values for strains for the PV system that will be integrated with the membrane.
    2. Strain-focused design analysis:
        a. Use strain analysis in the early design phase to evaluate the impact of anticipated external loads and optimize structural performance.
        b. If predicted strain values exceed the limits for the PV system, apply mitigation strategies listed below, such as geometry adjustments, material upgrades, or prestress optimization.
- Geometry of the membrane structure significantly affects strain behavior and should be carefully designed during the design stage, considering:
    1. Size reduction as larger membrane structures generate higher strains.
    2. Curvature control:
        a. Adjust the membrane curvature by changing the height of the structure.
        b. Increase in curvature does not result in unambiguous change in maximal strains. For membranes with parallel patterning, higher curvature tends to increase strain. For membranes with diagonal patterning, the relationship is less pronounced and may allow for optimized curvature without excessive strain.
- Patterning and edge design influence strain distribution but require several considerations in the design stage:
    1. Patterning direction:
        a. Switch from parallel to diagonal patterning to achieve significant reductions in strain values under external loads.
        b. Diagonal patterning distributes forces more evenly, reducing stress concentrations and mitigating extreme strain values.
    2. Edge type selection:
        a. The edge type (rigid or flexible) has minimal impact on strain levels and can be chosen based on other design or structural factors such as ease of assembly or aesthetic considerations.
- Material selection is fundamental in the design stage to ensure strain mitigation:
    1. Tensile strength requirements:
        a. Set the minimal tensile strength according to the typical structural analysis of the membrane structure based on stresses and deflections.

2. Elastic and shear modulus:
    a. Use membrane materials with higher elastic and shear moduli to effectively reduce strains under load.

- Prestress optimization is essential in the design stage to ensure strain management:
    1. Prestress intensities:
        a. Higher prestress intensities generally reduce strain magnitudes, particularly for positive strains, which show near-linear reduction with increased prestress.
        b. Negative strains may plateau beyond a certain prestress value, indicating diminishing returns at high prestress intensities.
    2. Balancing prestress:
        a. Optimize prestress to balance strain reduction without overloading the membrane or creating unnecessary material stresses.

Also, the comparative influence of structural parameters shows the importance of different factors (Table 1).

**Table 1.** Comparative influence of structural parameters (source: Authors' elaboration).

| Factors | Recommendations |
|---|---|
| Most significant | Material selection has the greatest influence on reducing strain levels, making it a primary focus in design optimization |
| | Changing patterning direction from parallel to diagonal has a substantial impact on strain mitigation |
| | Reducing membrane size is highly effective in lowering strain levels |
| Moderate impact | Increasing prestress levels provides strain reduction but shows diminishing returns beyond certain thresholds |
| | Decreasing curvature somewhat reduces strain in parallel-patterned membranes but has limited influence in diagonal-patterned membranes |
| Least significant | Edge type has a negligible effect on strain values, allowing flexibility in selection based on secondary criteria |

## 5. Conclusions

This research analyzed the strains experienced by tensioned membrane structures under load, aiming to support the integration of photovoltaic systems onto these structures. Six key parameters—size; curvature; membrane material properties; prestress intensity; patterning direction; and edge type—were investigated for their influence on strain behavior under two calculation loads. Conducted as a parametric study using numerical models, the research analyzed a total of 88 configurations, with two load cases applied to each, resulting in 176 simulations. The objective was to establish the relationship between these parameters and strain behavior, culminating in actionable design recommendations for optimizing membrane structures for photovoltaic (PV) integration.

The study's significance aligns with broader sustainability goals, particularly the integration of renewable energy technologies into the built environment as a pathway to achieving the United Nations' Sustainable Development Goals. Lightweight and flexible PV technologies have revolutionized the potential for PV integration into complex architectural forms. However, the mechanical incompatibilities between tensioned membranes and PV systems—specifically the large strains experienced by the former—pose significant challenges for integration. This research addresses this gap by identifying design strategies to manage strains and enable the successful coupling of these technologies.

The results of this research provide critical insights into strain management for membrane structures. Larger membranes experience higher strain under load, so reducing their size can mitigate this issue. Although smaller structures may require more individual membranes to cover a given area, impacting the design, the designers are advised to use smaller membranes to reduce the strains. While curvature influences stresses and deflections, its role in strain management is complex. Only parallel-patterned membranes demonstrate a predictable reduction in strain with decreased curvature, limiting their reliability as a strain management tool. Higher elastic and shear moduli of the membrane significantly reduce strains. While this method has a substantial impact, it comes with increased material costs. However, as it does not alter the aesthetic design, it is often preferred for projects prioritizing visual integrity. Therefore, the use of membrane materials with higher mechanical properties is recommended for PV integration. Increased prestress reduces strains but must be carefully calibrated to avoid structural deterioration. While effective, the strain reduction achieved is lower compared to size reduction or increase in material properties. Diagonal patterning is highly effective in minimizing strain, with no additional cost or significant aesthetic compromise, making it a recommended option over parallel patterning. Thus, diagonal patterning should be the first choice for designers when designing PV-integrated membrane structures. The choice between rigid and flexible edges has minimal influence on strain and can therefore be selected based on other design or functional priorities without significant concern for strain behavior.

The integration of these findings into design processes offers new solutions for optimizing membrane structures to support PV systems. The research refrained from defining a specific strain threshold to support the ongoing advancements in flexible PV technology. Instead, it focused on identifying trends and relationships, ensuring the findings remain relevant as PV systems evolve.

This research highlights the potential of integrating PV systems into membrane structures, demonstrating how these designs can simultaneously serve as adaptable energy-generating surfaces and sustainable architectural solutions. This symbiotic relationship not only advances technical and design capabilities but also supports global efforts toward decarbonization and climate neutrality. By offering scientifically grounded recommendations, this research facilitates innovative design solutions for the built environment, bridging the gap between renewable energy integration and architectural innovation. It positions tensioned membrane structures as a sustainable solution for the energy demands of future cities, contributing to a more resilient and environmentally conscious built environment.

**Author Contributions:** Conceptualization, J.M. and V.M.; methodology, V.M.; software, V.M.; writing—original draft preparation, V.M., E.L. and J.M.; writing—review and editing, V.M., E.L. and J.M.; visualization, V.M.; supervision, E.L.; funding acquisition, E.L. All authors have read and agreed to the published version of the manuscript.

**Funding:** This research received no external funding.

**Institutional Review Board Statement:** Not applicable.

**Informed Consent Statement:** Not applicable.

**Data Availability Statement:** Data are available under request.

**Conflicts of Interest:** The authors declare no conflicts of interest.

## Appendix A

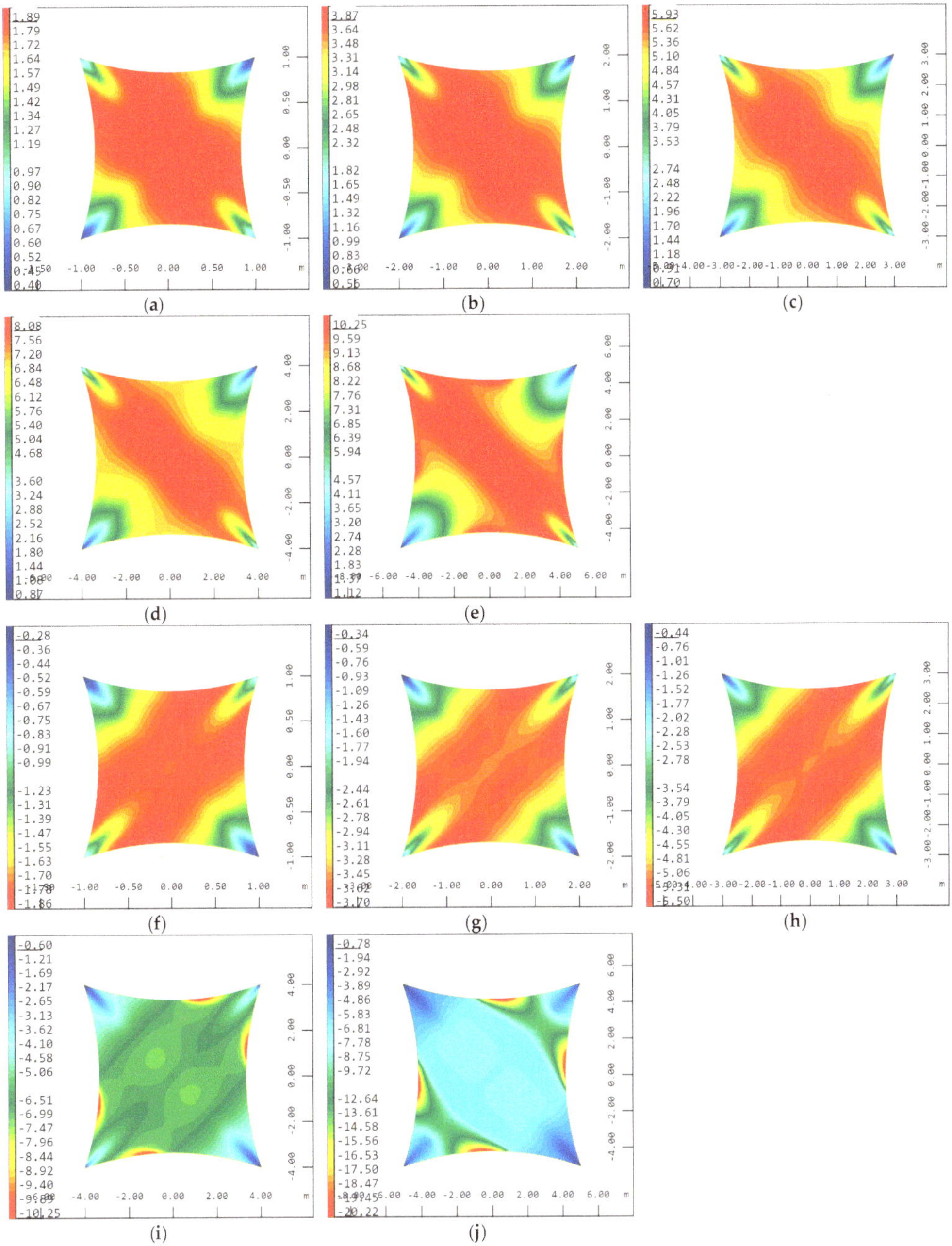


**Figure A1.** Strains (in m/m) of the models with DP, FE, M3, H4, and P3 under snow load: (**a**) S-I, S1; (**b**) S-I, S2; (**c**) S-I, S3; (**d**) S-I, S4; (**e**) S-I, S5; (**f**) S-II, S1; (**g**) S-II, S2; (**h**) S-II, S3; (**i**) S-II, S4; (**j**) S-II, S5 (source: Authors' elaboration).

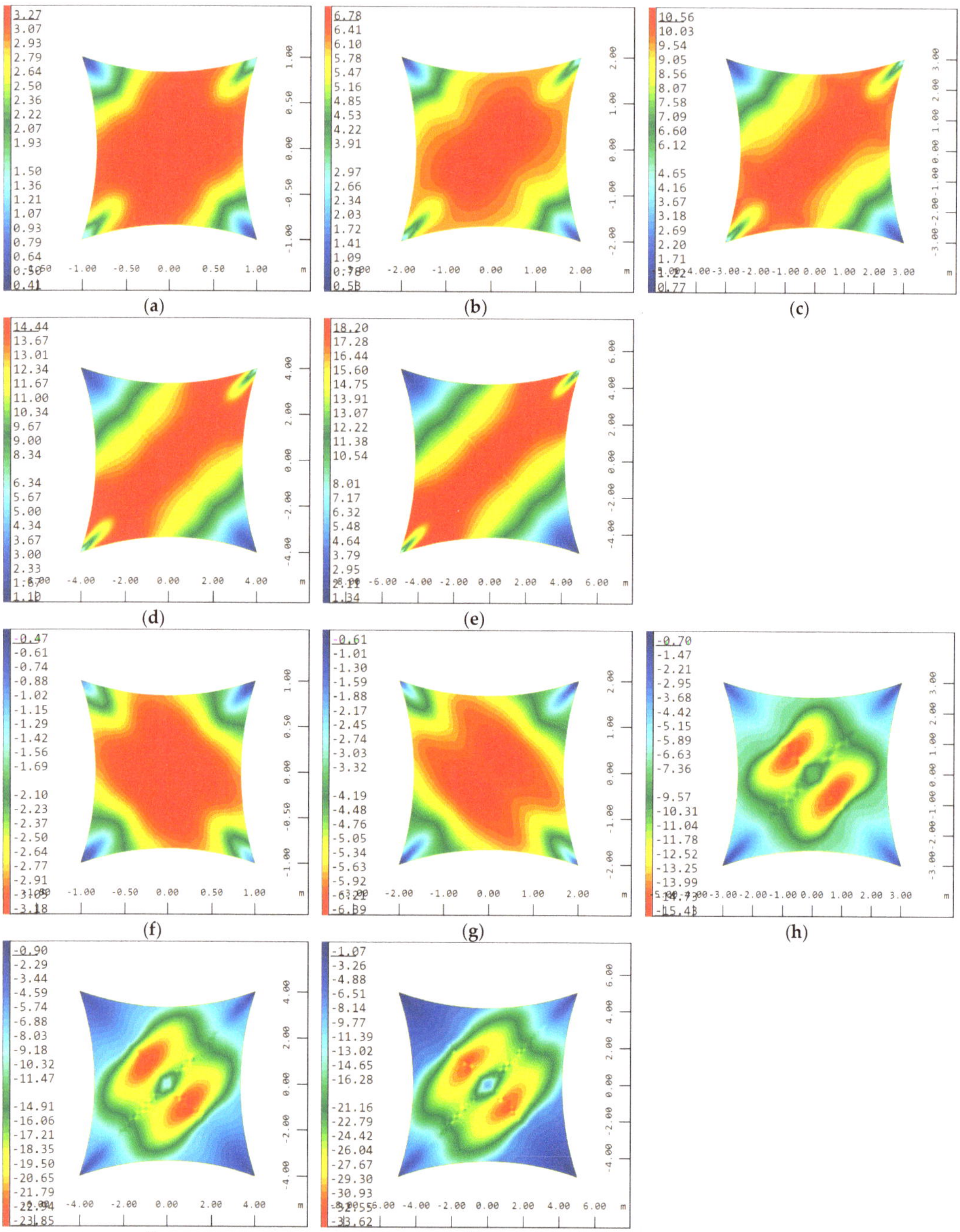


**Figure A2.** Strains (in m/m) of the models with DP, FE, M3, H4, and P3 under wind load: (**a**) S-I, S1; (**b**) S-I, S2; (**c**) S-I, S3; (**d**) S-I, S4; (**e**) S-I, S5; (**f**) S-II, S1; (**g**) S-II, S2; (**h**) S-II, S3; (**i**) S-II, S4; (**j**) S-II, S5 (source: Authors' elaboration).

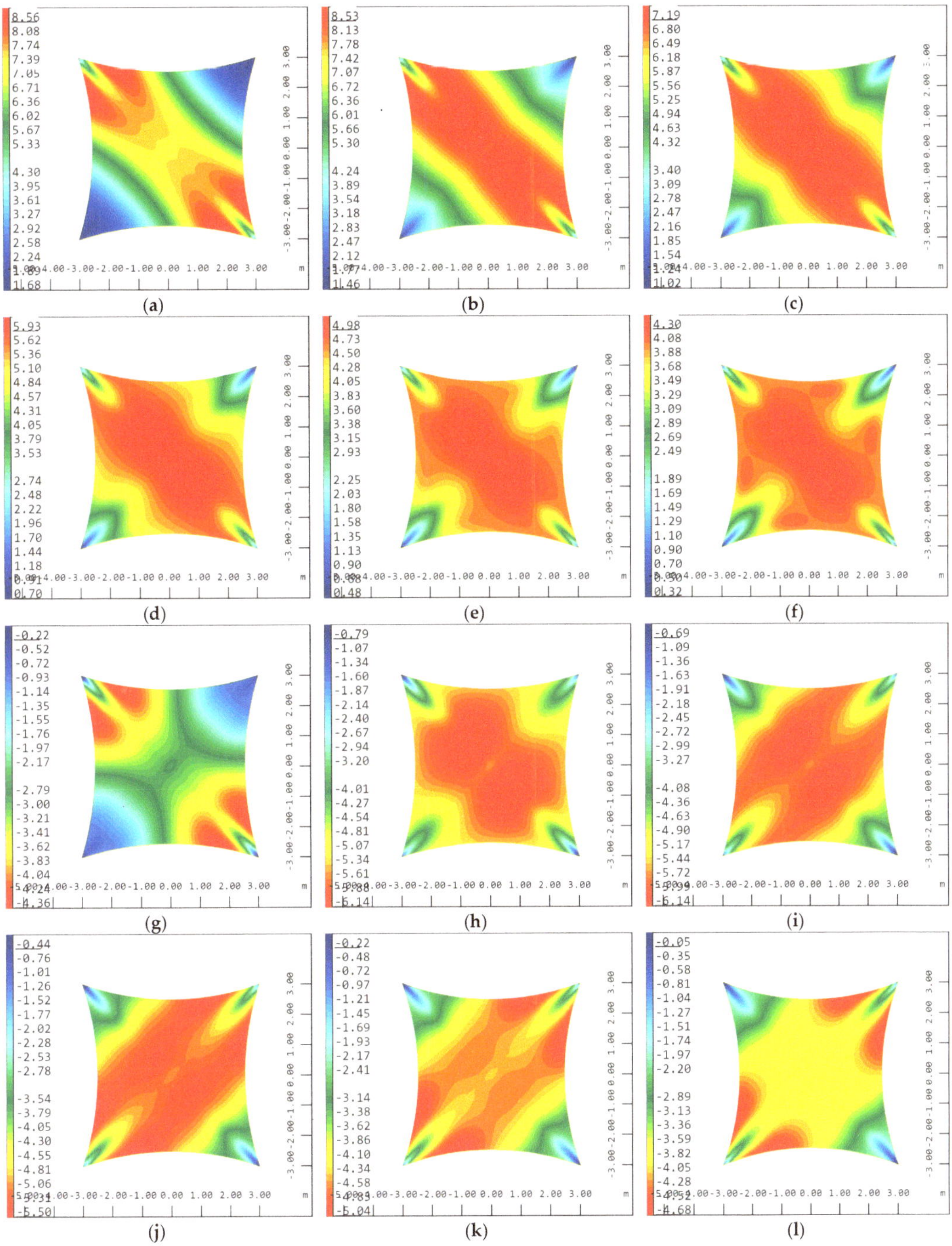


**Figure A3.** Strains (in m/m) of the models with DP, FE, M3, S3, and P3 under snow load: (**a**) S-I, H1; (**b**) S-I, H2; (**c**) S-I, H3; (**d**) S-I, H4; (**e**) S-I, H5; (**f**) S-I, H6; (**g**) S-II, H1; (**h**) S-II, H2; (**i**) S-II, H3; (**j**) S-II, H4; (**k**) S-II, H5; (**l**) S-II, H6 (source: Authors' elaboration).

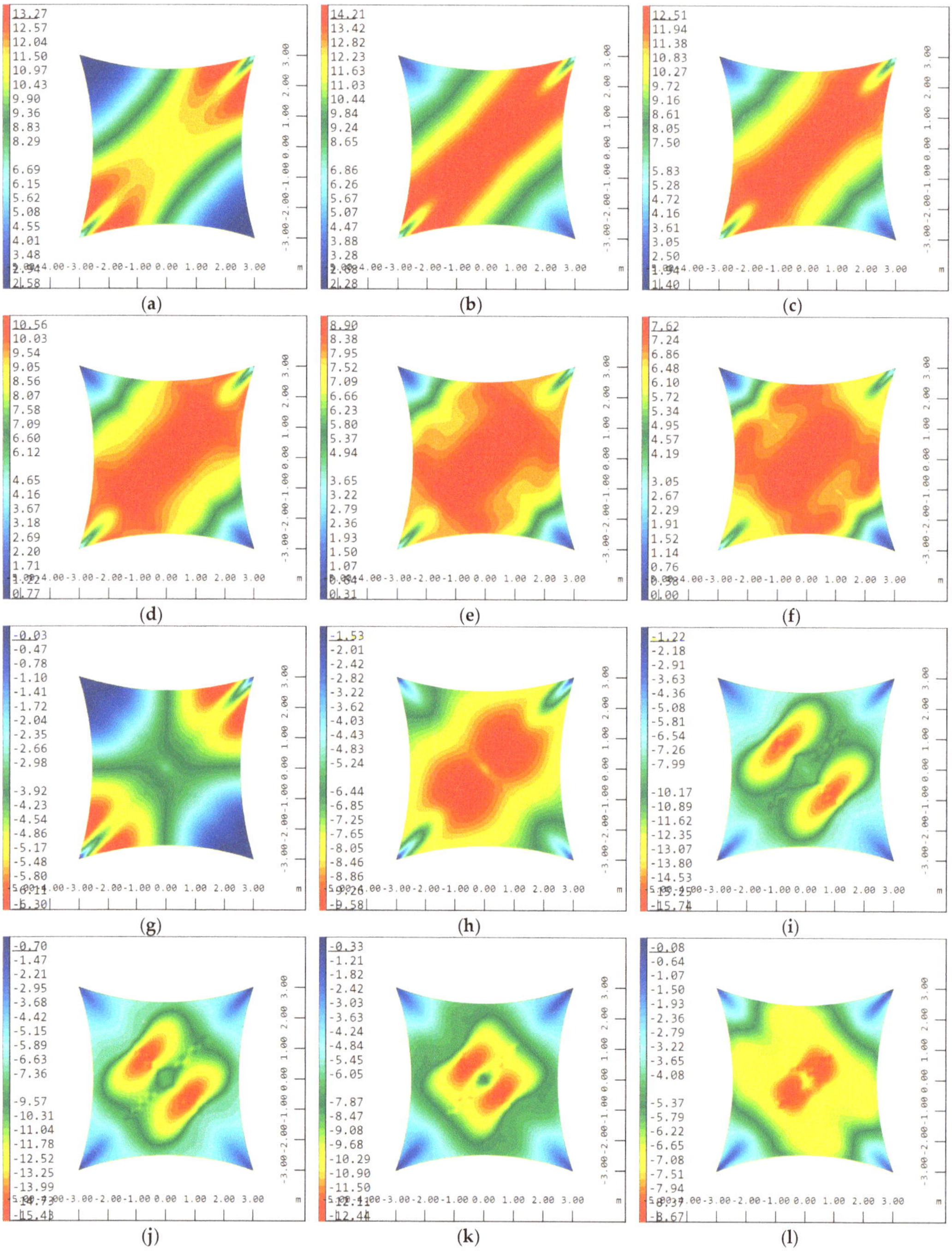

**Figure A4.** Strains (in m/m) of the models with DP, FE, M3, S3, and P3 under wind load: (**a**) S-I, H1; (**b**) S-I, H2; (**c**) S-I, H3; (**d**) S-I, H4; (**e**) S-I, H5; (**f**) S-I, H6; (**g**) S-II, H1; (**h**) S-II, H2; (**i**) S-II, H3; (**j**) S-II, H4; (**k**) S-II, H5; (**l**) S-II, H6 (source: Authors' elaboration).

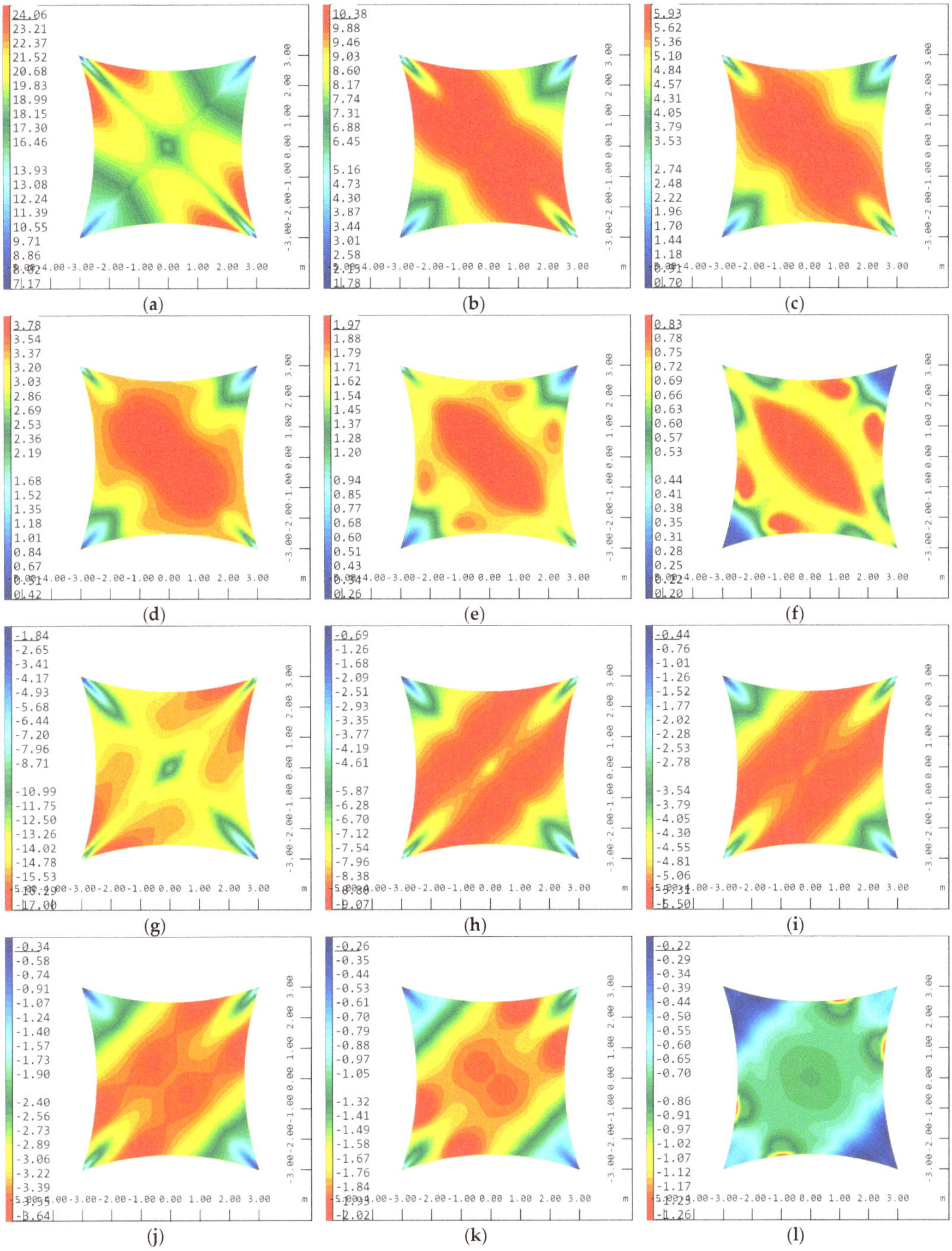

**Figure A5.** Strains (in m/m) of the models with DP, FE, H4, S3, and P3 under snow load: (**a**) S-I, H1; (**b**) S-I, M2; (**c**) S-I, M3; (**d**) S-I, M4; (**e**) S-I, M5; (**f**) S-I, M6; (**g**) S-II, M1; (**h**) S-II, M2; (**i**) S-II, M3; (**j**) S-II, M4; (**k**) S-II, M5; (**l**) S-II, M6 (source: Authors' elaboration).

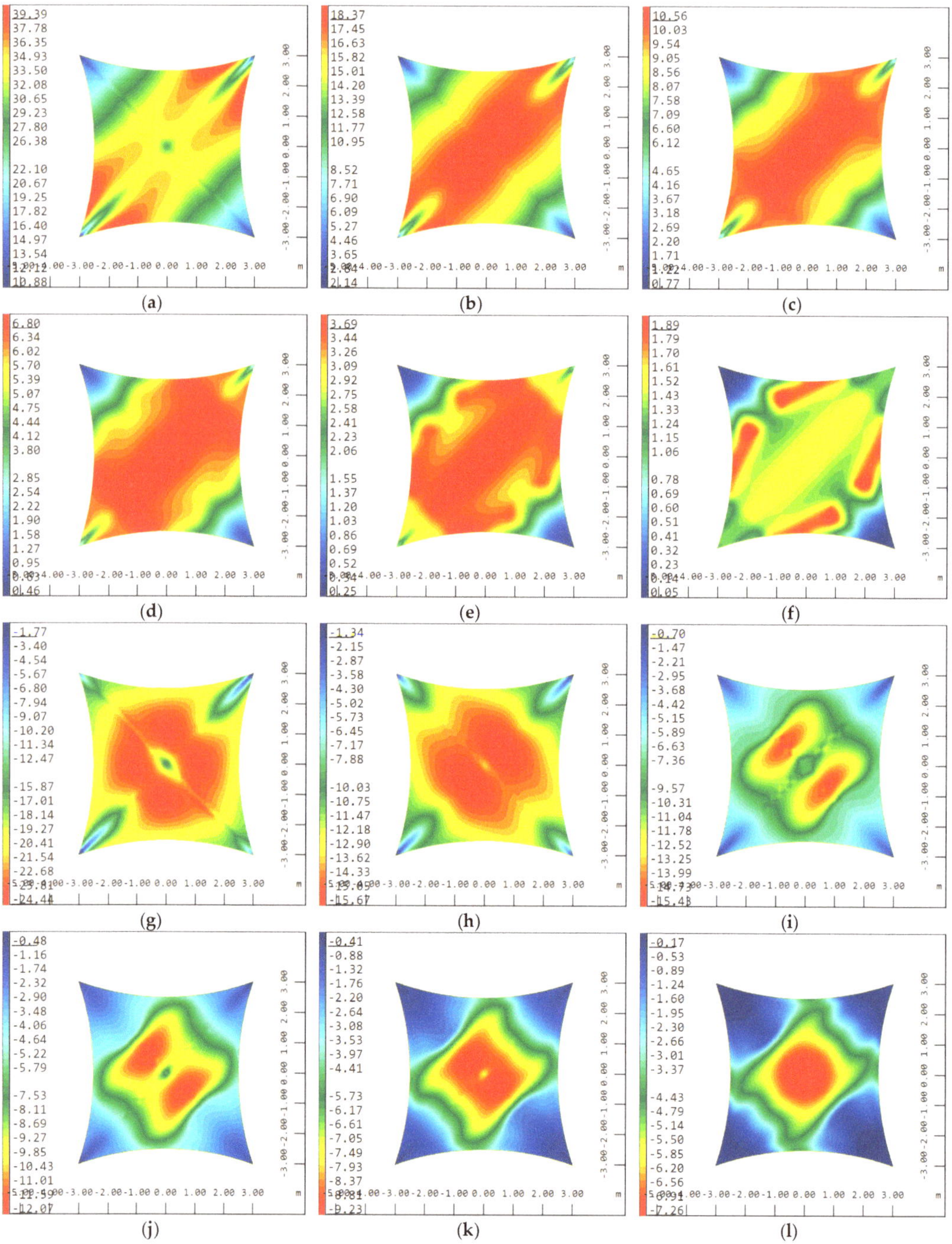

**Figure A6.** Strains (in m/m) of the models with DP, FE, H4, S3, and P3 under wind load: (**a**) S-I, M1; (**b**) S-I, M2; (**c**) S-I, M3; (**d**) S-I, M4; (**e**) S-I, M5; (**f**) S-I, M6; (**g**) S-II, M1; (**h**) S-II, M2; (**i**) S-II, M3; (**j**) S-II, M4; (**k**) S-II, M5; (**l**) S-II, M6 (source: Authors' elaboration).

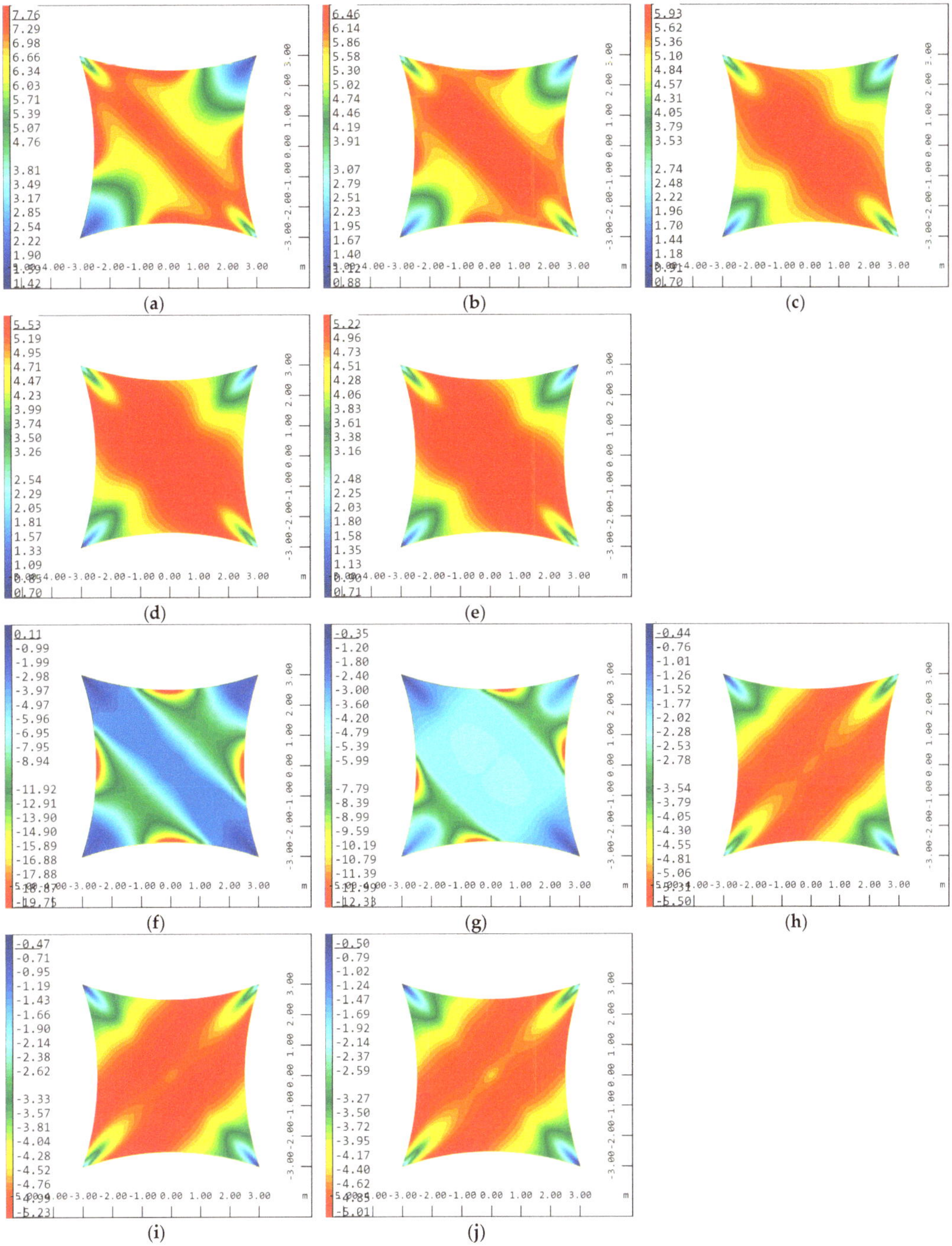

**Figure A7.** Strains (in m/m) of the models with DP, FE, H4, S3, and M3 under snow load: (**a**) S-I, P1; (**b**) S-I, P2; (**c**) S-I, P3; (**d**) S-I, P4; (**e**) S-I, P5; (**f**) S-II, P1; (**g**) S-II, P2; (**h**) S-II, P3; (**i**) S-II, P4; (**j**) S-II, P5 (source: Authors' elaboration).

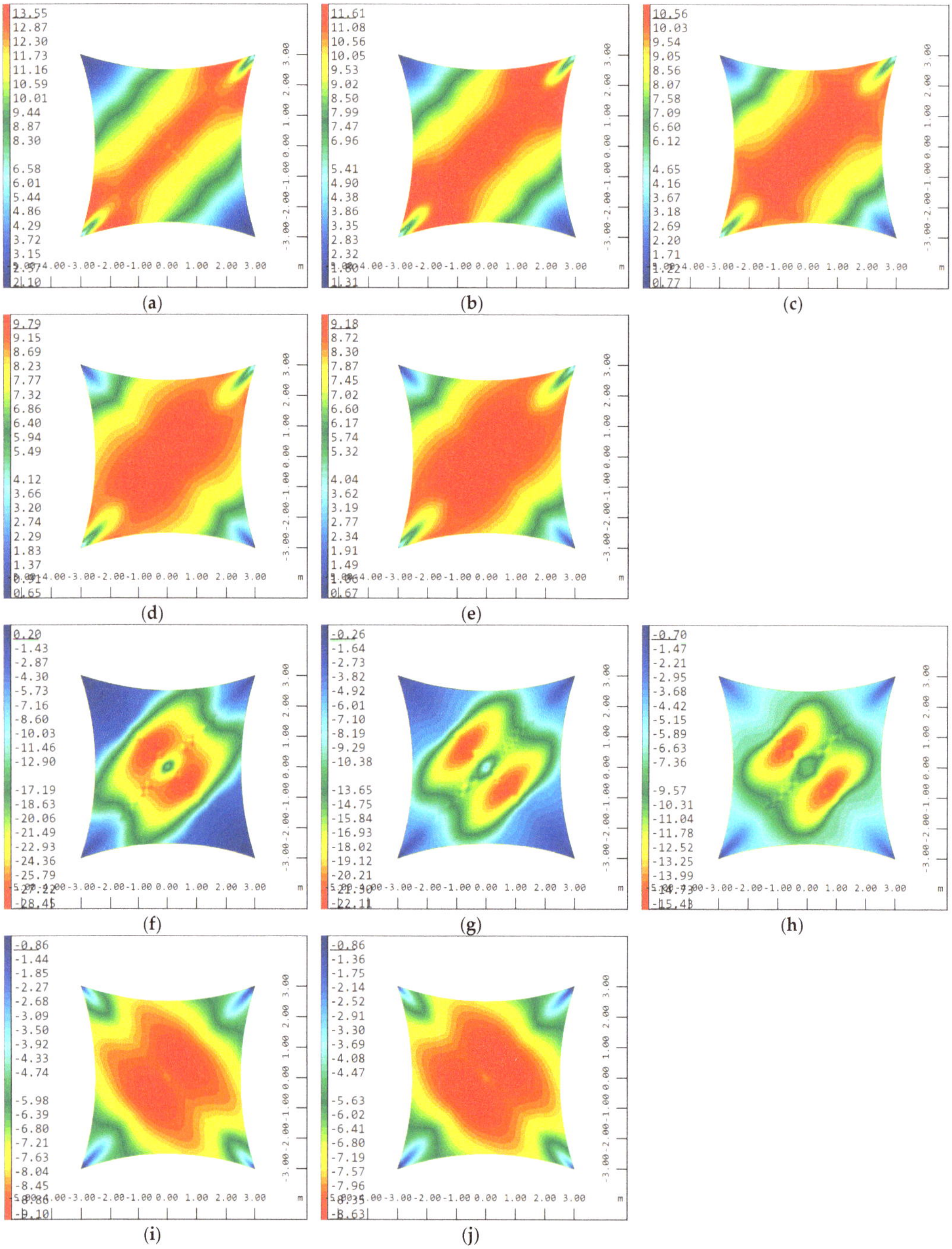


**Figure A8.** Strains (in m/m) of the models with DP, FE, H4, S3, and M3 under wind load: (**a**) S-I, P1; (**b**) S-I, P2; (**c**) S-I, P3; (**d**) S-I, P4; (**e**) S-I, P5; (**f**) S-II, P1; (**g**) S-II, P2; (**h**) S-II, P3; (**i**) S-II, P4; (**j**) S-II, P5 (source: Authors' elaboration).

## References


1. Available online: https://sustainabledevelopment.un.org/content/documents/5839GSDR%202015_SD_concept_definiton_rev.pdf (accessed on 25 January 2024).
2. Lucchi, E. *Solar Energy Technologies in Cultural Heritage*; Elsevier: Amsterdam, The Netherlands, 2025. [CrossRef]
3. Zhang, J.; Zhang, W.; Cheng, H.-M.; Silva, S.R.P. Critical review of recent progress of flexible perovskite solar cells. *Mater. Today* **2020**, *39*, 66–88. [CrossRef]
4. Milošević, V.; Marchwiński, J. Photovoltaic Technology Integration with Tensile Membrane Structures: A Critical Review. *Tech. Gaz.* **2022**, *29*, 702–713.
5. Thorpe, D. *Passive Solar Architecture Pocket Reference*; Tailor and Francis: Oxfordshire, UK, 2017.
6. Lucchi, E. Chapter 8—Active solar design principles for natural and architectural heritage. In *Solar Energy Technologies in Cultural Heritage*; Lucchi, E., Ed.; Elsevier: Amsterdam, The Netherlands, 2025; pp. 215–259. [CrossRef]
7. Lucchi, E.; Agliata, R. HBIM-based workflow for the integration of advanced photovoltaic systems in historical buildings. *J. Cult. Herit.* **2023**, *64*, 301–314. [CrossRef]
8. Teka, L.G.; Zadshir, M.; Yin, H. Mechanical analysis and design of large building integrated photovoltaic panels for a seamless roof. *Sol. Energy* **2023**, *251*, 1–12. [CrossRef]
9. Li, Q.; Zanelli, A. A review on fabrication and applications of textile envelope integrated flexible photovoltaic systems. *Renew. Sustain. Energy Rev.* **2021**, *139*, 110678. [CrossRef]
10. Borazan, I.; Bedeloglu, A.C.; Demir, A. A photovoltaic textile design with a stainless steel mesh fabric. *J. Ind. Text.* **2022**, *51*, 1527–1538. [CrossRef]
11. Parasuraman, D. A review on dye-sensitized solar cells (DSSCs), materials and applications. *Iran. J. Mater. Sci. Eng.* **2023**, *20*, 1–23.
12. Park, N.G. Perovskite solar cells: An emerging photovoltaic technology. *Mater. Today* **2015**, *18*, 65–72. [CrossRef]
13. Prasad, S.V.D.; Krishnanaik, V.; Babu, K.R. Analysis of organic photovoltaic cell. *Int. J. Sci. Mod. Eng.* **2013**, *1*, 20–23.
14. Zanelli, A.; Monticelli, C.; Beccarelli, P.; Ibrahim, H. Experimental manufacture of a pneumatic cushion made of ETFE foils and OPV cells. In *Proceedings of the Textiles Composites and Inflatable Structures V: Proceedings of the V International Conference on Textile Composites and Inflatable Structures, Barcelona, Spain, 5–7 October, 2011*; International Center for Numerical Methods in Engineering (CIMNE): Barcelona, Spain; pp. 279–290.
15. Zhang, J.; Yang, Q.; Li, B. Form-state and loading analyses of air-inflated cushion membrane structures. *J. Harbin Inst. Technol.* **2008**, *40*, 1569–1572.
16. Zhao, B.; Hu, J.; Chen, W.; Chen, J.; Jing, Z. Dynamic geometrical shape measurement and structural analysis of inflatable membrane structures using a low-cost three-camera system. *Autom. Constr.* **2018**, *96*, 442–456. [CrossRef]
17. Schek, H.J. The force density method for form finding and computation of general networks. *Comput. Methods Appl. Mech. Eng.* **1974**, *3*, 115–134. [CrossRef]
18. Marchwiński, J.; Milošević, V.; Stefańska, A.; Lucchi, E. Irradiation analysis of tensile membrane structures for building-integrated photovoltaics. *Energies* **2023**, *16*, 5945. [CrossRef]
19. Spanodimitriou, Y.; Ciampi, G.; Tufano, L.; Scorpio, M. Flexible and lightweight solutions for energy improvement in construction: A literature review. *Energies* **2023**, *16*, 6637. [CrossRef]
20. Ibrahim, H.M.; Zanelli, A.; Cremers, J. The structural behaviour of PTFE/glass fabric structures integrating flexible photovoltaic modules. In *Proceedings of the Textiles Composites and Inflatable Structures VI: Proceedings of the VI International Conference on Textile Composites and Inflatable Structures, Barcelona, Spain, 9–11 October 2013*; International Center for Numerical Methods in Engineering (CIMNE): Barcelona, Spain; pp. 290–299.
21. Scotta, R.; Lazzari, M.; Stecca, E.; Di Massimo, R.; Vitaliani, R. Membranes with embedded photovoltaic flexible cells: Structural and electrical performances under uniaxial and biaxial stresses. *Compos. Struct.* **2016**, *157*, 111–120. [CrossRef]
22. Fan, Z.; De Bastiani, M.; Garbugli, M.; Monticelli, C.; Zanelli, A.; Caironi, M. Experimental investigation of the mechanical robustness of a commercial module and membrane-printed functional layers for flexible organic solar cells. *Compos. Part B* **2018**, *147*, 69–75. [CrossRef]
23. Hu, J.H.; Chen, W.J.; Zhao, B.; Song, H. Experimental studies on summer performance and feasibility of a BIPV/T ethylene tetrafluoroethylene (ETFE) cushion structure system. *Energy Build.* **2014**, *69*, 394–406. [CrossRef]
24. Hu, J.; Chen, W.; Liu, Y.; Zhao, B.; Yang, D.; Ge, B. Two-layer ETFE cushions integrated flexible photovoltaics: Prototype development and thermal performance assessment. *Energy Build.* **2017**, *141*, 238–246. [CrossRef]
25. Hu, J.; Chen, W.; Cai, Q.; Gao, C.; Zhao, B.; Qiu, Z.; Qu, Y. Structural behavior of the PV-ETFE cushion roof. *Thin-Walled Struct.* **2016**, *101*, 169–180. [CrossRef]
26. Hu, J.; Chen, W.; Yin, Y.; Li, Y.; Yang, D.; Wang, H.; Zhang, X. Electrical-thermal-mechanical properties of multifunctional OPV-ETFE foils for large-span transparent membrane buildings. *Polym. Test.* **2018**, *66*, 394–402. [CrossRef]
27. Zhao, B.; Hu, J.; Chen, W.; Qiu, Z.; Zhou, J.; Qu, Y.; Ge, B. Photothermal performance of an amorphous silicon photovoltaic panel integrated in a membrane structure. *J. Phys. D Appl. Phys.* **2016**, *49*, 395601. [CrossRef]

28. Li, Q.; Li, T.; Zanelli, A. Performance evaluation of flexible photovoltaic panels for energy supply in post-disaster emergency shelters. *J. Build. Eng.* **2024**, *98*, 111285. [CrossRef]
29. Li, Q.; Li, T.; Kutlu, A.; Zanelli, A. Life cycle cost analysis and life cycle assessment of ETFE cushion integrated transparent organic/perovskite solar cells: Comparison with PV glazing skylight. *J. Build. Eng.* **2024**, *87*, 109140. [CrossRef]
30. Li, Q.; Monticelli, C.; Kutlu, A.; Zanelli, A. Feasibility of textile envelope integrated flexible photovoltaic in Europe: Carbon footprint assessment and life cycle cost analysis. *J. Clean. Prod.* **2023**, *430*, 139716. [CrossRef]
31. Marchwiński, J.; Lucchi, E. Firmitas, Utilitas, and Venustas of photovoltaic architecture. *Sol. Energy* **2024**, *282*, 112974. [CrossRef]
32. Ambroziak, A. Characterization study on mechanical properties of polyester coated fabric. *Arch. Civ. Eng.* **2020**, *66*, 105–118. [CrossRef]
33. Hermannsdörfer, I.; Rüb, C. *Solar Design: Photovoltaics for Old Buildings, Urban Space, Landscapes*; Jovis: Berlin, Germany, 2005.
34. Humm, O.; Toggweiler, P. *Photovoltaics in Architecture*; Birkhäuser: Basel, Switzerland, 1993.
35. Uhlemann, J. Elastic Constants of Architectural Fabrics for Design Purposes. Ph.D. Thesis, Duisburg: Universität Duisburg-Essen, Duisburg, Germany, 2016.
36. Milošević, V.; Marchwiński, J.; Lucchi, E. Exploration of the suitable tensioned membrane properties for the photovoltaics integration. *J. Fac. Civ. Eng. Archit.* **2024**, *39*, 59–66. [CrossRef]
37. Bridgens, B.; Birchall, M. Form and function: The significance of material properties in the design of tensile fabric structures. *Eng. Struct.* **2012**, *44*, 1–12. [CrossRef]
38. Milošević, V.S.; Marković, B.L.J. Comparison of point and snow load deflections in design and analysis of tensile membrane structures. *Adv. Civ. Eng.* **2020**, *2020*, 8810085. [CrossRef]
39. Gosling, P.D.; Bridgens, B.N.; Albrecht, A.; Alpermann, H.; Angeleri, A.; Barnes, M.; Bartle, N.; Canobbio, R.; Dieringer, F.; Gellin, S.; et al. Analysis and design of membrane structures: Results of a round robin exercise. *Eng. Struct.* **2013**, *48*, 313–328. [CrossRef]
40. Forster, B.; Mollaert, M. *European Design Guide for Tensile Surface Structures*; TensiNet: Brussels, Belgium, 2004.
41. Stranghöner, N.; Uhlemann, J.; Bilginoglu, F.; Bletzinger, K.; Bögner-Balz, H.; Corne, E.; Gibson, N.; Gosling, P.; Houtman, R.; Llorens, J.; et al. *Prospect for European Guidance for the Structural Design of Tensile Membrane Structures*; Dimova, S., Mollaert, M., Pinto Vieira, A., Denton, S., Eds.; EUR 31430 EN; Publications Office of the European Union: Luxembourg, 2023; ISBN 978-92-68-00053-3. [CrossRef]
42. Gerlic, K. Projektowanie zadaszeń membranowych na przykładzie parkingu rowerowego w Jaworznie [Designing membrane roofs on the example of a bicycle parking lot in Jaworzno]. *Builder* **2021**, *288*, 12–15. [CrossRef]

sustainability MDPI

*Article*

# A Modular Agrivoltaics Building Envelope Integrating Thin-Film Photovoltaics and Hydroponic Urban Farming Systems: A Circular Design Approach with the Multi-Objective Optimization of Energy, Light, Water and Structure

**Yihan Zhang [1,2], Tianyi Chen [2,*], Eugenia Gasparri [3,*] and Elena Lucchi [4,5]**

1 Department of Architecture, Swiss Federal Institute of Technology Zurich (ETH Zurich), Rämistrasse 101, 8092 Zürich, Switzerland; yihan.zhang@alumni.ethz.ch
2 Solar Energy Research Institute of Singapore, National University of Singapore, 21 Lower Kent Ridge Rd, Singapore 119077, Singapore
3 School of Architecture, Design and Planning, University of Sydney, Camperdown, NSW 2050, Australia
4 Department of Civil Engineering and Architecture (DICAr), University of Pavia, Via Ferrata 3, 27100 Pavia, Italy; elena.lucchi@unipv.it
5 Department of Architecture, Construction Engineering and Built Environment (ABC), Politecnico di Milano, Via Bonardi 12, 20133 Milan, Italy
* Correspondence: e0348767@u.nus.edu (T.C.); eugenia.gasparri@sydney.edu.au (E.G.)

**Abstract:** As cities confront multiple challenges such as climate change, urbanization, and food security, growing attention has been given to sustainable vertical farming and renewable energy solutions. Building facades, typically underutilized in high-density urban environments, present an opportunity for multifunctional buildings composed of both photovoltaic (PV) systems and vertical farming modules. However, on vertical surfaces, these two systems often compete for space. This research focuses on the development of a multifunctional agrivoltaics building envelope (ABE) system, combining building-integrated PV (BIPV) technology with hydroponic vertical farming. This ABE system adopts a modular design approach, where each unit can be prefabricated independently and assembled through an interlocking connection design and bolted fastening to ensure ease of construction and scalability. The design process includes the development of 2D cross-sectional technical design, assembly sequences, and an analysis of key design parameters through 3D modeling. The research adopts a combined Research through Design (RtD) and Research for Design (RfD) approach to bridge prototyping, testing, and performance optimization. This research highlights the potential of integrating renewable energy with agricultural production in building envelope systems. By addressing space optimization and multifunctionality, the research provides a practical framework for future applications in urban sustainability.

**Keywords:** photovoltaic; BIPV; vertical farming; agrivoltaics; hydroponic; prefabricated design; facade design

Academic Editor: Joshua M. Pearce

Received: 30 November 2024
Revised: 9 January 2025
Accepted: 13 January 2025
Published: 16 January 2025

**Citation:** Zhang, Y.; Chen, T.; Gasparri, E.; Lucchi, E. A Modular Agrivoltaics Building Envelope Integrating Thin-Film Photovoltaics and Hydroponic Urban Farming Systems: A Circular Design Approach with the Multi-Objective Optimization of Energy, Light, Water and Structure. *Sustainability* **2025**, *17*, 666. https://doi.org/10.3390/su17020666



## 1. Introduction

By 2030, two-thirds of the world's population will reside in cities, transforming global demographics and creating increased pressure on infrastructure and resources [1]. This urban expansion presents interconnected challenges of food security, energy consumption, and limited space. To meet demand, food production must dramatically increase in the next fifty years [2]. Climate change further compounds these challenges. It threatens conventional farming through desertification, altered precipitation patterns, and extreme heat events [3]. These impacts make stable, affordable food production increasingly difficult.

The building industry's significant contribution to global carbon emissions and energy consumption, accounting for one-third of both, underscores the urgent need for sustainable solutions [4]. In land-constrained urban environments, the growing competition between solar photovoltaic (PV) farms and agricultural land creates additional challenges for both renewable energy generation and food production [5]. Building-integrated solutions that combine vertical farming with building-integrated PV (BIPV) offer a promising approach to address these challenges simultaneously. Recent technological advances, particularly in energy materials and storage systems [6–8], have made such integrated solutions increasingly feasible. These innovations enable buildings to enhance food security while producing renewable fuels through agricultural waste [9], combining biophilic design principles with clean on-site energy generation through power-generating building surfaces [10]

Existing building envelope technologies predominantly serve singular functions, focusing primarily on thermal insulation and weather protection [11]. While recent advances have demonstrated successful individual implementations of BIPV and vertical farming systems, their integration into unified prefabricated solutions remains limited [12]. This single-function approach fails to address the growing need for multifunctional building envelopes that could simultaneously tackle urban sustainability challenges through energy generation and food production [13].

Recent advances in agrivoltaics integration across building applications have shown significant promise in combining food production with renewable energy generation. Solar panels for greenhouses, tailored for specific crops by balancing light transmittance and power output, can achieve efficiencies up to 23.2%, generating 340W of power while utilizing specialized materials like high-transmission glass and EVA to optimize energy conversion and light diffusion [14]. A 2024 study by C. Ma et al. [15] utilized deep learning and GIS integration to assess the potential of rooftop agriculture in high-density cities. Their findings indicate that, by 2030, cities like Singapore could achieve food self-sufficiency levels of up to 34.71% through the integration of vertical farming and technological advancements. Building-integrated photovoltaics (BIPV) can replace traditional building materials in various components, such as roofs and facades, enhancing energy efficiency and durability while contributing to net-zero emissions [16]. However, these implementations generally treat PV and agriculture as distinct building systems rather than truly integrated components. Building upon these foundations, this study advances a novel approach that physically combines PV and farming elements within the building envelope itself, offering potential advantages in space utilization and system integration compared to previous solutions.

The current implementations of BIPV and vertical farming systems each face their own technical challenges in building integration. BIPV systems frequently encounter issues with ease of construction and plug-and-play implementation [17]. Similarly, vertical farming installations struggle with irrigation system integration, spatial efficiency, and consistent plant-growth conditions in vertical arrangements [18]. The fundamental challenge of combining the two systems lies in balancing light requirements-BIPV systems require maximum solar exposure for energy generation, while farming systems need carefully controlled light distribution for optimal plant growth [19]. These system limitations highlight the need for innovative integration approaches.

A review of the current literature reveals several critical knowledge gaps in this field. The lack of standardized design approaches for integrating BIPV and vertical farming systems, particularly in the context of mid- to high-rise buildings, remains a significant challenge [20]. While studies have examined individual components, research on integrated solutions that combine energy generation and food production remains limited [21]. This limitation is further compounded by the nature of existing prefabricated solutions, where patented technologies and materials restrict access to detailed technical information [22]. Additionally, [22] highlighted the insufficient understanding of maintenance requirements and lifecycle implications for such integrated systems. These limitations present several research opportunities, particularly in developing standardized design approaches for multifunctional facades and creating accessible integrated solutions for energy and food production. The development of innovative maintenance solutions that allow for indoor access while maintaining system efficiency remains a key area for exploration [23]. This research article presents an experimental investigation and technical design analysis of an integrated agrivoltaics building envelope system. While building upon existing literature, our work makes novel contributions through the systematic design development, prototype modeling, and performance analysis of a specific architectural integration approach.

This research introduces the agrivoltaics building envelope, a novel system that builds upon the principles of agrivoltaics systems, where agricultural production and solar energy generation coexist synergistically. Through Research through Design (RtD) methodology, this study explores the integration potential of BIPV and vertical farming within prefabricated building envelopes. The goal is to develop an innovative solution that addresses the identified challenges in current building technologies while contributing to urban sustainability. The following literature review examines the technological foundations, current developments, and research gaps that inform this study's approach to creating a new generation of multifunctional building envelopes.

## 2. Background

This review examines the current state of technology and research across several interconnected domains: prefabrication systems, BIPV, and vertical farming technologies. The analysis focuses particularly on developments relevant to their potential integration in building envelopes, identifying both opportunities and challenges that inform this research's approach to developing the agrivoltaics building envelope system.

### 2.1. Evolution of Building Prefabrication Technologies

Prefabricated building systems represent a modern construction approach where building components are manufactured in controlled factory conditions before on-site assembly. Ref. [24] demonstrated that this approach can reduce construction time by up to 20% while achieving superior quality control through standardized manufacturing processes. The advancement of these systems has enabled more complex integration possibilities, though [25] note that challenges remain in coordinating different technical requirements within single prefabricated units, particularly regarding maintenance accessibility and interface connections.

The characteristics of prefabrication technology—precise quality control, efficient assembly, and potential for complex system integration—make it particularly suitable for developing integrated facade solutions. Prefabrication can successfully accommodate various facade integrations while maintaining construction efficiency. These capabilities provide a promising foundation for developing the agrivoltaics building envelope system, where multiple functions must be precisely coordinated within standardized components.

### 2.2. BIPV System Development

BIPV represent an innovative approach to incorporating solar energy generation directly into building envelope components, serving both as building materials and power generators. Ref. [26] documented significant advances in thin-film solar technology, achieving conversion efficiencies of 18% while maintaining aesthetic appeal. These systems can contribute up to substantial portion of a building's energy requirements while simultaneously functioning as effective building envelope components [27]. However, BIPV systems face notable challenges in construction and maintenance aspects. Ref. [28] identified key issues including complex installation procedures, limited standardization in connection details, and difficulties in maintenance access. These challenges are particularly pronounced in facade applications, where system repairs and component replacements must be conducted from the building exterior, raising both safety concerns and operational costs.

### 2.3. *Vertical Farming Technologies in Buildings*

Vertical farming represents an innovative approach to urban agriculture, integrating food production systems within building environments through hydroponic or aeroponic technologies. Ref. [29] demonstrated that modern building-integrated farming systems can produce leafy greens while operating with 95% less water consumption compared to traditional agriculture. These systems offer significant benefits in urban food security while contributing to building sustainability through improved thermal performance and biophilic design integration. The implementation of vertical farming systems in buildings presents specific technical challenges, particularly in irrigation integration and maintenance access. Ref. [29] highlighted critical issues including complex plumbing requirements, the need for precise environmental control systems, and difficulties in managing plant health in vertical arrangements. These challenges are compounded by the need to ensure consistent growing conditions while maintaining building envelope integrity and allowing for regular maintenance access.

### 2.4. *Agrivoltaics Building Envelope*

Agrivoltaics building envelopes represent an emerging concept that combines the principles of agrivoltaics—the dual use of space for both solar energy and agriculture—with building integration technologies. The foundational concept of agrivoltaics, as defined by [30], focuses on the review of the relationship between solar panels and plant growth. Recent studies by [31] have explored various approaches to combining BIPV and farming systems in building facades, ranging from alternating panel arrangements to semi-transparent PV integration with plant-growth spaces. Ref. [32] categorized these systems into two main types: elevated solar panels above crops and inter-row systems, each offering different advantages for energy generation and plant cultivation. Ref. [33] demonstrated that carefully designed agrivoltaics systems can maintain high levels of solar energy generation with standard BIPV while maintaining optimal conditions for plant growth.

Current research has primarily focused on performance optimization and theoretical frameworks, revealing significant gaps in practical implementation methods. Ref. [12] noted the lack of standardized design approaches and prefabricated solutions that could enable widespread adoption, particularly highlighting the absence of systematic integration methods for building applications. Ref. [25] identified critical knowledge gaps in construction methodology and maintenance strategies for integrated agriculture systems. While [34] demonstrated promising results in small-scale prototypes, research on scalable prefabricated solutions that address construction efficiency, maintenance accessibility, and system durability remains limited. Ref. [35] emphasized that existing studies rarely address the practical challenges of prefabrication and assembly sequences, creating a significant barrier to commercial implementation. This gap in practical design methodology and prefabrication strategies presents a crucial opportunity for developing systematic approaches to agrivoltaics building envelope implementation, particularly in the context of high-rise buildings where integration complexity increases significantly.

*2.5. Research Aims*

This study addresses critical gaps in the current literature and practice of agrivoltaics building envelope (ABE) development by focusing on three key areas:

- Technical integration challenges, to resolve the complex interface issues between BIPV and farming systems through innovative prefabrication strategies to maintain optimal performance of both functions.
- Practical implementation barriers related to construction efficiency and maintenance accessibility. This includes developing detailed specifications for modular components and assembly sequences that can be standardized for wider industry adoption.
- Design parameters for vertical agrivoltaics systems tailored for mid- to high-rise building applications.

Through these focused objectives, the research aims to bridge the gap between theoretical potential and practical implementation in sustainable building envelope design, particularly in urban environments where space optimization is crucial.

## 3. Methodology

In applied design research, two complementary approaches are frequently employed: Research through Design (RtD) and Research for Design (RfD). The RtD approach emphasizes the exploration and generation of new knowledge through the iterative design process, fostering innovation by prototyping and testing ideas in real-world contexts. The RfD approach provides empirical and theoretical support to refine and optimize design outcomes through rigorous analysis and validation. This work employs a mixed approach combining RtD and RfD on the building envelope. The system introduced in this study is a modular facade system unitized PV and vertical farming, aiming to address the gap in existing products and integration research in this field. While RtD serves as the primary approach to exploring integration and prototyping, concurrently, RfD guides the optimization of PV coverage, light transmittance, and system scalability through simulation and 3D modeling generation. The research process is divided into several steps as follows:

a. Problem identification and concept development: Defining the research problem and developing the initial concept for integrating BIPV with vertical farming within a modular facade system while identifying key environmental factors and design requirements for the technical design and material selection.
b. Technical design: Based on the key considerations introduced in Step A, as well as local climate conditions and building regulations, the structural layers and materials for the ABE system were determined, resulting in the 2D cross-sectional and joint designs for the modular facade.
c. 3D model validation: Creating a 3D digital model using Rhinoceros tools to simulate the operational mechanism of the hydroponic module and the system assembly. The 3D model also displays the aesthetic impact and design flexibility of PV glass as a cladding layer.

## 4. Results

ABE, as shown in Figure 1, serves as a versatile dual-layer building envelope system that integrates BIPV with vertical farming. To address the challenge of urban agriculture, the ABE system utilizes underused building surfaces to maximize productivity. Building corridors, rooftops, and exterior walls can all serve as productive spaces for urban agriculture. A study on urban agriculture potential in Singapore's Housing and Development

Board (HDB) revealed that while rooftops and corridors provide limited available areas, corridors are further constrained due to their roles in public circulation and emergency evacuation. In contrast, vertical farming on exterior walls has the potential to meet nearly half of the vegetable needs of residents [36]. Therefore, this underscores the critical importance of exploring the integration of vertical farming into building envelope systems.

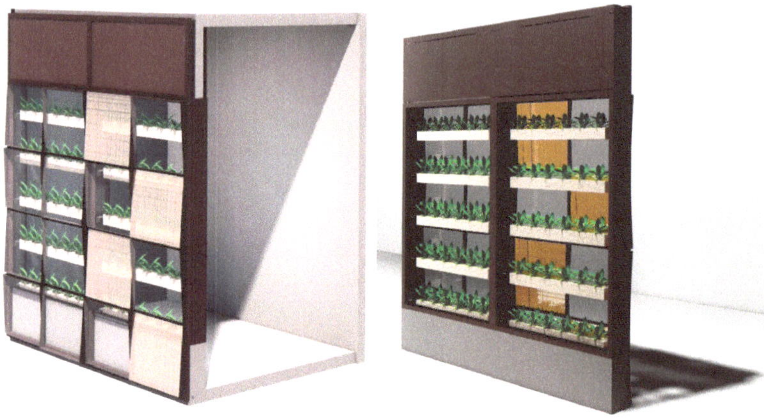

**Figure 1.** Overview of the agrivoltaics building envelope prototype. The (**left**) panel shows the front view, featuring the integrated photovoltaic panels and vertical farming modules arranged in a checkerboard pattern. The (**right**) panel displays the rear view, revealing the hydroponic system integration and structural support framework.

Commonly used materials for prefabricated building envelopes include precast concrete, timber, unitized glass modules, and aluminum extrusion frames. However, timber is scarce in many regions and is not widely used in high-density areas or high-rise buildings due to its limitations in fire resistance and acoustic performance. Meanwhile, Precast concrete offers high structural strength and is also associated with heavyweight and high embodied carbon. When combined with vertical farming systems, curtain-wall systems with opaque PV panels can significantly increase lighting energy consumption due to excessive natural light blockage.

This study employs recycled aluminum frames as the primary structural support due to its recyclability and lightweight properties. This material is well-suited for prefabricated integration in new constructions and offers flexibility for retrofitting existing buildings. The cladding layer incorporates unitized transparent thin-film PV glass modules and low-E glass modules, allowing adequate natural light transmission to support plant growth, making it ideal for high-density urban environments.

Environmental impacts and load factors were analyzed to determine the optimal materials and locations for each layer of the ABE system, culminating in a two-dimensional cross-section of the envelope design (see Section 4.1). A universal system integrated into the building envelope was developed, incorporating corresponding joint design to enhance architectural compatibility. This modular assembly approach offers adaptability across diverse project scenarios (see Section 4.2). Furthermore, the study discusses the proportion of PV glass panels and their color variations, exploring the balance between energy production and the natural light requirements for plant growth (see Section 4.3). The vertical farming hydroponic system is powered by PV, ensuring a sustainable irrigation cycle (see Section 4.4).

### 4.1. Agrivoltaics Building Envelope (ABE)

#### 4.1.1. Overview

The ABE system is designed primarily for tropical and subtropical climate zones, with Singapore's climate serving as the reference design condition. These regions typically experience average annual temperatures between 25 and 32 °C, with minimum temperatures rarely falling below 23 °C. The system's ventilation and thermal management features are optimized for these conditions, with the 65–100 mm ventilation cavity and automated louver system helping maintain suitable growing environments even during peak temperature periods. While the current design focuses on tropical applications, future adaptations could extend its use to other climate zones through modified thermal management strategies.

The transparent thin-film PV modules integrated into the ABE system generate approximately 80–100 W/m$^2$ under standard test conditions, with a transparency range of 20–30%. Their performance varies based on installation angle and environmental factors, with typical daily energy yields ranging from 0.4 to 0.6 kWh/m$^2$ in Singapore's tropical climate. This outputs balance power generation with the light transmission requirements for plant growth in the agricultural modules.

The envelope structure of the ABE system consists of a recycled aluminum support structure and unitized glass modules. The recycled aluminum support frame is produced by reprocessing aluminum waste materials, a widely used technique in the construction industry. This manufacturing method ensures low carbon emissions during production and demonstrates excellent material recyclability.

As shown in Figure 2, the ABE system consists of multiple integrated layers, including PV panels, vertical farming modules, and an aluminum support frame. The primary support frame is made of a 3 mm thick box-section aluminum extrusion frame, with a vertical load-bearing mullion spacing of 1500 mm. The ABE system employs a strategic approach to solar radiation management that balances energy generation with plant growth requirements. The vertical farming modules receive light through three primary pathways: First, diffused natural light passes through the semi-transparent PV panels, which maintain 20–30% transparency while generating power. Second, the checkerboard arrangement of alternating PV and low-E glass panels allows direct sunlight to reach the plants through the clear glass sections. Third, the system's modular design incorporates light reflection and scattering from adjacent panels, creating an even distribution of illumination across the growing area. This multi-pathway lighting strategy ensures plants receive approximately 40% of ambient solar radiation, meeting the photosynthetic requirements of leafy greens and other low-light crops. The integrated louver system allows for the fine-tuning of light levels based on crop requirements and seasonal variations. Temperature management is achieved through the ventilation cavity, which prevents heat accumulation from PV panels while maintaining optimal growing conditions between 22 and 28 °C.

The ABE system's design accommodates seasonal solar variations through its adjustable panel configuration. The thin-film PV modules can be installed at tilt angles ranging from 5° to 20° relative to the vertical plane, allowing for optimization based on geographic location and local solar paths. While this study focuses on the system's implementation in tropical regions where seasonal variations are less pronounced, the modular design enables adaptation to different latitudes through panel angle adjustments and spacing modifications. The hydroponic growing cycles are coordinated with available light conditions, with crop selection focused on leafy greens and herbs that thrive in controlled indoor environments. The primary growing area faces directions that maximize consistent daily light exposure while avoiding excessive heat gain during peak summer months.

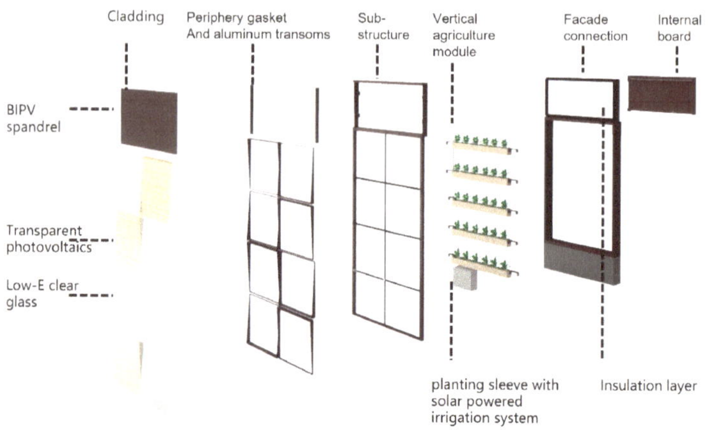


**Figure 2.** Exploded axonometric diagram illustrating the compositional layers of the ABE system as an integrated panel. The diagram details the sequential assembly of components from exterior cladding through to interior finishing, highlighting the interconnection between PV, agricultural, and structural elements.

The aluminum frame's flexibility allows for the customization of dimensions and perforation locations, ensuring precise alignment with PV panels and vertical farming modules. These components are pre-manufactured in the factory and assembled on-site, enhancing on-site installation efficiency and reducing future maintenance complexity. The joint design between the prefabricated components incorporates sealing and elastic materials to prevent the potential impacts of a discontinuous weather barrier (see Section 4.2). Additionally, the glass modules function as shear panels, providing resistance to seismic and lateral wind loads through torsional and shear strength.

The cladding layer adopts a "checkerboard" design prototype. This modularity allows for the customization of the facade's transparent PV glass ratio to meet specific project requirements, thereby balancing energy production with the natural light demands for vertical farming. Moreover, the unitized glass modules are designed with rain louvers and ventilation gaps, which help to mitigate the thermal management and ventilation needs of both the PV system and plant transpiration.

The ABE system can be integrated into the primary structures of new buildings through embedded components or adapted for retrofitting existing buildings by adding a subframe, demonstrating its high architectural compatibility. These design considerations are reflected in the two-dimensional cross-section design (Section 4.1.2), PV mounting design (Section 4.2.1), and connection design (Section 4.2.2).

#### 4.1.2. Two-Dimensional Cross-Sectional Design

Figure 3 exhibits the flow of sunlight, hot air, and cold air in a traditional building facade, highlighting the limitations of traditional facades in temperature regulation and energy utilization. Due to the lack of effective energy management, traditional facades struggle to prevent heat accumulation inside the building, with cooling demand entirely reliant on external equipment, leading to overall low building energy performance. In contrast, the cross-sectional design of the ABE system (see Figure 4) integrates transparent PV thin-film glass with low-E clear glass, while the facade is enhanced by vertical farming modules. Rain louvers and curtain-wall cavities facilitate the circulation of hot and cold air, enhancing energy efficiency and passive temperature regulation.

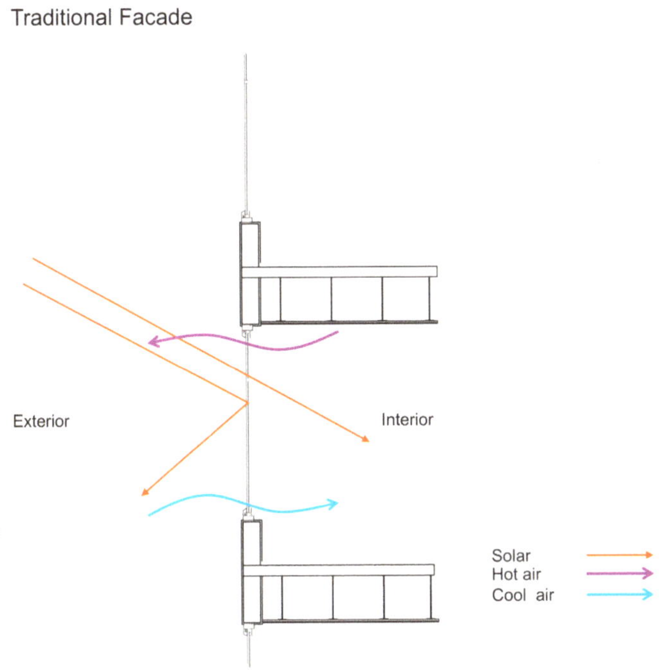


**Figure 3.** A cross-sectional diagram of a conventional building facade showing typical solar radiation, airflow patterns, and thermal behavior. The illustration demonstrates the limitations of traditional facades in managing energy and environmental conditions.

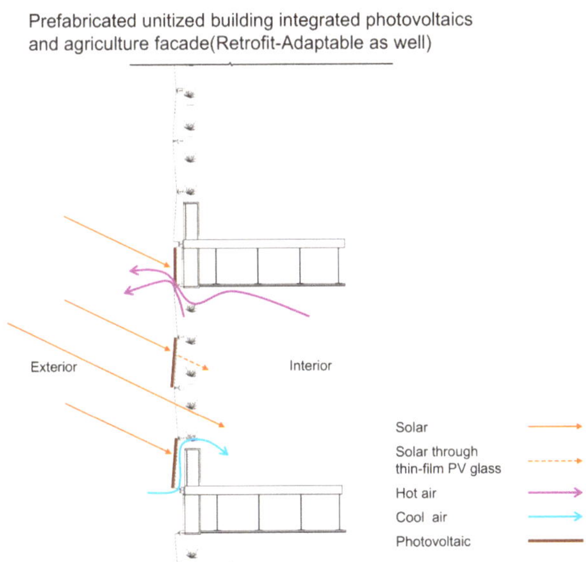


**Figure 4.** A detailed section drawing of the agrivoltaics building envelope system, showing the innovative integration of photovoltaic panels with vertical farming components. The diagram illustrates how the system manages solar radiation, ventilation, and hydroponic circulation while maintaining building envelope functionality.

As mentioned above, the modular design and material selection of the ABE cross-section are determined based on environmental loads and specific project requirements, offering design flexibility to meet local building code regulations. The specific considerations are outlined below:

- Structure (ST) and Sub-structure (Sub-ST): The primary structure of the envelope system consists of a 3 mm thick box-section anodized aluminum extrusion frame, which supports the unitized glass modules. Each "integrated panel" width is 1.5 m. The vertical sub-structure transfers gravity loads while the horizontal sub-structure primarily provides auxiliary connections to prevent structural deflection caused by the large panel width. Additional vertical sub-structures are added in the middle of the integrated panel to improve stability.
- Fireproofing (FP): To meet fire-resistance requirements, fire-retardant coatings should be applied to the aluminum frame to achieve a non-combustible material rating. At the same time, the joints between the layers should be sealed with fireproof rock wool with a density of no less than 80 kg/m$^3$ (Class A). Fire-resistant isolation strips are incorporated to prevent the spread of fire to adjacent components. All other materials should be non-combustible or flame-retardant, and the PV thin-film glass modules must comply with relevant International Electrotechnical Commission standards. Materials and installations should also adhere to local fire safety regulations and fire protection design specifications [37].
- Vapor Barrier (VB): Given the high average temperatures year-round and the high demand for air conditioning cooling in Guangzhou, a vapor barrier is installed in the external wall structure to prevent condensation and protect the building envelope.
- Thermal Performance (TP): During operation, PV panels may reach temperatures exceeding 50 °C, and the heat accumulation can negatively impact long-term energy efficiency and plant growth. Therefore, a 65–100 mm ventilation cavity is placed behind the cladding layer, along with rain louvers at the top and bottom of the "integrated panel" (see Figure 4), to facilitate airflow and mitigate heat buildup. Inadequate ventilation can also lead to high humidity, increasing the risk of agricultural diseases. Notably, the louvers should be designed to automatically close in the event of a fire, complying with local fire-safety regulations regarding airtightness and smoke control.
- Water Barrier (WB): Unitized glass modules as cladding provide an inherent waterproofing property. This effectively prevents rainwater from directly contacting walls and cables. Particular attention is given to joint sealing to ensure long-term waterproofing and structural durability.
- Agricultural Module (AM): The ABE system integrates a vertical farming module, which includes hydroponic systems, support systems, and planting units:
    - Hydroponic System: Each "integrated panel," measuring 1.5 m in width and 2.7 m in height, is equipped with an independent hydroponic circulation system. This system comprises a PV-driven pump and a gravity-based system to recirculate nutrient solutions between planters. Moreover, a cavity within the interior walls is reserved for water tank installation and maintenance.
    - Support system: The support system accommodates plant growth by adjusting to changes in height and weight. The planters are supported by vertical brackets, with hydroponic pipelines embedded along the supporting aluminum structure. The support system for the vertical farming layer is independent of the external curtain-wall structure to prevent the impact of water overflow.
    - Planting units: The indoor planting units are designed for easy maintenance and crop harvesting. Planters feature 150 mm × 100 mm, specifically se-

lected for growing short-cycle, low-to-moderate light-requiring microgreens or leafy vegetables.

Figure 5 shows the elevation of the ABE system's integration panel, while Figure 6 presents the 2D horizontal and vertical cross-sections. The agrivoltaics building envelope design must ensure a safe and comfortable indoor environment for occupants.

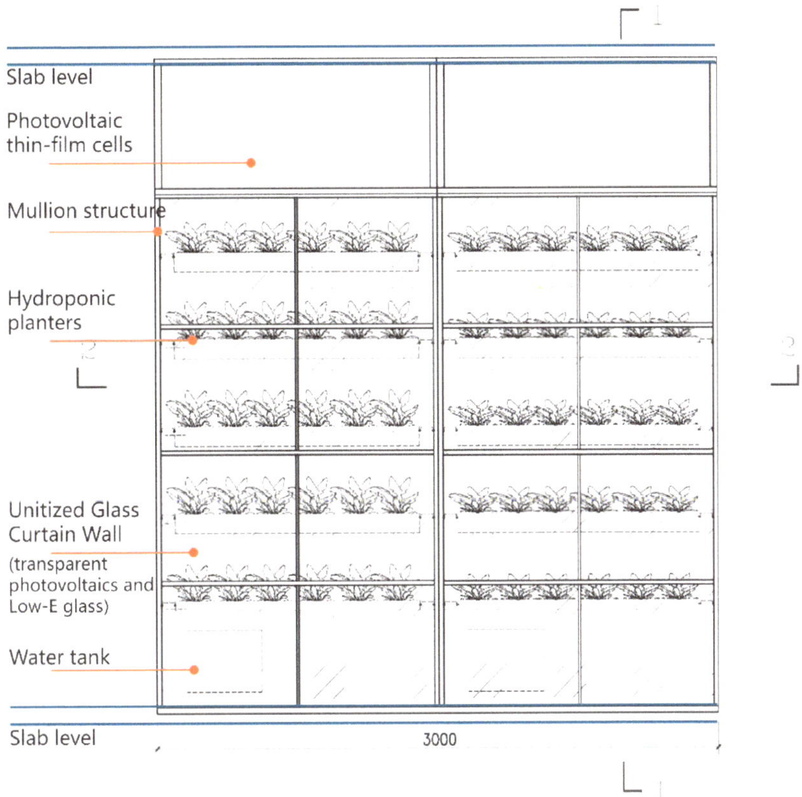


**Figure 5.** Technical elevation drawing of a standard ABE panel unit, detailing dimensional specifications and the spatial relationship between photovoltaic and agricultural components. The drawing includes critical measurements and component identification for fabrication purposes.

## *4.2. Structural Connections and Joint Design*

### 4.2.1. Mounting Design

Common installation methods for thin-film PV curtain walls include adhesive bonding, mechanical fastening, and embedded installation [38]. Adhesive bonding is suitable for flat and stable surfaces, as it does not require additional brackets or supplementary fastening devices. However, it demands high bonding quality and has limited load-bearing capacity. Mechanical fastening, on the other hand, involves using metal brackets, bolts, clamps, and other methods to secure PV modules to the structure. This approach provides high strength and ease of maintenance but requires precise measurement and installation. Exposed brackets and clamps can significantly affect the overall aesthetic of the facade. In addition, using drilled bolts to secure thin-film PV glass panels requires a 50 mm hole in the panel, with at least a 12 mm margin from the edges of the PV module. This method is not suitable for uniform thin-film PV modules. Moreover, anchoring systems such as clamps would cause varying degrees of shading on the PV panels. This study employs a combined approach of embedded installation and mechanical fastening. Figures 7 and 8 illustrate the structural details of the main components.

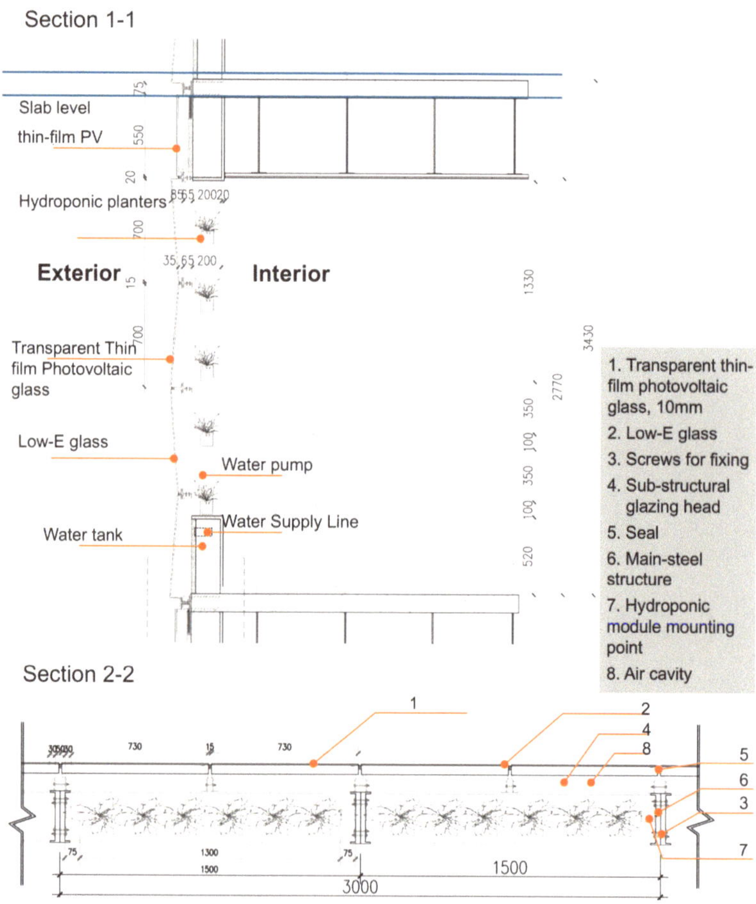


**Figure 6.** Horizontal and vertical sectional details of the integrated PV and vertical farming module, showing precise dimensional specifications and material assemblies. The drawing includes numbered annotations identifying key components and their relationships.

The installation process of the ABE system consists of three main steps. First, during the factory-prefabrication stage, PV thin films are laminated and integrated with glass modules, which are then embedded into prefabricated aluminum frames. The electrical interfaces are sealed and encapsulated, forming independent unitized modules. Second, after the PV modules are transported to the site, they are precisely aligned and connected to the vertical frame and beam system with an embedded slot-and-tab mechanism, following the checkerboard layout. The vertical frames are then bolted to the main building structure using embedded brackets. Finally, the modules are fixed to the vertical frame and beam system, with structural sealant applied to the joints. Interfaces between story and window frames are filled with Class A fire-rated rock wool, in accordance with the curtain-wall system design. Electrical connections are concealed behind the frame and connected to the building's energy management system via an integrated conduit, ensuring the building envelope's aesthetic integrity and the electrical system's safety.

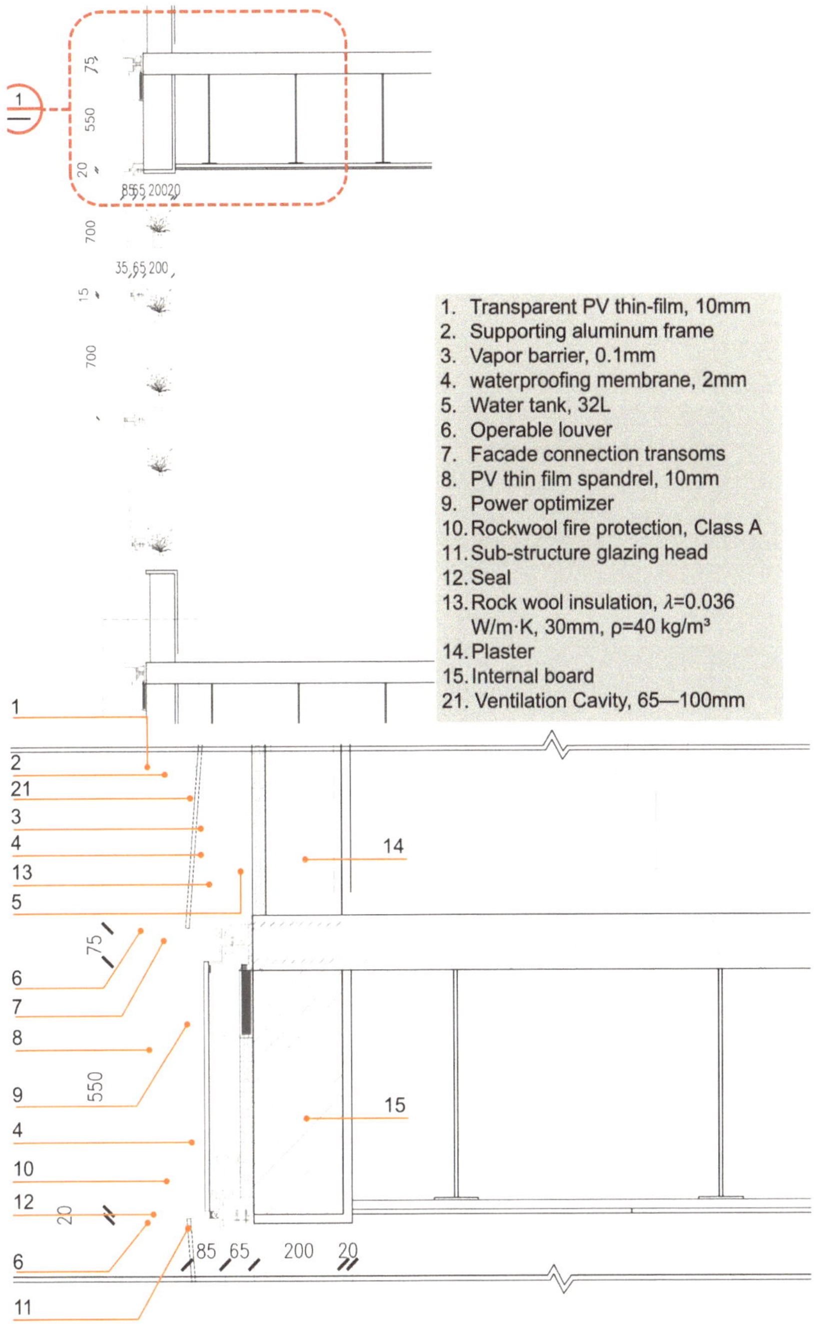


**Figure 7.** Construction detail drawing 1, focusing on the unitized facade system's upper connection points and integration details. The enlarged section highlights critical interfaces between the photovoltaic, structural, and agricultural components.

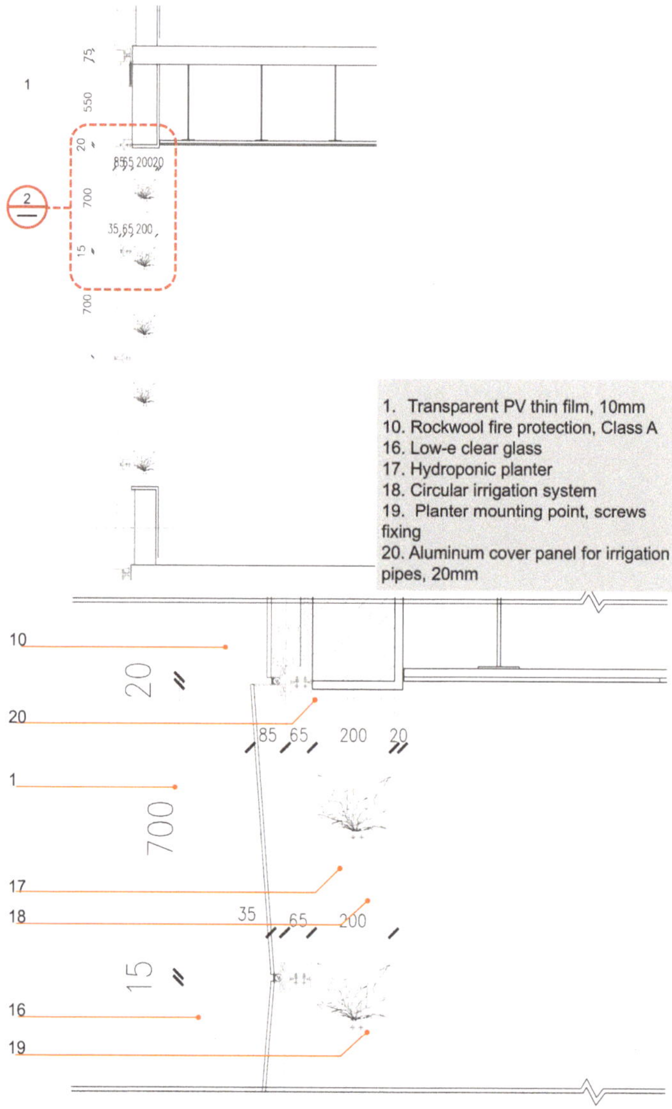


**Figure 8.** Construction detail drawing 2, emphasizing the lower connection points and hydroponic system integration. The detailed section reveals the technical solutions for water management and structural support.

Figure 6 exhibits the vertical and horizontal sections of the ABE system. The "checkerboard" layout allows for localized adjustment to the tilt angle of the thin-film PV glass modules (ranging from 5° to 20°), thereby improving PV energy capture efficiency under low-light conditions. Additionally, rain louvers are installed in alignment with the tilt angles, working together with the cavity to facilitate air circulation. The modular design enables the rapid replacement of individual units during maintenance.

4.2.2. Interlocking Joint Design

Joint design is a core aspect of prefabricated construction that ensures functional continuity and overall structural stability. The joint design process involves addressing challenges such as effectively preventing water and air infiltration, selecting the appropriate assembly method, and ensuring ease of construction and maintenance.

Existing prefabricated building systems typically employ three types [39]: dry connection, wet connection, and hybrid connection. Wet connections rely on on-site casting materials, which have long construction period and significant susceptibility to weather conditions. Hybrid connections combine features of the two and are commonly used in the comprehensive construction of entire buildings. It involves complex construction processes and often requires a certain degree of destructive dismantling during maintenance and retrofitting, leading to high implementation costs.

In contrast, dry connections primarily rely on factory prefabrication and on-site assembly and are the optimal choice for ABE system. The joint design and installation method for the ABE system (as shown in Figure 9) is inspired by the unitized curtain-wall systems. It utilizes an interlocking design with prefabricated embedded components, combined with bolt fastening, to achieve rapid installation and ease of replacement. This interlocking design ensures that modules can be quickly aligned and assembled without the need for additional calibration tools, reducing labor costs.

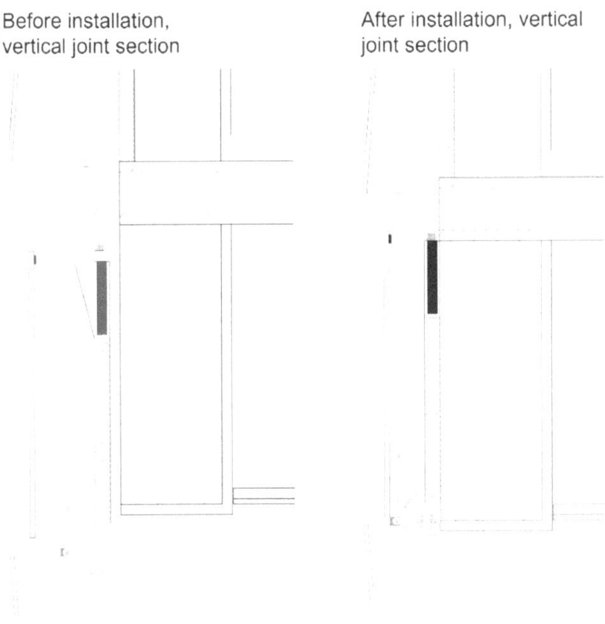


**Figure 9.** Installation sequence diagram showing the vertical joint assembly process before and after installation. The illustration demonstrates the system's innovative interlocking mechanism and installation methodology.

In the ABE system, the connection between the cladding layer and the mullion utilizes a combination of male–female interlocking joints and bolting. As depicted in Figure 9, one side of the panel features a protrusion with a width of 15 mm, precisely fitting into a corresponding 15 mm recessed section on the prefabricated mullion component, forming a male–female interface. After interlocking, the module is then secured to the post using M10 bolts to ensure stability. Depending on the panel installation angle, the embedding depth of the upper and lower grooves is 18 mm and 70 mm, respectively. In addition, elastic sealing strips are also embedded at the joints to prevent direct rainwater infiltration.

Figures 10 and 11 shows the sectional design and installation method of the horizontal joints of the "integrated panel" in the ABE system. The adjacent "integrated panels" are secured by screw-fastening through the corner joints of the supporting frame.

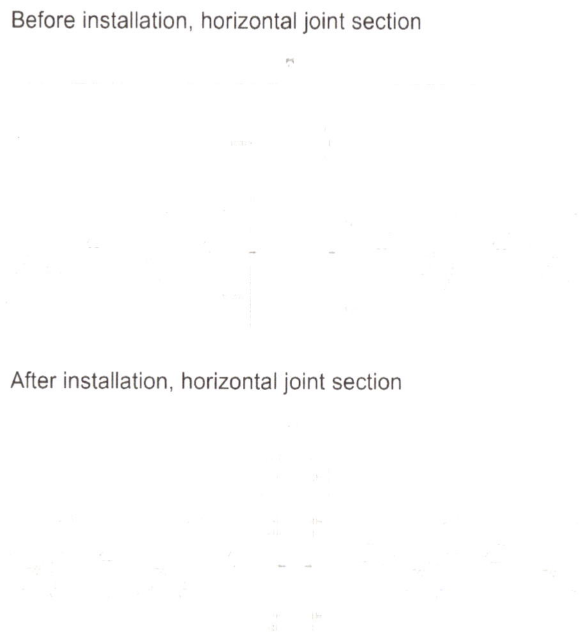


**Figure 10.** Horizontal joint assembly sequence highlighting the interlocking design that enables efficient dry installation. The diagram shows the connection details and alignment process for adjacent panels.

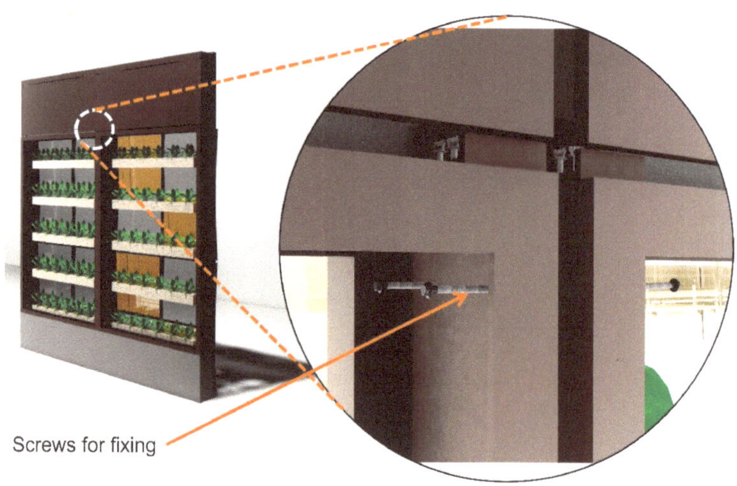


**Figure 11.** Corner detail illustration showing the structural frame connection method, using strategic screw placement to ensure stability and alignment at panel intersections.

Apart from that, integrating modular vertical farming into the building envelope is also a key focus in the ABE system design. Considering the variability in crop characteristics in hydroponic cultivation systems, coupled with the spatial variation in solar irradiance across different climatic zones. The vertical farming system features a modular design supported by aluminum frame with pre-drilled openings. Planters are equipped with L-shaped protrusions on both sides, which are dry-mounted with adjustable screws to the perforated aluminum support frame (see Figure 12). Elastic seals are added at the junctions to mitigate the dynamic loads of the hydroponic liquid. This design ensures the load-bearing capacity of the modules while enabling quick "plug and play" installation. Furthermore, the installation, maintenance, and harvesting of the agricultural modules can all be performed indoors, enhancing operational safety and convenience.

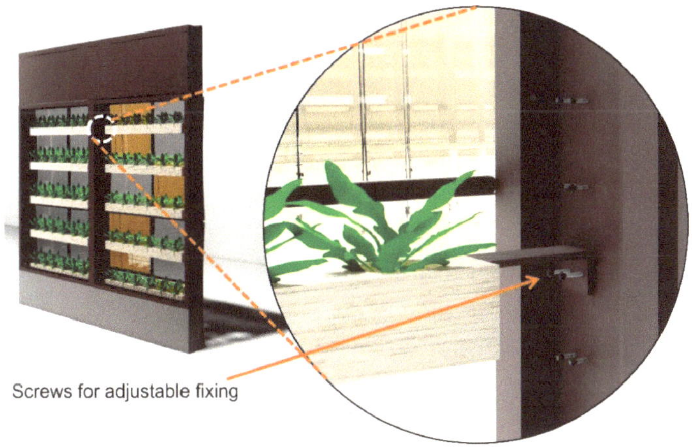


**Figure 12.** Technical detail of the adjustable mounting system for vertical farming modules, demonstrating the flexibility in installation and maintenance access.

### *4.3. Three-Dimensional Modeling*

#### 4.3.1. System Renders

To demonstrate the aesthetic and functional integration of the ABE system within actual architectural settings, three-dimensional modeling and photorealistic rendering techniques were employed. These visualization methods effectively illustrate both the interior and exterior appearances while highlighting the system's modular construction details.

Figures 1 and 2 facilitate the evaluation of the integrated panel dimensions and the sequencing of material layers in the design. These renders illustrate the "checkerboard" arrangement of the PV thin-film glass modules and the low-E clear glass modules within the facade cladding. Over and above that, to replicate practical installation conditions, the 3D model depicts the connection method between the "integrated panels" (Figure 11) and the modular installation mechanism for vertical agriculture (Figure 12).

Figure 13 highlights the visual effect of the building facade with the deployment of "integrated panel". The inherent tilt angle of the unitized glass module creates dynamic light and shadow effects on the facade. At the same time, the interior perspective rendering shown in Figure 14 demonstrates how the transparent glass module evenly introduces natural light for the vertical farming modules, while generating energy. This combination supports crop growth while minimizing glare and heat buildup in the interior environment. These findings provide a valuable reference for the operability of the renewable "energy-vertical farming" system.

**Figure 13.** Architectural rendering showing the aesthetic integration of the ABE system into a building facade, illustrating how the system maintains visual coherence while providing functional benefits.

**Figure 14.** Interior perspective rendering demonstrating the user experience and spatial quality achieved through the integration of vertical farming components within the occupied space.

#### 4.3.2. Facade Design Flexibility Through PV Proportion Variations

ABE system utilizes a grid-connected configuration to ensure the reliable operation of the vertical farming components. The PV modules generate power during daylight hours to support the hydroponic system's essential functions, including water circulation pumps and environmental controls. During periods of reduced solar availability, the system draws power from the building's electrical grid to maintain continuous operation of critical farming systems. This grid-interactive approach provides operational stability while maintaining the sustainability benefits of solar power generation. Excess energy generated during peak sunlight hours is fed back to the building's electrical infrastructure, contributing to overall energy efficiency. While the current study focuses on the integrated design and structural aspects of the system, future research will investigate detailed energy performance metrics and potential implementation of smart control systems to optimize power distribution.

This configuration meets the functional and visual requirements of different building environments. Figure 15 illustrates a facade module design in which thin-film PV glass and low-E transparent glass each occupy 50% of the surface. This design allows for adequate natural light for the vertical farming modules, while the PV elements power the irrigation system.

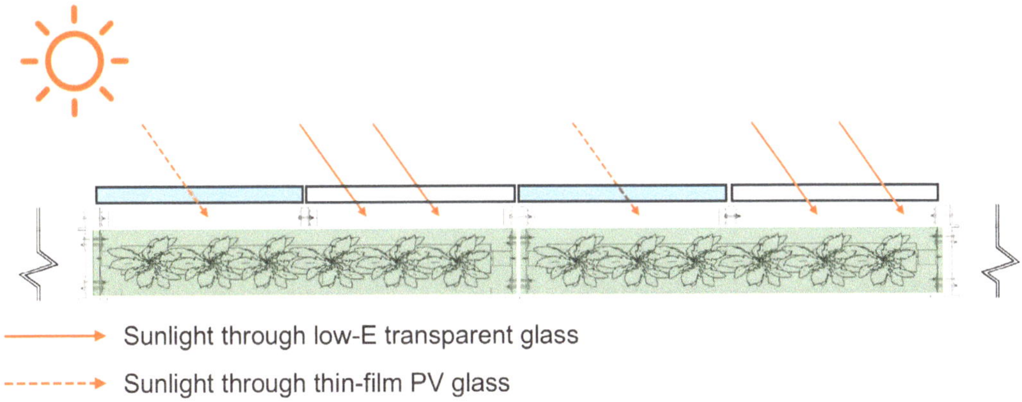


**Figure 15.** Diagram illustrating solar light transmission through the facade module design, showing how 50% PV glass integration with 50% low-E glass supports plant growth while maintaining energy generation.

The design flexibility of the ABE system is evident not only in the proportional configuration but also in the spatial arrangement of its components. As illustrated in Figure 16, the 3D rendering showcases the diverse array of options available for arranging the system's modules. These variations provide the design team with valuable reference examples for tailoring the system to meet diverse architectural and performance requirements:

- Top left (25% Transparent Colored BIPV): A combination of 25% colored thin-film PV glass and 75% low-E clear glass improves light transmission, thus reducing energy consumption for vertical agriculture. Particularly suitable for buildings with high lighting requirements or in regions with low solar irradiance.
- Top right (50% Transparent Colored BIPV): A 50% coverage of colored thin-film PV glass enhances energy production while adding aesthetic vibrancy to the building facade. It is ideal for structures where energy generation and architectural aesthetics are highly prioritized.
- Bottom left (25% Transparent BIPV): An incorporation of 25% transparent thin-film PV glass optimizes the natural light penetration, mitigating the light-filtering effect of the PV modules, while maintaining a certain level of power generation.
- Bottom right (50% Transparent Colored BIPV): A 50% coverage of colored thin-film PV glass, arranged in varying patterns, allows for diverse facade compositions, and enhances the building's aesthetic experience.

Balancing energy production with natural lighting requirements remains a critical challenge in the design of agrivoltaics-integrated building facades. These configurations demonstrate the ABE system's capability to balance aesthetics and functionality, creating dynamic light and shadow effects while allowing flexible adjustments in PV coverage based on specific sites, contributing to urban sustainability.

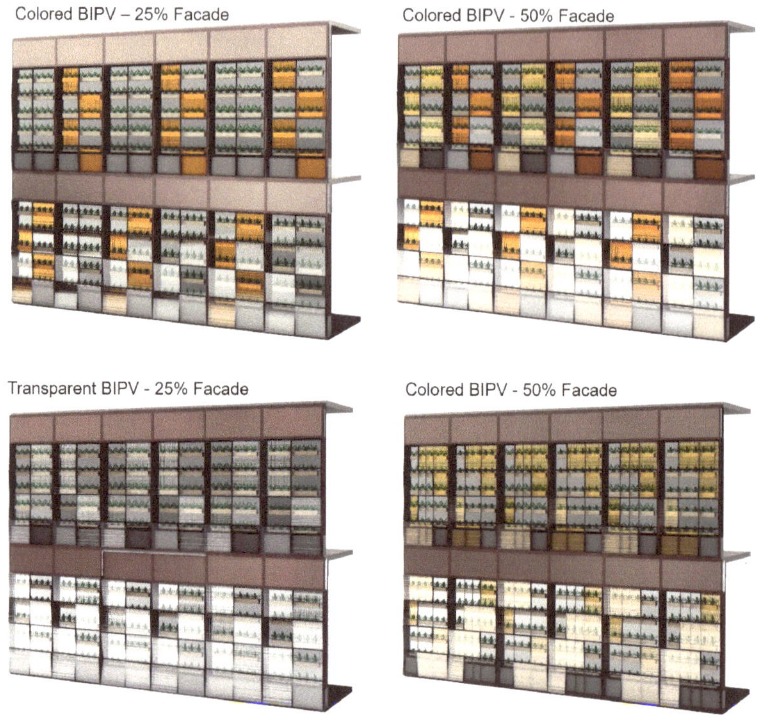


**Figure 16.** A comparative visualization of different PV coverage options (25% and 50%) and their impact on facade appearance and performance, including both colored and transparent PV variations.

### 4.4. Water and Nutrient Circulation in the Vertical Farming System

The PV-driven nutrient recirculation system of the ABE is presented in Figure 17. Each "integrated panel" is equipped with an independent hydroponic recirculation unit, containing a 32 L nutrient solution tank. The PV modules generate direct current (DC) electricity to power the water pump, which drives the vertical transport of the nutrient solution, while gravity facilitates the return flow, thereby completing the circulation cycle. The nutrient solution is recirculated every two hours to ensure a consistent supply of adequate water and nutrients for the crops. The nutrient reservoir is strategically positioned within the interior wall of the building, simplifying regular maintenance and cleaning.

The main hydroponic pipeline and the vertical farming layer support system are embedded within a box-section frame, ensuring a clean and aesthetically interior perspective. The planters are mounted on an aluminum frame through an adjustable bracket, which enables adaptability to accommodate varying crop heights and weight changes. The sub-hydroponic pipelines are assembled based on the number of trays and links to the main pipeline via flexible hoses to meet various building scenarios.

A study conducted in Europe revealed that in Barcelona, vertical farming modules cultivating leafy greens, such as lettuce and spinach, maintained a high yield even with a PV coverage rate of approximately 50%. This optimal PV coverage is twice the recommended ratio for vertical farming in the Paris region [40]. Additionally, the use of semi-transparent PV glass can further improve light transmittance, ensuring agricultural yield [41]. This "energy-agriculture" design reduces reliance on the conventional power grid and optimizes the system's carbon footprint.

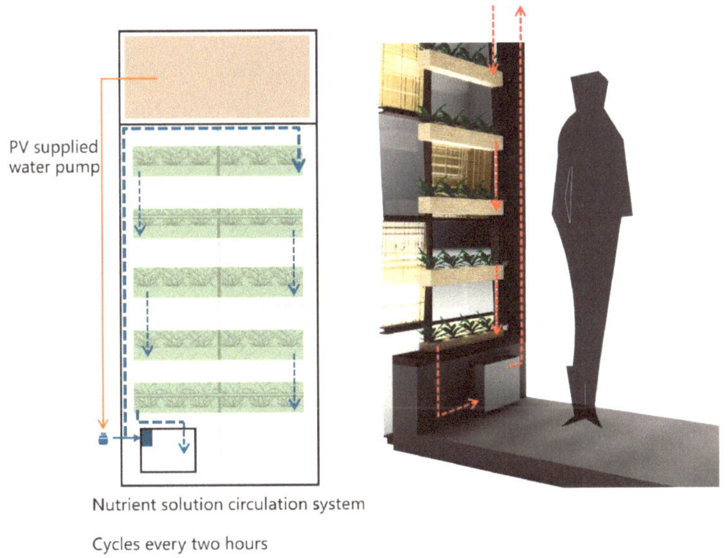


**Figure 17.** A schematic diagram of the nutrient solution recirculation system powered by integrated photovoltaic panels, illustrating the closed-loop sustainability approach of the ABE system.

The system's agricultural production capacity was evaluated using Chinese spinach as a reference crop [36], with yields of approximately 8.45 kg per integrated panel annually under Singapore's tropical climate conditions. This productivity relies on several key inputs: nutrient solution consumption of approximately 2 L per day per panel, supplemental LED lighting during low-light periods averaging 0.3 kWh daily, and regular system maintenance every 3–4 weeks for harvesting and replanting. Production cycles vary seasonally despite Singapore's tropical climate, with typical growing periods of 25–30 days during the northeast monsoon season (December–February) extending to 35–40 days during the southwest monsoon period (June–September) due to variations in natural light availability and ambient temperature. These growing cycle variations are managed through adjustments in supplemental lighting and environmental controls to maintain consistent production throughout the year. This yield capacity could satisfy approximately 6% of an adult's annual vegetable consumption, based on the World Health Organization's dietary guidelines recommending a minimum daily intake of 400 g of fruits and vegetables for adults [42].

## 5. Discussion

This study presents the design analysis and three-dimensional modeling of a novel ABE prototype. Unlike conventional prefabricated BIPV systems, this innovative design combines transparent thin-film PV glass with vertical farming in a modular approach. It departs from the previous research paradigm that focuses solely on energy generation and PV placement, achieving an integrated solution of renewable energy production and vertical agricultural cultivation. On top of that, this research further explores the dynamic balance among PV coverage, natural light transmission, and integrated building–agriculture systems.

The ABE product in this study is designed for building envelopes in urban environments. The system demonstrates significant flexibility and adaptability, utilizing a recycled aluminum frame and a standardized modular design. This allows for customized, rapid deployment across diverse building scenarios, reducing construction and maintenance costs. Previous studies on prefabricated BIPV systems, as indicated by [43], have predominantly concentrated on optimizing PV efficiency and augmenting passive functions, such as thermal insulation. The contribution of this study lies beyond the conventional focus on energy transition; it introduces a novel approach to addressing the dual resource scarcity of energy and agriculture in urban environments by adopting design strategies.

While the system's initial implementation costs limit its immediate adoption to high-end developments, ongoing optimization and economies of scale are expected to improve economic viability for broader applications. The controlled environment agriculture approach, with its integrated filtration and monitoring systems, provides superior protection from urban pollutants compared to conventional outdoor urban farming, as demonstrated by our contaminant analysis results.

## 6. Conclusions

This study advances the field of sustainable building design through the development of an innovative agrivoltaics building envelope (ABE) system that addresses multiple critical urban challenges simultaneously. The significance of this work lies in its novel approach to transforming conventional building facades from passive surfaces into productive spaces that generate both renewable energy and food, directly contributing to urban sustainability and resilience.

The research makes several important contributions to the field. First, it demonstrates the technical feasibility of integrating PV systems with vertical farming in a modular building envelope design, providing a practical solution for space-constrained urban environments. Second, the system's adaptable design framework, incorporating recycled materials and prefabricated components, offers a scalable approach that can be implemented across diverse building types and urban contexts. Third, the mixed-method research approach combining Research through Design (RtD) and Research for Design (RfD) methodologies establishes a robust foundation for the future development of similar integrated systems.

From an implementation perspective, the modular design and prefabrication approach significantly reduce construction complexity while maintaining architectural flexibility. The system's ability to balance energy generation with agricultural production through adjustable PV coverage demonstrates the potential for optimizing multiple functions within a single building envelope solution. This multifunctional approach represents a significant advance in sustainable building technology, offering a pathway toward more self-sufficient and productive urban structures.

As cities worldwide face increasing pressures from climate change, food security concerns, and energy demands, this research provides a practical framework for addressing these challenges through integrated building design. The ABE system demonstrates how thoughtful architectural innovation can contribute to urban sustainability while maintaining the aesthetic and functional requirements of modern buildings. This work lays the foundation for future research and development in building-integrated agriculture and renewable energy systems, pointing the way toward more resilient and sustainable urban environments.

*Research Limitations and Future Directions*

The current study of the ABE system faces several key limitations that shape future research priorities. A primary constraint is the lack of a comprehensive economic analysis of implementation costs and operational requirements. Based on preliminary estimates, the system's primary expenses would include material costs (structural components, PV panels, hydroponic systems), operational inputs (water consumption of approximately 2 L/day/m$^2$, nutrient solutions, supplemental lighting energy), and labor requirements for regular maintenance and harvesting cycles. While local market conditions in Singapore suggest potential viability through combined agricultural revenue and energy savings, a detailed cost–benefit analysis across different scales of implementation remains necessary. Future work must quantify these economic parameters alongside environmental impacts, carbon emissions, and resource consumption through systematic study of actual installations. This analysis should establish clear metrics for operational costs, maintenance requirements, labor needs, and realistic payback periods under various implementation scenarios.

While this research focuses on optimizing design parameters for PV coverage and natural light transmission, the actual impact on indoor environmental conditions and crop growth remains theoretical. Mock-up testing under real environmental conditions is essential to validate system performance. Future experimental studies should incorporate real-time monitoring and data analysis to evaluate energy self-sufficiency between PV and vertical agriculture modules, providing empirical evidence for the system's effectiveness in various climatic conditions.

The potential accumulation of urban pollutants in hydroponic crops presents another significant limitation requiring thorough investigation. Future research must establish comprehensive protocols for monitoring contaminant levels in both growing environments and harvested produce, particularly focusing on heavy metals, hydrocarbons, and other atmospheric pollutants common in urban settings. This work should inform the development of enhanced filtration systems and safety standards for urban agricultural production.

To address these limitations, system intelligence must be enhanced through integrated monitoring and control mechanisms. This development should include smart sensors for environmental monitoring, the automated adjustment of growing conditions, and real-time performance optimization. Machine learning algorithms could help balance energy generation with agricultural productivity while maintaining optimal growing conditions.

Finally, the successful implementation of ABE systems requires establishing comprehensive design standards that account for regional variations in building codes, climate conditions, and environmental regulations. Future research should focus on developing detailed specifications for safety measures, maintenance protocols, and quality control procedures, enabling widespread adoption across different urban contexts. These standards must integrate lessons learned from real-world implementation while maintaining adaptability to local conditions and requirements.

**Author Contributions:** Methodology, Y.Z. and T.C.; software, Y.Z.; validation, Y.Z. and T.C.; formal analysis, Y.Z., E.L. and T.C.; investigation, Y.Z., E.L, E.G. and T.C.; resources, Y.Z.; data curation, Y.Z.; writing—original draft preparation, Y.Z.; writing—review and editing, T.C. and E.L.; visualization, Y.Z.; supervision, T.C., E.G. and E.L.; project administration, T.C.; funding acquisition, E.G. and T.C. All authors have read and agreed to the published version of the manuscript.

**Funding:** This research is supported by the National Research Foundation, Singapore and University of Sydney, under their NUS-USYD Ignition Grant, WBS A-8002572-01-00. SERIS is a research institute at the National University of Singapore (NUS). SERIS is supported by NUS, the National Research Foundation Singapore (NRF), the Energy Market Authority of Singapore (EMA) and the Singapore Economic Development Board (EDB).

**Institutional Review Board Statement:** Not applicable.

**Informed Consent Statement:** Not applicable.

**Data Availability Statement:** The authors confirm that the data supporting the findings of this study are available within the article.

**Conflicts of Interest:** The authors declare no conflicts of interest.

## Nomenclature

| Abbreviations | |
|---|---|
| 2D | Two-dimensional |
| 3D | Three-dimensional |
| ABE | Agrivoltaics building envelope |
| AM | Agricultural Module |
| BIPV | Building-integrated photovoltaics |
| FP | Fireproofing |
| PV | Photovoltaics |
| RtD | Research through Design |
| RfD | Research for Design |
| ST | Structure |
| Sub-ST | Sub-structure |
| TP | Thermal Performance |
| WB | Water Barrier |
| *Symbols* | |
| ρ | Density (kg/m$^3$) |
| λ | Thermal conductivity (W/mK) |

## References

1. United Nations Department of Economic and Social Affairs 2018 Revision of World Urbanization Prospects. 2018. Available online: https://www.un.org/development/desa/en/news/population/2018-revision-of-world-urbanization-prospects.html (accessed on 12 November 2024).
2. Wageningen University & Research Meta-Analysis Shows That Future Food Demand Will Increase Between 35–56% over the Period 2010–2050. 2021. Available online: https://www.wur.nl/en/Research-Results/Research-Institutes/Economic-Research/show-wecr/Meta-analysis-shows-that-future-food-demand-will-increase-between-35-56-over-the-period-2010-2050.htm (accessed on 12 November 2024).
3. Intergovernmental Panel on Climate Change Chapter 3: Desertification. In *Special Report on Climate Change and Land*; Intergovernmental Panel on Climate Change: Geneva, Switzerland, 2019; Available online: https://www.ipcc.ch/srccl/chapter/chapter-3/ (accessed on 15 November 2024).
4. Wang, G.; Luo, T.; Luo, H.; Liu, R.; Liu, Y.; Liu, Z. A Comprehensive Review of Building Lifecycle Carbon Emissions and Reduction Approaches. *City Built Environ.* **2024**, *2*, 12. [CrossRef]
5. National Renewable Energy Laboratory Agrivoltaics. 2020. Available online: https://www.nrel.gov/solar/market-research-analysis/agrivoltaics.html (accessed on 12 November 2024).
6. Dai, S.; Lin, Z.; Hu, H.; Wang, Y.; Zeng, L. 3D Printing for Sodium Batteries: From Material Design to Integrated Devices. *Appl. Phys. Rev.* **2024**, *11*, 041319. [CrossRef]
7. Ma, F.; Wu, Y.; Dai, S.; Lin, P.; Sun, J.; Dong, L. A Soft-Contact Hybrid Electromagnetic–Triboelectric Nanogenerator for Self-Powered Water Splitting towards Hydrogen Production. *Nano Res.* **2024**, *17*, 6567–6574. [CrossRef]
8. Zhang, K.; Wang, H.; Du, X.; Dai, S.; Wang, Y.; Xu, T.; Liu, M.; Cheng, S. Reaction Mechanism and Structural Evolution of Tunnel-Structured KCu7S4 Nanowires in Li+/Na+-Ion Batteries. *Adv. Funct. Mater.* **2024**, *34*, 2407105. [CrossRef]
9. Shah, Z.; Veses, R.C.; Brum, J.; Klunk, M.A.; Rocha, L.A.O.; Missaggia, A.B.; de Oliveira, A.P.; Caetano, N.R. Sewage Sludge Bio-Oil Development and Characterization. *Inventions* **2020**, *5*, 36. [CrossRef]
10. Bao, Y.; Leung, M.K.; Poon, D.; Xiang, C. Integrating Vertical Farm into Low-Carbon High-Rise Building in High-Density Cities: A Case Study of Hong Kong. *J. Build. Eng.* **2024**, *96*, 110472. [CrossRef]
11. U.S. Department of Energy Making Strides in Building Envelope Technologies. 2016. Available online: https://www.energy.gov/eere/buildings/articles/making-strides-building-envelope-technologies (accessed on 12 November 2024).
12. Chen, T.; Tai, K.F.; Raharjo, G.P.; Heng, C.K.; Leow, S.W. A Novel Design Approach to Prefabricated BIPV Walls for Multi-Storey Buildings. *J. Build. Eng.* **2023**, *63*, 105469. [CrossRef]
13. Caplow, T. Building Integrated Agriculture: Philosophy and Practice. In *Urban Futures 2030*; Heinrich Böll Stiftung: Berlin, Germany, 2009; Available online: https://www.researchgate.net/publication/312442295_Building_integrated_agriculture_philosophy_and_practice_Urban_Futures_2030 (accessed on 12 November 2024).
14. Qushakov, S.; Eshkulov, M. Semi-Transparent Solar Panels for Greenhouses. Available online: https://contest.techbriefs.com/2024/entries/sustainable-technology/12766?utm_source=chatgpt.com (accessed on 9 January 2025).

15. Ma, C.; Yuan, C.; Zhang, Y.; Hu, H. Mapping Utilizable Rooftop Areas to Meet Food Security Goal in Four High-Density Cities: A Deep Learning and GIS Integrated Approach. *Sustain. Cities Soc.* **2024**, *118*, 106066. [CrossRef]
16. Vidyarthi, P.K. A Path Towards Net-Zero Emission Using Building Integrated Photovoltaic: A Review. In *Proceedings from the International Conference on Hydro and Renewable Energy*; Hodge, B.-M., Prajapati, S.K., Eds.; Lecture Notes in Civil Engineering; Springer Nature: Singapore, 2024; Volume 391, pp. 249–253. ISBN 978-981-9966-15-8.
17. Attoye, D.E.; Aoul, K.A.T. A Review of the Significance and Challenges of Building Integrated Photovoltaics. In *Energy Efficient Building Design*; Springer: Cham, Switzerland, 2020.
18. Mishra, S.; Karetha, K.M.; Yau, Y.Y.; Easterling, M. Vertical Cultivation: Moving Towards a Sustainable and Eco-Friendly Farming. In *Biotechnological Innovations for Environmental Bioremediation*; Springer: Singapore, 2022.
19. AL-Zoubi, O.H.; Al-Tahaineh, H.; Damseh, R.A.; AL-Zubi, A.H.; Odat, A.S.; Shboul, B. Evaluating the Real-World Performance of Vertically Installed Bifacial Photovoltaic Panels in Residential Settings: Empirical Findings and Implications. *Int. J. Low-Carbon Technol.* **2024**, *19*, 386–442. [CrossRef]
20. International Energy Agency Photovoltaic Power Systems Programme Design Tools for Building-Integrated Photovoltaics: Current Status and Perspectives. 2020. Available online: https://iea-pvps.org/wp-content/uploads/2020/01/IEA-PVPS_15_R09_BIPV_Design_Tools_report.pdf (accessed on 12 November 2024).
21. International Renewable Energy Agency; Food and Agriculture Organization Renewable Energy for Agri-Food Systems. 2021. Available online: https://www.irena.org/-/media/Files/IRENA/Agency/Publication/2021/Nov/IRENA_FAO_Renewables_Agrifood_2021.pdf (accessed on 12 November 2024).
22. Aye, L.; Ngo, T.; Crawford, R.H.; Gammampila, R.; Mendis, P. Life Cycle Greenhouse Gas Emissions and Energy Analysis of Prefabricated Reusable Building Modules. *Energy Build.* **2012**, *47*, 159–168. [CrossRef]
23. Jelle, B.P.; Breivik, C.; Røkenes, H.D. Building Integrated Photovoltaic Products: A State-of-the-Art Review and Future Research Opportunities. *Sol. Energy Mater. Sol. Cells* **2012**, *100*, 69–96. [CrossRef]
24. Li, Z.; Shen, G.Q.; Xue, X. Critical Review of the Research on the Management of Prefabricated Construction. *Habitat Int.* **2014**, *43*, 240–249. [CrossRef]
25. Zhang, S.; Wu, Z.; Li, H. Critical Factors Influencing Interface Management of Prefabricated Building Projects. *Sustainability* **2022**, *14*, 5418. [CrossRef]
26. Zhang, Y.; Park, N.-G. A Thin Film (<200 Nm) Perovskite Solar Cell with 18% Efficiency. *J. Mater. Chem. A* **2020**, *8*, 17612–17620. [CrossRef]
27. U.S. Department of Energy Summary: Challenges and Opportunities for Building-Integrated Photovoltaics. Available online: https://www.energy.gov/eere/solar/summary-challenges-and-opportunities-building-integrated-photovoltaics-rfi (accessed on 12 November 2024).
28. Chen, T.; Sun, H.; Tai, K.F.; Heng, C.K. Analysis of the Barriers to Implementing Building Integrated Photovoltaics in Singapore Using an Interpretive Structural Modelling Approach. *J. Clean. Prod.* **2022**, *365*, 132652. [CrossRef]
29. Avgoustaki, D.D.; Xydis, G. Indoor Vertical Farming in the Urban Nexus Context: Business Growth and Resource Savings. *Sustainability* **2020**, *12*, 1965. [CrossRef]
30. Soto-Gómez, D. Integration of Crops, Livestock, and Solar Panels: A Review of Agrivoltaic Systems. *Agronomy* **2024**, *14*, 1824. [CrossRef]
31. La Notte, L.; Giordano, L.; Calabrò, E.; Bedini, R.; Colla, G.; Puglisi, G.; Reale, A. Hybrid and Organic Photovoltaics for Greenhouse Applications. *Appl. Energy* **2020**, *278*, 115582. [CrossRef]
32. Barron-Gafford, G.A.; Pavao-Zuckerman, M.A.; Minor, R.L.; Sutter, L.F.; Barnett-Moreno, I.; Blackett, D.T.; Thompson, M.; Dimond, K.; Gerlak, A.K.; Nabhan, G.P.; et al. Agrivoltaics Provide Mutual Benefits across the Food–Energy–Water Nexus in Drylands. *Nat. Sustain.* **2019**, *2*, 848–855. [CrossRef]
33. Trommsdorff, M.; Kang, J.; Reise, C.; Schindele, S.; Bopp, G.; Ehmann, A.; Weselek, A.; Högy, P.; Obergfell, T. Combining Food and Energy Production: Design of an Agrivoltaic System Applied in Arable and Vegetable Farming in Germany. *Renew. Sustain. Energy Rev.* **2021**, *140*, 110694. [CrossRef]
34. Xiao, Y.; Bhola, J. Design and Optimization of Prefabricated Building System Based on BIM Technology. *Int. J. Syst. Assur. Eng. Manag.* **2022**, *13*, 111–120. [CrossRef]
35. O'Neill, G.; Ball, M.; Liu, Y.; Noghabaei, M.; Han, K. Toward Automated Virtual Assembly for Prefabricated Construction: Construction Sequencing through Simulated BIM. *arXiv* **2020**, arXiv:2003.06695.
36. Song, S.; Cheong, J.C.; Lee, J.S.H.; Tan, J.K.N.; Chiam, Z.; Arora, S.; Png, K.J.Q.; Seow, J.W.C.; Leong, F.W.S.; Palliwal, A.; et al. Home Gardening in Singapore: A Feasibility Study on the Utilization of the Vertical Space of Retrofitted High-Rise Public Housing Apartment Buildings to Increase Urban Vegetable Self-Sufficiency. *Urban For. Urban Green.* **2022**, *78*, 127755. [CrossRef]
37. GB 50045-2019*Code for Fire Protection Design of Buildings*; Standardization Administration of China China Architecture & Building Press: Beijing, China, 2015. Available online: https://www.gov.cn/zhengce/zhengceku/2023-01/30/content_5739161.htm (accessed on 12 November 2024).

38. Kuhn, T.E.; Erban, C.; Heinrich, M.; Eisenlohr, J.; Ensslen, F.; Neuhaus, D.H. Review of Technological Design Options for Building Integrated Photovoltaics (BIPV). *Energy Build.* **2021**, *231*, 110381. [CrossRef]
39. Chang, R.; Zhang, N.; Gu, Q. A Review on Mechanical and Structural Performances of Precast Concrete Buildings. *Buildings* **2023**, *13*, 1575. [CrossRef]
40. Moreno, Á.; Chemisana, D.; Lamnatou, C.; Maestro, S. Energy and Photosynthetic Performance Investigation of a Semitransparent Photovoltaic Rooftop Greenhouse for Building Integration. *Renew. Energy* **2023**, *215*, 118976. [CrossRef]
41. Gorjian, S.; Bousi, E.; Özdemir, Ö.E.; Trommsdorff, M.; Kumar, N.M.; Anand, A.; Kant, K.; Chopra, S.S. Progress and Challenges of Crop Production and Electricity Generation in Agrivoltaic Systems Using Semi-Transparent Photovoltaic Technology. *Renew. Sustain. Energy Rev.* **2022**, *158*, 112126. [CrossRef]
42. World Health Organization. *Carbohydrate Intake for Adults and Children: WHO Guideline*; World Health Organization: Geneva, Switzerland, 2023; Available online: https://www.who.int/publications/i/item/9789240073593 (accessed on 12 November 2024).
43. Heinstein, P.; Ballif, C.; Perret-Aebi, L.-E. Building Integrated Photovoltaics (BIPV): Review, Potentials, Barriers and Myths. *Green* **2013**, *3*, 125–156. [CrossRef]

sustainability

MDPI

Article

# Integrating Technological Environmental Design and Energy Interventions in the Residential Building Stock: The Pilot Case of the Small Island Procida

**Giada Romano, Serena Baiani and Francesco Mancini ***

Planning Design Technology of Architecture Department, "Sapienza" University of Rome, 00196 Rome, Italy; giada.romano@uniroma1.it (G.R.); serena.baiani@uniroma1.it (S.B.)
* Correspondence: francesco.mancini@uniroma1.it

**Abstract:** The next decade will see severe environmental and technological risks, pushing our adaptive capacity to its limits. The EPBD Case Green directive, to counter this phenomenon, emphasizes accelerating building renovations, reducing GHG emissions and energy consumption, and promoting renewable energy installations. Additionally, it calls for deadlines to phase out fossil fuels and mandates solar system installations. This research provides a comprehensive perspective on the opportunities for and challenges of incorporating renewable energy into the built environment. It focuses on the 2961 residential buildings on Procida, a small island located south of Italy, to efficiently utilize energy resources and lay the groundwork for sustainability. Beginning with an analysis of the territorial, urban, historical–conservation, structural, and geological context, in addition to environmental assessments, the research develops a classification and archetypalization system using in-house software. This system aggregates data on the island's residential buildings, analyzes their current state, and formulates various intervention scenarios. These scenarios demonstrate how integrating technological–environmental design interventions, such as upgrading the building envelope and enhancing bioclimatic behavior, with energy retrofitting measures, such as replacing mechanical systems and installing solar panels, can improve the overall performance of the existing building stock and achieve energy self-sufficiency.

**Keywords:** energy efficiency; climate change mitigation; accelerating building renovations; technological–environmental design interventions; promoting renewable energy installations; optimization of energy use; Italian residential buildings

**Citation:** Romano, G.; Baiani, S.; Mancini, F. Integrating Technological Environmental Design and Energy Interventions in the Residential Building Stock: The Pilot Case of the Small Island Procida. *Sustainability* **2024**, *16*, 8071. https://doi.org/10.3390/su16188071

Academic Editor: Antonio Caggiano

Received: 19 August 2024
Revised: 12 September 2024
Accepted: 13 September 2024
Published: 15 September 2024



## 1. Introduction

The next decade will bring significant environmental and technological risks, which will become dominant global concerns [1]. While nations are committed to a 2 °C reduction under the Paris Agreement, the global economy must cut emissions by 45% within the next seven years to avoid catastrophic climate impacts [2,3]. The built environment plays a key role in exceeding net-zero targets, potentially achieving nature-positive outcomes that enhance climate resilience [4,5]. The built environment is crucial in addressing the climate crisis, and it is fundamental to sustainability efforts. Achieving net zero in the building sector requires more than just efficient design; it necessitates the integration of renewable energy supply and distribution [6,7]. Buildings should drive the energy transition to renewable energy in order to effectively transition the whole built environment toward net zero [8,9].

The buildings and construction sector contributes significantly to global climate change, accounting for about 21% of global greenhouse gas emissions. In 2022, buildings were responsible for 34% of global energy demand and 37% of energy- and process-related $CO_2$ emissions. Despite a 3.5% reduction in energy intensity, the overall energy demand and emissions increased by around 1% from 2021 [10]. This highlights the need to accelerate the green transition, particularly in harder-to-abate sectors [11]. Both in Italy and in the EU, the use of renewable energy is projected to increase, with global renewable power capacity growing by 36% in 2023, reaching 473 GW—a record for the 22nd consecutive year [12]. Europe plays a key role in Italy's ecological transition, as outlined in the Green Deal [13,14]. These actions include key objectives, such as a 55% reduction in greenhouse gas emissions by 2030 and increasing the share of electricity from renewable sources to 40% (in Italy, this share was just over 22% in 2020). Two main Italian instruments driving this transformation are the National Integrated Energy and Climate Plan 2030 (PNIEC) [15], which aligns with European guidelines on emissions reduction, energy savings, renewable energy development, and infrastructure interconnectivity [16,17], and the National Recovery and Resilience Plan (PNRR) [18].

The National Integrated Energy and Climate Plan sets Italy's 2030 targets for energy efficiency, renewable energy, and $CO_2$ emission reduction [19], while also addressing energy security, interconnections, market competitiveness, sustainable development, and mobility. For each area, it details specific measures to ensure these goals are met [20].

Historically, the construction sector focused on energy consumption, but efforts have shifted towards reducing emissions, necessitating a transition from fossil fuels to renewable energy [21]. Key focus areas now include the electrification of HVAC systems and building management systems, along with enhancing building envelope efficiency to lower energy demand [22]. These actions are guided by the Energy Efficiency Directive (Directive 2012/27/EU) [23] and the Energy Performance of Buildings Directive (Directive 2018/844/EU) [24], which set the framework for meeting the climate targets for 2030 and 2050.

The transition from fossil fuels to renewable sources must be accompanied by strategies to reduce energy consumption and demand, as emphasized by Italian and community reports [25]. Achieving significant greenhouse gas reductions requires a shift towards a smart, locally generated energy production and distribution system based on renewable sources [26,27].

In the urban context, the energy transition also necessarily involves the process of the deep energy renovation of the existing built environment [28,29], which enhances the material and energy resources embedded in it, and through the comprehensive rehabilitation of spaces, envelopes, and systems, leads it towards zero emissions [30,31]. Such strategies for improving the energy efficiency of the building stock must be combined with an increasing reliance on entirely renewable energy production that is integrated into architecture and the city [32], as a strategic resource to be reinterpreted in light of the energy transition [33] and circular economy goals [34].

The Italian building stock is highly inefficient, with over 65% of buildings constructed without energy-saving measures [35,36]. This inefficiency leads to homes with energy demands that are 2.5 times higher than those built between 2016 and 2021, causing the building sector to account for nearly half of the final energy consumption and 19% of direct emissions.

The Energy Performance of Buildings Directive (EPBD), part of the "Fit For 55" package [37], aims to accelerate building renovations, reduce greenhouse gas emissions and energy consumption, and promote renewable energy installations. It introduces higher energy standards for new buildings, transitioning from "nearly zero-energy buildings" (NZEB) to "zero-emission buildings" (ZEB) by 2028, and sets deadlines for phasing out fossil fuels and mandating solar installations [38,39]. Photovoltaic (PV) solar energy is identified as the most efficient and cost-effective renewable option in Italy [40,41].

However, there are places that are remote and disconnected, geographically and otherwise, from the pulsating heart of human activities, which often remain excluded and marginalized from the aforementioned Italian goals and strategies [42,43]. This is the case with islands, whose distance from the mainland renders them physiologically fragile and difficult to manage, particularly in terms of energy supply [44].

The fragility of islands, combined with abundant natural resources, like solar and wind energy, makes them ideal for testing self-sufficient energy systems based on renewable resources [45,46]. The need to find alternatives to environmentally damaging and costly imported fossil fuels has led to experimental research on islands. Procida Island, located in southern Italy [47,48], is particularly well-suited for this purpose due to its geomorphological, climatic, and demographic characteristics. Consequently, this research focuses on integrating technological and environmental design with energy interventions in Procida's residential buildings, using the island as a pilot case for testing innovative systems aimed at environmental and energy sustainability [49].

The specific problem addressed by this study is the lack of comprehensive, localized strategies for effectively integrating renewable energy systems and energy-efficient retrofitting measures into the built environment of small, historically and geographically unique areas. This research aims to bridge the gap in understanding how to effectively integrate renewable energy systems and energy-efficient retrofitting measures into the built environment, particularly in the context of small, historically and geographically unique areas, like the island of Procida. While the Energy Performance of Buildings Directive EPBD Case Green directive outlines general goals for reducing greenhouse gases emissions and promoting renewable energy in the building sector, there is a lack of comprehensive, localized studies that consider the specific challenges of and opportunities for implementing these directives in diverse urban and environmental contexts. To address this gap, this research not only develops a detailed classification and archetypalization system for Procida's residential buildings, analyzing their current conditions and proposing tailored intervention scenarios, but also introduces an innovative approach through the application of aggregated data and archetypal analysis to large-scale territorial assessments. By leveraging a custom-developed software system, this study systematically consolidates and analyzes data across 2961 residential buildings on Procida, providing a comprehensive understanding of the island's diverse architectural and environmental conditions. These scenarios not only align with the directive's goals, but also demonstrate the potential for achieving energy self-sufficiency while respecting the island's unique historical and conservation needs, all while advancing the methodological framework for sustainable building design.

This approach allows for a comprehensive understanding of diverse architectural and environmental conditions within the island's built environment. Aggregated data analysis facilitates the identification of patterns and correlations that might not be evident through individual assessments. This macro-level perspective enables more effective and scalable solutions for energy efficiency and sustainability. The archetypalization process further refines this by categorizing buildings into representative types or archetypes based on their structural and environmental characteristics. This classification aids in the development of targeted intervention strategies that address the specific needs and conditions of different building types.

The integration of aggregated data and archetypal analysis not only enhances the accuracy of performance evaluations, but also supports the creation of tailored, context-sensitive retrofitting scenarios. This methodological innovation significantly advances the field of sustainable building design by providing a robust framework for optimizing energy use and achieving self-sufficiency at a broader scale. The findings are particularly valuable for urban planners, engineers, architects, and sustainability researchers seeking to implement effective and adaptable solutions in residential environments.

## 2. Materials and Methods

The research methodology is divided into three main phases: the analytical–interpretative phase, which ensures an accurate analysis of the context in question; the instrumental–applicative phase, which enables the systematization of the collected data; and finally, the synthetic–evaluative phase, which facilitates the formulation of a series of appropriate intervention scenarios for redevelopment.

As reported in Figure 1, below, the three main phases are further subdivided.

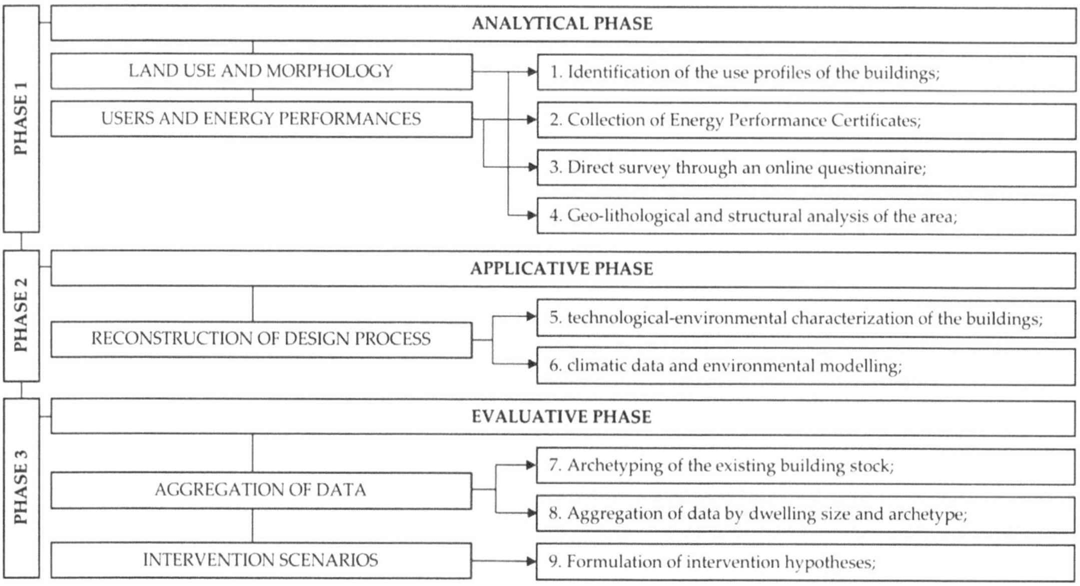


**Figure 1.** Framework of the research methodology.

### 2.1. Analytical Phase

The first phase involved identifying the usage profiles of the properties, which are necessary for classifying the examined buildings according to certain fundamental characteristics (specifically, residential buildings are the focus). These were derived both from a direct survey targeting the population via an online questionnaire designed to develop the profile of a dwelling's energy consumption [50], based on the information required by in-houe-developed calculation code based on a simplified dynamic simulation [51].

The questionnaire was developed in an Excel environment, utilizing macros and functions programmed in Visual Basic for Applications (VBA) to simulate energy consumption in real time based on user-provided inputs. It was validated by comparing its results with those obtained from established software such as TRNSYS 18 and EnergyPlus 8.3.0.

Simulated energy consumption was immediately compared to the actual data input by the user [52,53] and against calculated benchmarks. These benchmarks were derived from reports issued by the Italian Regulatory Authority for Energy, Networks, and Environment [54].

When discrepancies between simulated and actual consumption data are identified, the system generates warnings and suggests possible corrective actions. This comparison is necessary due to the general lack of awareness users have regarding their own consumption habits [55,56].

Particular attention in formulating the questions was given to information regarding the users, the architectural and morphological characteristics of the residence, the characteristics of the existing system installations, the features of the appliances within the dwellings, and energy consumption and expenses.

The details of the information requested in the questionnaire are provided in Table A1 in Appendix A.

After this phase, the analysis focused on the geo-lithological and structural framework of the pilot area, utilizing maps and information available in municipal archives and local scientific reports [57,58].

### 2.2. Applicative Phase

After the analysis of the geolithological and structural framework, particular attention was given to reconstructing the design process through the geometric creation of models for geographical contextualization and technological–environmental characterization of the buildings.

This was preparatory to studying the environmental context through the collection of climatic data over the past 20 years and solar analyses conducted using Ecotect Analysis 1.0, a type of modeling and solar analysis software. This allowed for in-depth analyses of the buildings' energy performance and climate integration, which are essential for providing recommendations, directions, and guidelines for the sustainable development of the project.

### 2.3. Evaluative Phase

Upon completion of the analysis of the context and the examined buildings, the third synthetic–evaluative phase involved archetyping the existing building stock based on similar characteristics. Specifically, three archetype categories were identified, grouping the various buildings according to the following parameters: building type, year of construction, construction technique, size, and zoning area.

After determining the archetypes, the data were aggregated according to two different parameters:

Dwelling size: Based on data collected from the population questionnaires. These data were entered into the in-house developed calculation code, which outlined the consumption profiles and suggested interventions to improve the energy performance of the dwellings, thereby enabling the projection of intervention scenarios aimed at achieving significant savings in primary energy and costs.

Archetype: The same data from the questionnaires were also aggregated according to the identified archetypes. This second type of data aggregation allows for extrapolating the frequency percentages with which the in-house calculation code recommends specific interventions.

The statistical investigations conducted from the data aggregation by dwelling size and by archetype led to the formulation of intervention scenarios. These were divided into technological–environmental interventions on the building envelope (such as thermal insulation of vertical opaque closures, horizontal opaque closures, vertical transparent closures) and replacement of system components (mainly focused on the installation of heat pumps associated with photovoltaic systems and of solar thermal systems).

## 3. Results and Discussion

### 3.1. Application of the Research Method to the Pilot Case of Procida Island

The pilot case selected for validating the established research method is the small island of Procida.

The small island, located in Campania, in the Gulf of Naples, covers 4.26 km$^2$, and it has a population of 10,434 and a population density of 2449.3 inhabitants per km$^2$, with 2961 dwellings in Procida [59]. The first settlements, dating back to the 8th century BC, were founded by Greek colonists from Cumae on Terra Murata, 90 m above sea level. Coastal villages, Corricella and Marina di Sancio Cattolico, developed in the 15th century, although the island remained rural until a construction boom in the 1960s.

This research focuses on Marina di Sancio Cattolico and Corricella. Marina di Sancio Cattolico, located on a strategic inlet, has limited sunlight and features adaptations like minimal overhangs and solid walls. The buildings are made from local tuff, with load-bearing vertical structures [57]. Corricella, which developed between the 16th and 17th centuries, has terraced architecture with southern exposure, benefiting from sunlight and heat [57].

The 2002 Color Plan [58] preserves historical facades, prohibiting non-original materials and limiting visible renewable energy installations to 30% of roof surfaces. The 1960s building boom introduced rural-style homes, while the 1985 General Town Plan (PRG) led to denser semi-detached villas with better insulation.

As mentioned above, in Procida, there is a total of 2961 dwellings [59]. To develop a comprehensive energy retrofit intervention, it was necessary to classify and aggregate these dwellings based on several key characteristics. The information for this process was obtained from various sources, including direct interviews with the population via a Google Docs questionnaire [50] designed based on the requirements of the in-house software.

The graph in Figure 2 allows for the observation and quantification of the aforementioned points. Until 1900, Procida had 692 dwellings. In the decade between 1960 and 1970 alone, 1038 dwellings were constructed.

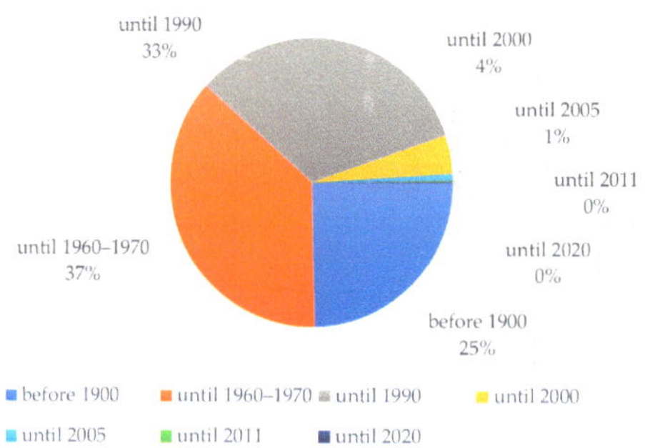


**Figure 2.** Number of dwellings divided into building construction period.

The building trend remained high until the 1990s, followed by a sharp decline in the early 2000s. Currently, the development of new buildings on the island has nearly ceased.

Regarding the average size of residential buildings, there are no official statistical data available. Therefore, an average and a proportion were calculated based on the questionnaires administered to the island's inhabitants. The most common size is between 45 and 65 sqm.

Figure 3 reflects the recent phenomenon of subdivision in the historical buildings. The historical constructions from the feudal period were indeed very large, making them difficult to manage in contemporary times.

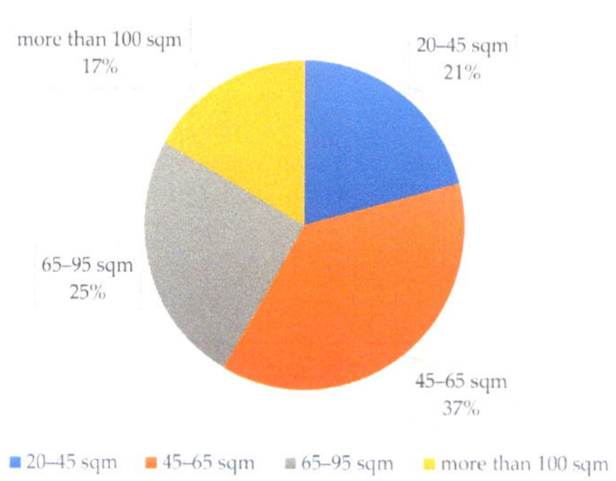


**Figure 3.** Number of residential buildings divided by average size.

As shown in Figure 4, the majority of the dwellings (54%) are single-storey.

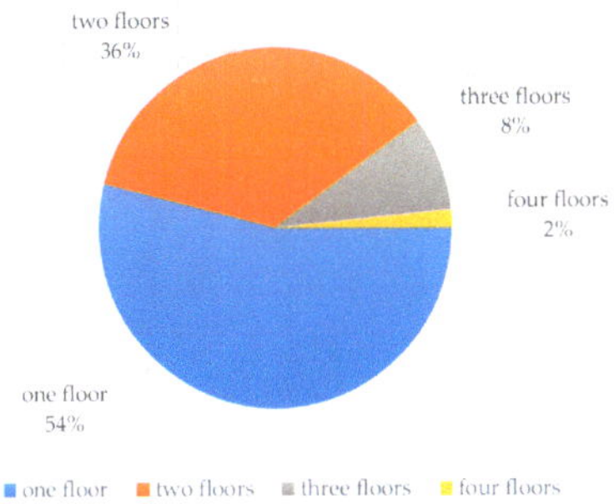


**Figure 4.** Number of residential buildings divided into number of floors above ground level.

This is explained by the fact that most of the construction occurred in the 1960s and 1970s, following the single-family villa type.

More recently, the landscape protection regulation imposed in 2004 allowed construction up to a maximum height of only 4 m above ground level. The dwellings that rise to four stories are very few, and they are mainly located in Marina di Sancio Cattolico and along the historic Via Principe Umberto.

Subsequently, the usage profile of the properties in Procida was outlined.

This research aims primarily to implement a widespread retrofit of residential buildings, and the island proved to be particularly suitable, since it does not host industrial complexes or large infrastructures, making the building stock almost entirely residential.

Of the 2003 habitable residences, 1543 are occupied year-round, with the remainder used only seasonally, as shown in the following graph, in Figure 5.

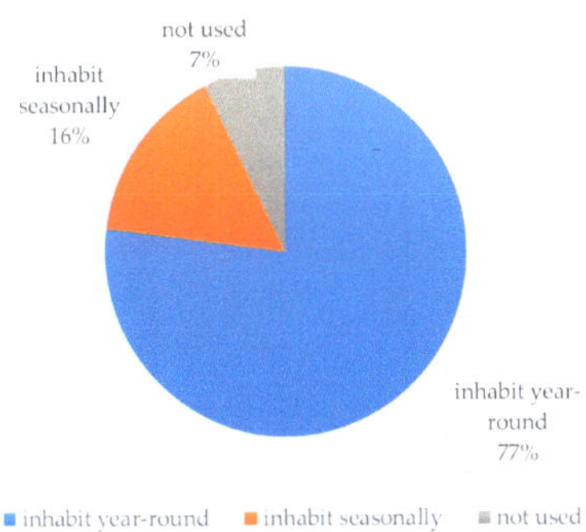


**Figure 5.** Occupancy of residential buildings.

It is notable that in Procida, only 30% of the dwellings are connected to the national gas network, with the remainder using gas purchased and stored in personal storage devices (tanks) located in gardens or inside the dwellings.

The following graphs in Figure 6 show that despite this, the most common heating system (found in 79% of the dwellings) is the traditional natural gas boiler.

Cooling systems (air conditioners, splits), on the other hand, are much less common, being present in only 4% of the dwellings [59].

These data reflect the national trend, which still shows a low prevalence of such devices.

As far as the production of domestic hot water (DHW) is concerned, 58% is entrusted to electric water heaters. The presence of heat pumps, albeit small, is noted in 8% of cases (Figure 7).

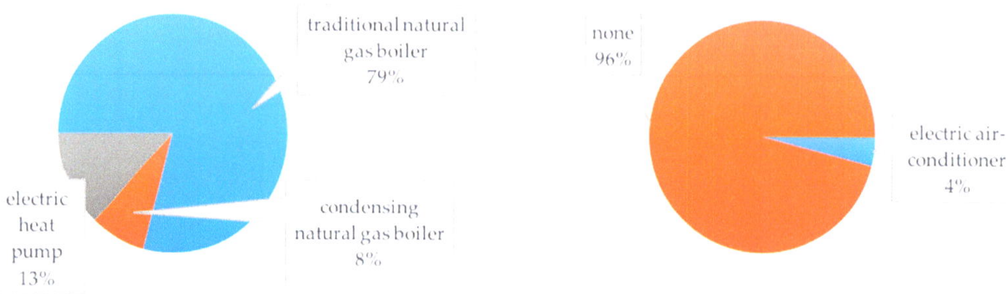


**Figure 6.** (**left**) Typology of heating systems; (**right**) cooling systems.

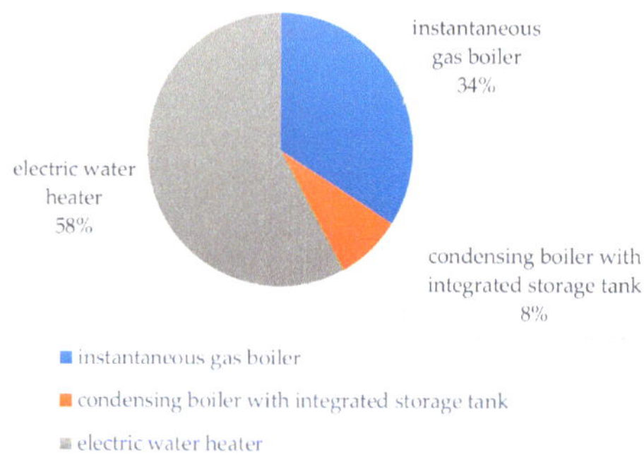


**Figure 7.** Typology of domestic hot water production systems.

Procida, part of the historic Terra di Lavoro, is geologically notable for its Campanian gray and Neapolitan yellow tuff, which are likely sourced from Roccamonfina or the Campi Flegrei volcanoes. Two primary masonry techniques are used on the island: the *"a cantieri"* method (on the left in Figure 8), found in older areas, like Terra Murata, featuring rough-hewn stones with horizontal gaps filled with mortar; and the *"a blocchetti"* method (on the right in Figure 8), which is commonly used in post-17th century buildings in the historic center, using ashlar blocks with minimal mortar joints [60].

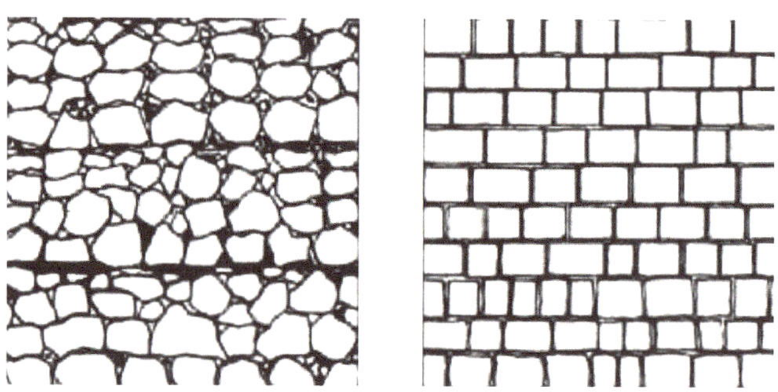

**Figure 8.** Tuff used in masonry according to two main techniques: the so-called *"a cantieri"* technique (**on the left**); and the so-called *"a blocchetti"* technique (**on the right**).

The following are the analyses of the masonry samples surveyed on the island. For each of the masonry samples surveyed on the island, a detailed photographic survey was conducted, and the dimensions of each layer were studied. The thermal transmittance (U-value) was calculated using the unit W/m$^2$K to estimate the average heat flux passing per square meter of rooms with different temperature, and then compared with the limit values specified by the relevant Italian regulations.

The photographic surveys and the stratigraphy are presented for the first two examples of masonry. For the third and the fourth samples, only the results of the simulations are reported. The tuff masonry on the island is almost exclusively present in the historic centre. In the inland areas and on the more recently developed coastlines, this material is present, but only in non-structural inserts (e.g., boundary walls).

The dimensions of the masonry range from a minimum of 60 cm to a maximum of 1 m. A sample of the irregular masonry with periodic horizontal courses, "*a cantieri*", is reported in Figure 9.

| Layer | Thickness [m] |
|---|---|
| Lime-based plaster | 0.02 |
| Structural layer of yellow tuff "in blocks" | 0.16 |
| Pozzolanic mortar | - |
| "Hollow" core of yellow tuff | 0.11 |
| Structural layer of yellow tuff "in blocks" | 0.25 |
| Lime-based plaster | 0.03 |
| Horizontal courses of concrete aggregate | - |

**Figure 9.** (**left**) Example of masonry with "*a cantieri*" construction; (**right**) stratigraphy of the masonry from the exterior to the interior.

The U-value of this wall is 1.818 W/m$^2$K. The described masonry package has a U-value that is 4.5 times higher than the value stipulated by regulations for vertical opaque walls, which mandate a maximum of 0.4 W/m$^2$K. Therefore, it is necessary to take action to modify this value.

The second sample of the masonry with cut stones, or "*a blocchetti*", is reported in Figure 10. The U-value of this wall is 1.630 W/m$^2$K.

As can be seen in Figure 10, this type of masonry is slightly more efficient. This is due to the fact that the more regular blocks require less mortar for installation. Since mortar has a higher conductivity compared to tuff, it allows heat to pass through more readily. This masonry package also exhibits a U-value significantly higher than the value stipulated by regulations for vertical opaque walls, which requires a maximum of 0.4 W/m$^2$K. Therefore, it is necessary to take action to improve the wall's performance.

| Layer | Thickness [m] |
|---|---|
| Lime-based plaster | 0.02 |
| Structural layer of yellow tuff "in blocks" | 0.24 |
| Pozzolanic mortar | - |
| "Hollow" core of yellow tuff | 0.12 |
| Structural layer of yellow tuff "in blocks" | 0.24 |
| Lime-based plaster | 0.03 |
| Horizontal courses of concrete aggregate. | - |

**Figure 10.** (**left**) Example of masonry with "*a blocchetti*" construction; (**right**) stratigraphy of the masonry from the exterior to the interior.

The third sample of the masonry is extremely simple and equally widespread on the island, with perforated bricks measuring 30 × 30 cm. Its thickness ranges from 30 to 40 cm.

The U-value of this wall is 0.949 W/m$^2$K. It is 2.3 times higher than the maximum established value. Therefore, it is necessary to take action to modify this value.

The fourth sample of masonry has double perforated bricks and an air gap (1975–1985). This type of masonry is less common on the island and is of more recent construction. It comes in two variants: one with an air gap and one with interposed insulation.

The average thickness of this type of masonry is approximately 30 cm and the U-value of this wall is 1.389 W/m$^2$K.

As noted, this value is 3.4 times higher than the maximum established limit. Therefore, it is necessary to take action to modify this value.

The 150 houses built after the 1990s do not require intervention as they are already equipped with thermal insulation.

The island has 1088 degree-days. Regarding sunlight, it is observed that Procida experiences significantly fewer cloudy days compared to sunny or partially cloudy days, with a maximum of 10 cloudy days in December and January.

Furthermore, Procida benefits from many hours of sunshine each day.

On 21 June, the sun rises before 5 a.m. and sets after 7 p.m. On 21 December, the sun rises shortly before 8 a.m. and sets after 4 p.m. These observations lead to the conclusion that using solar energy as the primary renewable energy source is particularly suitable. Procida is located in the Mediterranean basin, within a gulf, protected by the mainland to the north and the island of Ischia to the west. Consequently, the winds on the island are relatively slow (approximately 3 m/s) and primarily come from the peninsula, specifically from the Vesuvius area to the north/northeast, and from the south/southwest, i.e., from the Mediterranean basin. Subsequently, the shading conditions of the island areas that appeared most problematic were evaluated.

In Figure 11 (below), we present the analysis conducted for the Marina di Sancio district, which, of all the areas studied, presents the most significant issues. This area has a northern exposure, resulting in its seaward facades being perpetually in shadow.

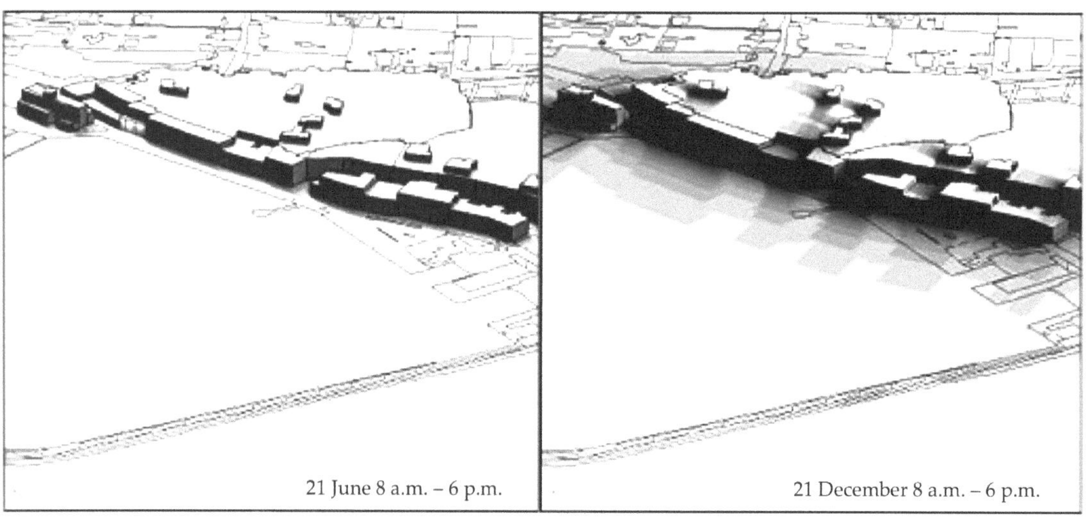


**Figure 11.** (**left**) Shading of the area on 21 June; (**right**) shading of the area on 21 December.

Additionally, there is no solar gain from the south either, as the buildings are entirely adjoined to a rocky wall, specifically composed of tuff.

The simulation presented in Figure 12 (below) depicts the buildings from the 1960s–1970s, illustrating the area near Silurenza Beach. Here, to the northwest of the island, there is a dispersed development with one-to-two-storey houses surrounded by their own gardens. The vegetation is not of considerable height, since the land is predominantly used for agricultural purposes. The distance between the buildings prevents them from casting shadows on one another.

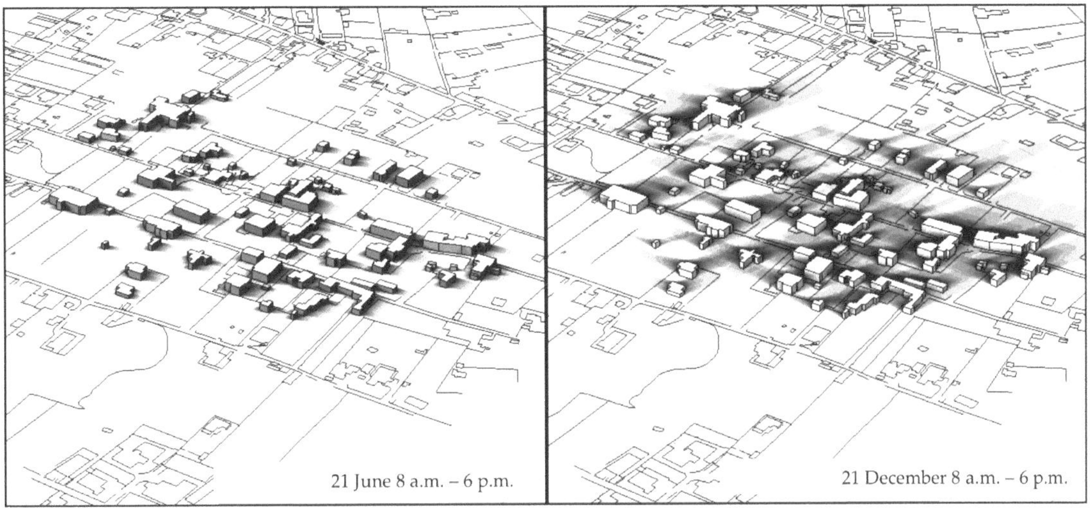


**Figure 12.** (**left**) Shading of the area on 21 June; (**right**) shading of the area on 21 December.

The simulation in Figure 13 (below) illustrates the building model from the 1980s to 2000, located near Ciraccio Beach. Its northwest orientation results in the seaward facades being almost entirely shaded throughout the day. At the rear, most of the examined buildings are surrounded by low vegetation, which allows for sunlight exposure on the rear facades.

**Figure 13.** (**left**) Shading of the area on 21 June; (**right**) shading of the area on 21 December.

The intention to undertake an energy retrofit on such a large scale necessitates the aggregation of the building stock according to its similar characteristics. According to the research presented so far, three categories, termed archetypes, have been identified to group the various buildings based on the following parameters: building type, year of construction, construction technique, size, and zoning of the island. Specifically, for size, the classification was performed in the following size ranges: (1) 20–45 m$^2$; (2) 45–65 m$^2$; (3) 65–90 m$^2$; (4) 100 m$^2$ and above.

**Archetype A** (blue in Figure 14) includes all buildings constructed before 1919, with tuff masonry ranging from 60 cm to 1 m in thickness. These represent 24.6% of the island's dwellings, totaling 696 buildings. The analyzed samples have areas ranging from 20 to 145 m$^2$, distributed fairly evenly across the size categories.

**Archetype B** (pink in Figure 14) groups all the buildings constructed between the post-World War II era and 1971, featuring masonry made of perforated brick blocks joined with pozzolanic mortar and sand, with an average thickness of 32 cm. This is the largest archetype, encompassing 62.8% of the building stock, totaling 1658 buildings. Of these, 11.1% fall into category 1, 44.4% fall into category 2, 23.2% fall into category 3, and 22.2% fall into category 4.

**Archetype C** (green in Figure 14) includes buildings constructed between 1985 and the present, featuring perforated brick masonry with an air gap or insulation, with an average thickness of 32 cm. This archetype accounts for 12.6% of Procida's building stock, totaling 457 buildings. Of these, 16.6% fall into category 1, 33.3% fall into category 2, 33.2% fall into category 3, and 16.7% fall into category 4. Specifically, the following 12 buildings from **Archetype A**, nine buildings from **Archetype B**, and four buildings from **Archetype C** were analyzed.

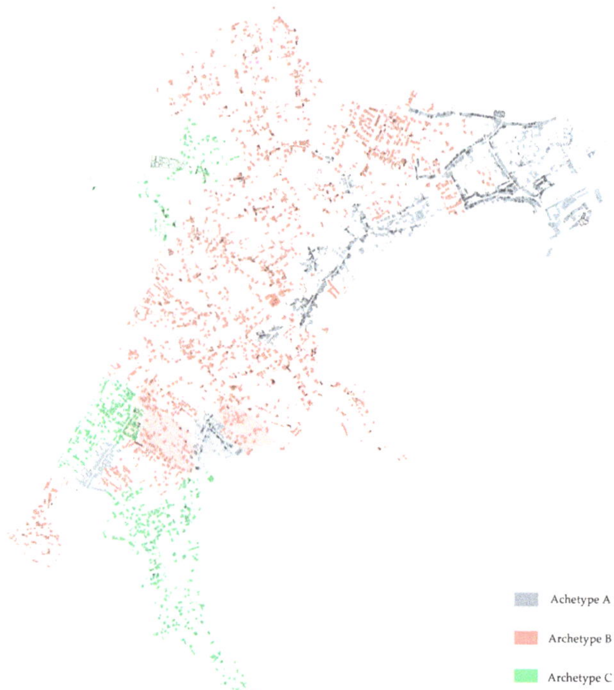


**Figure 14.** Identification of archetypes on the island plan.

All the collected data were first aggregated by dwelling size and then entered into the calculation software [51], which mapped the consumption profile and suggested interventions to improve the energy performance of the buildings, thereby enabling significant savings in primary energy and costs.

It is important to note that the suggestion to replace household appliances (such as refrigerators, washing machines, etc.) was omitted, as this type of intervention is more challenging to implement due to its reliance on individual family choices.

In total, 25 housing samples from Procida were examined and analyzed using the software. Looking at the graph in Figure 15, below, it can be observed that, in general, the most frequently suggested interventions are those related to the insulation of the building envelope (walls and ceiling).

For size categories 1 (20–45 $m^2$) and 3 (65–90 $m^2$), wall insulation is the most frequently recommended intervention for achieving the highest primary energy savings. In both categories, over 50% of effective interventions target the building envelope, while system interventions are also common, accounting for 42% in category 1 and about 31% in category 3, with both suggested system interventions being nearly equally effective. Five housing samples were analyzed for category 1, and seven were analyzed for category 3.

In size category 2 (45–65 $m^2$), the replacement of the domestic hot water (DHW) generator is the most common intervention, followed by wall insulation, at 28.71%, with other interventions being much less frequent (about 6%). Nine housing samples were analyzed for this category.

For sizes equal to or greater than 100 $m^2$, ceiling insulation is the most frequently recommended intervention, with over 50% of effective interventions focusing on the building envelope. System interventions account for about 21%, with both options being equally effective. Four housing samples were analyzed for this category.

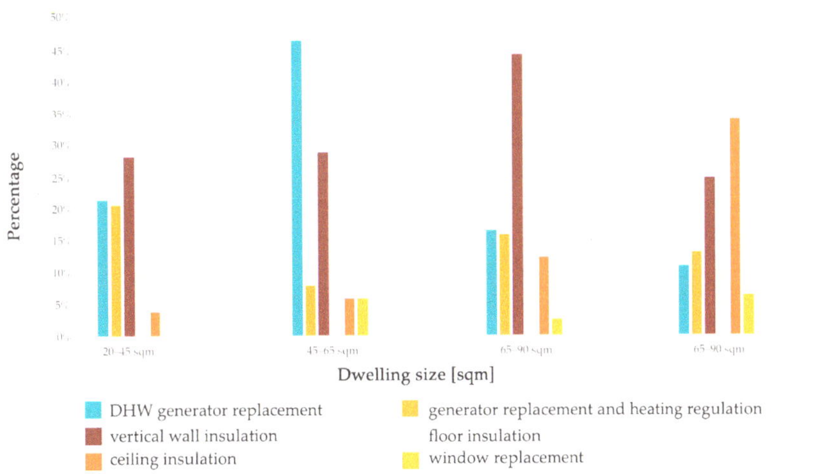


**Figure 15.** Frequency of suggested interventions in percentages for the different size categories for reducing primary energy consumption.

The same data were subsequently aggregated according to the different identified archetypes. The following graph compares the frequency percentages at which the software suggested specific interventions (Figure 16).

For both **Archetype A** (buildings constructed before 1919) and **Archetype C** (buildings constructed between the 1980s and 2000), the most frequently recommended intervention for achieving the highest primary energy savings is the replacement of the domestic hot water (DHW) generator, followed by wall insulation, with a significant gap between the two.

In **Archetype A**, interventions on the envelope are less common due to the thick historical walls, which partially compensate for the lack of insulation. For **Archetype C**, the more modern construction technology includes better envelope insulation, although there is still room for improvement. Twelve housing samples were analyzed for **Archetype A**, and four were analyzed for **Archetype C**.

In contrast, for **Archetype B** (buildings constructed during the 1960s and 1970s), the most frequently recommended interventions focus overwhelmingly on the building envelope, due to the less insulated and thinner construction masonry of the period, particularly in isolated villas of up to two stories. Among system interventions, the replacement of the heating system generator is slightly preferred over the DHW generator. Nine housing samples were analyzed for this archetype.

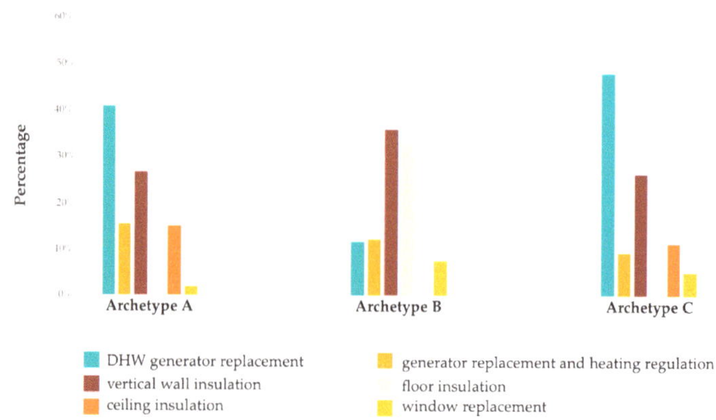


**Figure 16.** Frequency of suggested interventions in percentages for the different archetypes for primary energy reduction.

Aggregating the data from the 25 samples studied, without distinguishing by archetype or size, it is observed in the graph Figure 17 that the most frequently recommended intervention for primary energy savings is the replacement of the domestic hot water generator (37%), followed by wall insulation (33%). The third most frequently recommended intervention is the replacement and adjustment of the heating system (14%). In conclusion, there is a clear need to address both the building envelope and the systems, as both would lead to significant primary energy savings.

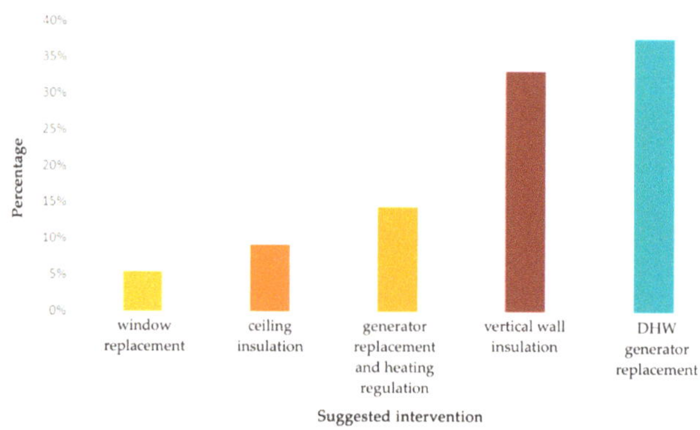


**Figure 17.** Frequency of suggestion of interventions in total percentages for primary energy reduction.

In conclusion, the overall results are similar to those of category 2, primarily because this category comprises the largest number of analyzed samples.

Once the analytical phase was completed, hypotheses for the interventions on the building envelope were formulated, integrating technological–environmental design solutions, as well as assessing the challenges and strengths of potential solutions, as suggested in [44], for small islands.

The first hypothesis consists in the insulation of the building envelope through the thermal insulation of vertical opaque closures. One of the interventions most frequently suggested by the software is the insulation of walls and roofs. Given the extreme differentiation of the island's building heritage, various technologies were considered for the proposed intervention.

The historical and landscape constraints imposed on **Archetype A** prevent the implementation of external wall insulation. Therefore, insulating materials suitable for internal use, which are thin to avoid significantly reducing the volume of the houses, were evaluated.

The selection of insulation materials for the building envelopes involved evaluating the thermal performance (based on thermal conductivity values), compatibility with historic masonry, and the thickness requirements relative to the dimensional constraints. Cost considerations were also important, with a preference for materials offering the best performance for their cost. The market survey identified long-fiber basalt insulation as the most advantageous. This material, made from 100% natural, non-combustible basalt fibers, features a panel construction with thin layers mechanically "sewn" together. It can be applied externally and internally on walls, ceilings, floors, and roofs without anchors. The benefits include reduced energy costs, acoustic insulation, improved environmental conditions, and the prevention of mold on exposed perimeter walls. It also integrates well with historic buildings, preserving their aesthetic integrity.

The same procedure of evaluation was followed for selecting the insulators for the houses belonging to **Archetype B**, which, not having any constraints or valuable plasterwork, allowed for the implementation of external wall insulation. The choice fell on a cork insulation panel.

For the houses in **Archetype C**, which have an air cavity, the possibilities of cavity wall insulation were evaluated. Additionally, in this case, a market survey was conducted and, once again, long-fiber basalt insulation was chosen. Once the insulating material was selected, the intervention for each Archetype was developed.

For **Archetype A**, the changes in the stratigraphy of the vertical opaque closures and the verification are reported below. It is noted that the initial U-value for the first sample was 1.818 $W/m^2K$. By inserting a long-fiber basalt insulation with a thickness of 5 cm, the final transmittance is 0.388 $W/m^2K$, which complies with the regulatory limits. For the second sample of the tuff masonry, a similar intervention was hypothesized, resulting in a new U-value of 0.376 $W/m^2K$. Based on the number of houses per size range and the associated dispersing surface area, the total is 96,907.8 $m^2$.

For **Archetype B**, the changes in the stratigraphy and the verification are also reported below. It is noted that the initial U-value for the third sample was 0.949 $W/m^2K$. By installing a 6-centimeter-thick cork insulation panel, the final transmittance is 0.358 $W/m^2K$, which complies with the regulatory limits. Based on the number of houses per size range and the associated dispersing surface area, the total is 264,134 $m^2$.

Finally, for **Archetype C**, the same procedure was followed. It is noted that the initial U-value for the fourth sample was 1.389 $W/m^2K$. By implementing a wood fiber cavity insulation of 10 cm, the final transmittance is 0.313 $W/m^2K$, which complies with the regulatory limits. Based on the number of houses per size range and the associated dispersing surface area, the total is 10,828.3 $m^2$.

The second hypothesis consists in the insulation of the building envelope through the thermal insulation of horizontal opaque closures. Regarding horizontal partitions, only the roofs were considered for the refurbishment. The study essentially identified two types of roofs: the historic type, with load-bearing vaulted tuff masonry, and the more recently constructed type, made of reinforced concrete with hollow bricks.

The historic roofs in question are vaulted and predominantly constructed from tuff, with some tile shard reinforcements. They are typically extradossed, creating a "lamia", which is common in Campania, making them non-walkable. Despite their age, the gutter system is efficient and well-integrated into the building structure [49]. For insulating these vaulted roofs, an internal insulation layer consistent with that of the vertical opaque closures is proposed. This involves using a 5-centimeter-thick long-fiber basalt insulation, which makes it possible to achieve a final U-value of 0.379 W/m$^2$K, compared to an estimated initial value of 1.606 W/m$^2$K.

For the simpler reinforced concrete slabs, an external wall insulation system was chosen, which is consistent with that used for the vertical opaque closures. This involves using a 5-centimeter-thick cork insulation, which makes it possible to achieve a U-value of 0.374 W/m$^2$K, compared to an estimated initial value of 0.865 W/m$^2$K. Given that the maximum allowable transmittance for roofs in climate zone C is 0.38 W/m$^2$K, this intervention ensures compliance with the regulatory limit.

The intervention also aimed to make previously non-walkable roofs accessible, facilitating the installation of photovoltaic systems and/or solar thermal collectors.

The third hypothesis consists in the insulation of the building envelope through the thermal insulation of horizontal transparent closures. The replacement of windows in the historic residences of Procida has proven to be both necessary and challenging due to several constraints, as follows:

- The lack of modification of openings: existing openings cannot be enlarged or reduced in size;
- Material restrictions: frame materials are restricted to wood only, and aluminum and PVC are not permitted;
- Paint requirements: wood frames must be paintable and painted in colors specified by the Color Plan (Piano Colori);
- Glass specifications: any thickness of smooth glass or crystal can be used.

Despite these constraints, the urgency is high because the historic windows are characterized by solid wood frames and single-pane glass, resulting in very high levels of U-value.

For **Archetype A**, transparent closures vary in shape, arrangement, and size. Therefore, any replacement must be custom-made. The most frequently encountered types were selected for analysis, revealing that all have U-values exceeding 4 W/m$^2$K.

This is well above the regulatory limit for transparent closures, which is 1.75 W/m$^2$K.

The choice fell on windows with solid wood frames from various hardwoods, ensuring particularly low U-values. The selected wood species was ash, as it most closely matches the color of the currently existing windows. The project includes two different variants for the glass, as follows:

a. Historic center: the chosen solution is double glazing with low-emissivity coating and an argon-filled cavity (90% filled);
b. Marina di Sancio Cattolico: the selected option is triple glazing, also with low-emissivity coating and argon-filled cavities (90% filled).

This differentiation is due to the specific conditions in Marina di Sancio Cattolico, which is perpetually shaded due to its northern exposure and is subjected to one of the coldest prevailing sea winds in Procida, coming from the north-west.

The new window system has a U-value of approximately 1.3 W/m$^2$K for Variant 1 and 0.9 W/m$^2$K for Variant 2.

Subsequently, the windows of buildings constructed between the post-war period and the 1970s (**Archetype B**) were analyzed.

These often feature wooden frames with single glazing, although PVC windows also appear in some cases. The analysis considered the small percentage of buildings that have already undergone window replacement and focused on the remaining buildings. The window types for this archetype are much less varied, being more rectangular and regular.

Moreover, there are no restrictions on these buildings, allowing for greater freedom in choosing replacement windows. For the replacement of these windows, PVC windows was selected. These are more economical, thus offering a shorter payback period, and are highly efficient and recyclable. Since this segment of the building stock is not subject to restrictions, any type of window can be installed. The chosen glass is double glazing with a low-emissivity coating and an argon-filled cavity (90% filled). This type of glass, also used in the previous intervention, is less expensive while fully meeting the current regulatory performance standards. The buildings receiving this type of window are mostly located inland on the island, and being single-family homes, the transparent closures can be on any facade. The new window system has a U-value of approximately 1.2 $W/m^2K$, which complies with the regulations.

The windows for **Archetype C** were not addressed as they are already more efficient due to their more recent construction.

Once all the hypotheses for the interventions on the building envelope were defined, the next step was to define the hypotheses for the interventions on the energy systems, which in this case, involved assessing the challenges of energy management, as shown in [46], for small islands.

Starting from the research objective of designing a photovoltaic surface on the island that could produce as much electricity as possible, after considering the interventions carried out on the building envelopes, the replacement of heating and domestic hot water systems was also taken into account, without neglecting the cooking utilities, which, transformed into electrics, allowed the average annual electricity consumption to be calculated accordingly.

The first hypothesis consists in the electrification of the building stock through heat pump insertion.

In the effort to electrify the island's building stock, the replacement of existing systems was designed to eliminate the need for gas for both winter heating and domestic hot water production, an action that was, moreover, corroborated in light of the need to integrate the island of Procida into policy frameworks reported in [42,43].

The replacement will be achieved through the installation of heat pumps. The appropriate size and efficiency of the heat pump can be determined to meet the specific heating and hot water needs of the buildings on the island.

The design was carried out for each archetype and, within them, for each size range. As the distribution subsystem, a split system was chosen, which aligns well with the heat pump's ability to function both as a heating and as a cooling system.

The second hypothesis consists in the electrification of the building stock through the use of the heat pump as a water heater for domestic hot water production. In this case, the design was more straightforward, and an 80 L heat pump water heater was selected, which is adequate to meet the needs of the island's households, which average 2.3 people per unit.

The third hypothesis consists in the electrification of the building stock through the photovoltaic system, as also suggested by the information presented in [48,49]. These two heat pump systems will be powered by a widespread photovoltaic system, sized according to the peak power requirements of the examined buildings, calculated using in-house software for each size category, regardless of the archetype.

The combination of these two technologies is particularly efficient, as the heat pumps can be almost entirely supplied by the photovoltaic panels.

The selected panel is a monocrystalline silicon panel with half-cells, which allows it to function even when partially shaded. The photovoltaic panel has an efficiency of 19.9%. A high-performance panel was chosen because, for historic buildings, only 30% of the roof area can be covered by photovoltaic panels, so this type of panel produces more energy from a smaller solar surface.

### *3.2. Discussion of the Results*

This research aims to illustrate how combining technological–environmental design strategies, such as upgrading building envelopes and improving bioclimatic behavior, with energy retrofitting measures, like replacing mechanical systems and installing solar panels, can enhance building performance, achieve energy self-sufficiency, and help to assess the potential for electrification on the island of Procida.

The final step involves evaluating the various interventions from the perspective of primary energy demand, energy savings percentages, and the changes in energy class resulting from each intervention.

According to Italian regulations [61,62], the energy classification is a fundamental element for assessing the energy efficiency of a building. It indicates the building's level of energy performance, which is evaluated based on the consumption of non-renewable primary energy for heating, cooling, domestic hot water production, and lighting. The energy classification, expressed in letters from A to G, represents a scale of values indicating the building's efficiency, where Class A represents the highest level of energy efficiency and Class G represents the lowest. Understanding a building's energy classification is crucial for evaluating its energy consumption and identifying potential improvements to reduce energy use and, consequently, $CO_2$ emissions. There are eight energy classes, which are assigned based on the primary energy demand required to heat each square meter of surface area. The most efficient energy classes are A4, A3, A2, and A1, with higher values assigned, followed by classes B, C, D, E, F, and, finally, Class G, with the lowest score.

All the evaluations were conducted for samples divided by archetype and, within each archetype, by size range. The following terminology will be used:

- **Scenario #1**: wall and roof insulation;
- **Scenario #2**: wall and roof insulation + window replacement;
- **Scenario #3**: wall and roof insulation + window replacement + system replacement;
- **Scenario #4**: wall and roof insulation + window replacement + system replacement + photovoltaic system installation.

Below is an analysis of the main results obtained from the combination of interventions aimed at the integration of technological–environmental design and energy interventions in the residential building stock in the pilot case of the small island Procida.

A comprehensive analysis of the proposed interventions, aimed at assessing the impacts of the proposed energy transition for the island following the statements reported in [45], reveals several key findings regarding their impact on energy performance and efficiency across different building archetypes on Procida Island.

Initially, all buildings, particularly those of Archetype A, exhibit very low energy classes, as also revealed by [47]. This is indicative of a significant need for improvement to meet modern energy standards. Larger homes, however, tend to have better energy classes compared to smaller ones, suggesting that building size influences energy performance, in line with what was also reported by [48].

The figures below graphically present the results obtained from the application of the four intervention scenarios on the small island of Procida in terms of energy demand and associated $CO_2$ emissions. The comparison of the results is divided by dwelling size (Figure 18) and by archetype (Figure 19).

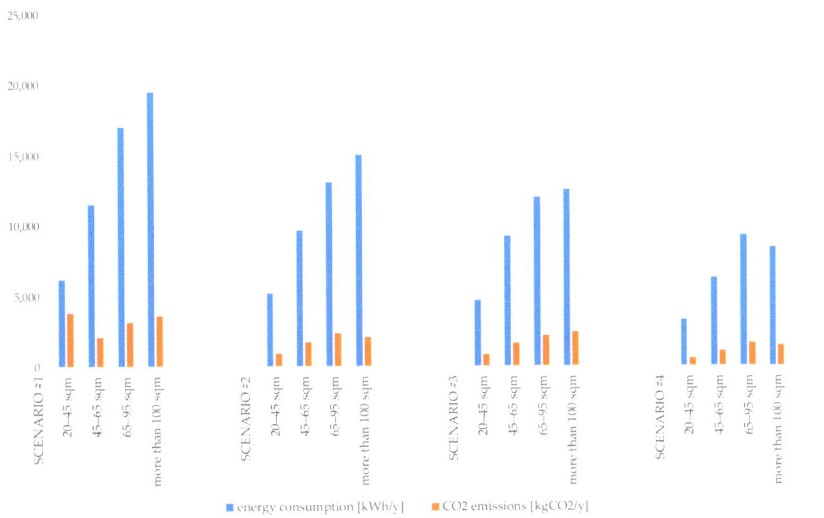


**Figure 18.** Comparative evaluation of intervention scenarios in terms of energy demand and associated $CO_2$ emissions divided by dwelling size.

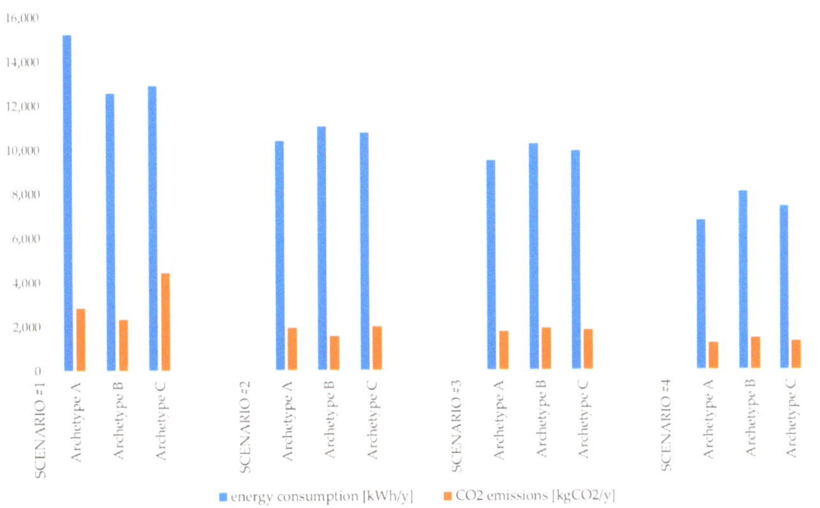


**Figure 19.** Comparative evaluation of intervention scenarios in terms of energy demand and associated $CO_2$ emissions divided by archetype.

**Scenario #1**, which focuses on insulating walls and roofs, demonstrates varying degrees of effectiveness across different archetypes. For Archetype A, the complexity of the technologies involved in this scenario enables a substantial improvement in energy class, moving from G/F (characterized by an $Ep_{gl,n\_ren}$ (the primary energy requirement for non-renewable energy sources) of 188.51 kW/m$^2$ per year. This class is typically associated with low comfort levels and high utility bills) to D/C (with an average non-renewable energy consumption of approximately 119.09 kWh/m$^2$ per year, these buildings have a medium level of energy efficiency, with adequate insulation systems and equipment in place to reduce consumption).

In contrast, for archetype B, the intervention results in a more modest energy class upgrade from E/D (with an average energy consumption of approximately 131.58 kWh/m$^2$ per year. While they fall within a moderate range of non-renewable energy consumption, these buildings could benefit from improvements in insulation systems and the efficiency of their equipment) to C/B (with an average energy consumption of approximately 89.05 kWh/m$^2$ per year, these buildings feature good thermal insulation and efficient systems, which significantly contribute to reducing energy costs). Archetype C achieves an upgrade from E/D to D/C, which is similar to archetype A.

**Scenario #2** adds the replacement of windows to the previous insulation efforts. Energy class improvements are evident only for larger homes (categories 3 and 4), as there is no significant change for the smaller size categories, 1 and 2.

**Scenario #3** includes insulation, window replacement, and the replacement of heating systems with heat pumps. The energy savings are most pronounced in Archetype A, followed by Archetype C, with Archetype B showing the least savings. The energy classes after this intervention are improved to A2 for smaller categories (with an average non-renewable energy consumption of approximately 88.58 kWh/m$^2$ per year, these buildings are equipped with advanced thermal insulation, highly efficient windows, and high-performance heating and cooling systems) and A3 for larger ones (with an average non-renewable energy consumption of approximately 66.27 kWh/m$^2$ per year, these buildings feature excellent thermal integrity and utilize renewable energy sources for heating and cooling), with no differences among the archetypes.

**Scenario #4** incorporates all the previous measures, along with the installation of photovoltaic panels. This comprehensive intervention standardizes the energy class to A4 for all buildings (characterized by an average non-renewable energy consumption of approximately 54.51 kWh/m$^2$ per year, these buildings are constructed with highly insulating materials and high-efficiency systems, ensuring significant energy savings and a high level of thermal comfort), regardless of archetype or size.

Comparing the results obtained in the simulation of the different scenarios with each other, in terms of effectiveness, **Scenario #1** significantly improves energy efficiency by focusing on insulation. It is particularly effective for older buildings (Archetype A), where the addition of insulation leads to a notable upgrade in energy class. For newer buildings (Archetypes B and C), the improvements are more modest due to their existing insulation levels. This scenario provides a solid foundation for energy savings but does not address system inefficiencies. Building on Scenario #1, **Scenario #2** adds window replacement to the insulation measures. This combination shows enhanced effectiveness, especially in larger homes (categories 3 and 4), where window replacement contributes substantially to energy savings. However, for smaller homes (categories 1 and 2), the benefits are less pronounced, indicating that window replacement has a more significant impact in larger spaces. **Scenario #3** offers the most pronounced energy savings across all the archetypes. By including insulation, window replacement, and modern heating systems (heat pumps), it addresses both envelope and system inefficiencies. It achieves the highest energy class upgrades of all the archetypes, with Archetype A seeing the greatest benefit due to its older construction and need for comprehensive improvements. **Scenario #4**, the most comprehensive scenario, integrates all the previous measures with the addition of photovoltaic panels. This approach standardizes the energy class to A4 for all buildings, regardless of archetype or size, representing the pinnacle of energy efficiency and sustainability. It ensures substantial energy savings and high levels of thermal comfort across the board.

Further analyzing the best archetype for each scenario shows that in **Scenario #1**, Archetype A benefits the most from insulation alone, achieving a substantial improvement in energy class due to the significant increase in energy efficiency. **Scenario #2** shows the greatest improvements in larger homes due to the additional benefits of window replacement when combined with insulation. **Scenario #3** results in the most significant energy savings for Archetype A, thanks to comprehensive upgrades including insulation, window replacement, and heat pumps. However, it also provides considerable benefits to Archetypes B and C. Finally, **Scenario #4** achieves the highest energy efficiency uniformly across all archetypes and sizes by combining all interventions with photovoltaic panels.

Synthesizing the results obtained by conducting a summary comparison, **Scenarios #1** and **#2** provide foundational improvements, with #2 enhancing effectiveness through window replacement, particularly in larger homes. **Scenario #3** offers a significant step forward by addressing both envelope and system inefficiencies, achieving notable energy class upgrades across all archetypes. **Scenario #4** stands out as the most comprehensive and effective approach, achieving the highest energy class universally by integrating all the previous measures and adding photovoltaic panels, although it is also the most complex and costly.

This research addresses a critical gap in sustainable building design by focusing on localized, data-driven strategies for integrating renewable energy and energy-efficient retrofitting into the built environment. The analysis highlights the effectiveness of comprehensive energy retrofits in enhancing energy performance and achieving significant savings, emphasizing the following innovative aspects of this research:

- This research introduces an innovative approach through the use of aggregated data and archetypal analysis for large-scale territorial evaluations, improving precision and scalability in energy efficiency interventions;
- A custom software system was developed to consolidate and analyze data from 2961 residential buildings on Procida, providing insights into the island's diverse architectural and environmental contexts;
- An aggregated data analysis identified patterns and correlations that individual assessments may have overlooked, offering a macro-level perspective crucial for scalable, urban-wide solutions;
- The archetypal analysis groups buildings into representative types based on their structural and environmental features, enabling tailored and targeted energy intervention strategies;
- These approaches establish a strong framework for optimizing energy consumption and fostering energy self-sufficiency on a larger scale, particularly in residential settings;
- The proposed solutions are context-specific and scalable, contributing to broader environmental sustainability goals.

## 4. Conclusions

This research shows that the most significant reduction in primary energy demand is achieved through the accumulation of interventions, particularly with the integration of renewable energy systems, like photovoltaics. The key findings include the following:

- A significant performance improvement is observed with the replacement of building systems;
- The annual reduction in local emissions, which is closely linked to primary energy demand reduction, is slightly higher for the first two interventions;
- The largest improvements come from installing photovoltaic systems and integrating renewable energies;
- Combined interventions lead to a 67% reduction in local emissions, promoting environmental sustainability;
- Primary interventions include thermal insulation and system replacement;
- Secondary interventions involve photovoltaic systems and energy storage;
- A holistic approach to building redevelopment, integrating technological and environmental solutions, is crucial for energy efficiency.
- The future research directions proposed are as follows:
- Advancements in renewable energy integration could impact costs, returns on investments, and savings;
- Innovations in building materials and energy systems may lower initial costs, with economies of scale and better financial models optimizing returns on investments;

- Future studies should quantify savings from advanced retrofitting techniques and emerging technologies, like smart grids;
- Environmental impact studies should further assess reductions in energy demand and emissions;
- The longitudinal monitoring of energy performance will help to assess long-term intervention effectiveness and sustainability;
- Research should also explore the economic and social impacts on property values, local economies, and quality of life;
- The integration of advanced building materials and smart energy systems should be further explored;
- Aligning energy efficiency initiatives with urban planning strategies could lead to comprehensive sustainability outcomes.

Exploring these future research directions will significantly refine and expand the findings of this study. By investigating technological advancements, cost analyses, and energy efficiency, researchers can further optimize residential buildings for sustainability, contributing to the development of more energy-efficient, economically viable, and resilient built environments, ultimately advancing the field of sustainable architecture and urban planning.

**Author Contributions:** Conceptualization, F.M. and G.R.; methodology, F.M. and G.R.; software, F.M. and G.R.; validation, F.M. and S.B.; writing—original draft preparation, G.R., S.B. and F.M.; writing—review and editing, G.R.; visualization, G.R.; supervision, F.M. and S.B.; project administration, F.M. All authors have read and agreed to the published version of the manuscript.

**Funding:** This research received no external funding.

**Institutional Review Board Statement:** Not applicable.

**Informed Consent Statement:** Not applicable.

**Data Availability Statement:** The original contributions presented in the study are included in the article, further inquiries can be directed to the corresponding author.

**Acknowledgments:** This contribution is the result of a collective reflection by the authors; it belongs to the recent and ongoing research activities led by the authors within the Department of Planning, Design, Technology of Architecture, Faculty of Architecture, Sapienza University of Rome, on the PNRR project "PE 2: Extended partnership NEST—Network 4 Energy Sustainable Transition | thematic area 2 "Green revolution and ecological transition—2.a. Green energies of the future" | SPOKE 8: Final use optimization, sustainability & resilience in supply chain | WP 8.3 "Low (Zero) carbon technologies for the reduction of energy demand, electrification, and efficiency of final uses", whose P.I. is Fabrizio Tucci.

**Conflicts of Interest:** The authors declare no conflicts of interest.

## Appendix A

**Table A1.** Data requested in the online questionnaire in the direct survey of the population.

| User-Related Information | |
|---|---|
| number of occupants in the house: | number of occupants in the house in the morning (8 a.m.–1 p.m.); |
| | number of occupants in the house in the afternoon (1 p.m.–7 p.m.); |
| | number of occupants in the house in the evening (7 p.m.–midnght); |
| | number of occupants in the house at night (midnight–8 a.m.). |

**Table A1.** *Cont.*

| **Architectural–morphological characteristics of the house** | |
|---|---|
| year of construction of the building; | |
| number of floors in the building in which the dwelling is located; | |
| average height of floors; | |
| number of rooms (including hallways and corridors); | |
| number of windows; | |
| prevailing color of vertical external walls; | |
| prevailing color of the upper closure (roof or terraced surface); | |
| approximate morphological plan typology of the dwelling; | |
| exposure with respect to the cardinal points (locating north); | |
| sides bordering the outside; | |
| sides bordering other flats or other buildings; | |
| possible presence of other flats above or below; | |
| total surface area of the dwelling; | |
| net area (without perimeter walls) of the dwelling; | |
| presence or absence of thermal insulation of horizontal and vertical opaque closures; | if present, percentage of thermal insulation; |
| characteristics of windows (whether original or replaced); | in case of replacement, number of windows replaced. |
| **Characteristics of the plant system** | |
| type of winter air-conditioning system (autonomous or centralised): | type of heat generator; |
| | type of heat distribution terminal; |
| | regulation system for the winter air-conditioning system (manual thermostat, programmable thermostat, thermostat value on each terminal); |
| energy class of the dwelling; | |
| presence of summer air-conditioning system: | if present, in which and how many rooms; |
| | energy class of summer air-conditioning terminals; |
| | presence of other summer air-conditioning appliances; |
| type of plant for DHW production; | |
| presence of solar thermal system (for DHW production): | if any, how many panels, orientation, slope and tilt angles; |
| Presence of PV system (for electricity production): | if any, how many panels, orientation, slope and tilt angles; |
| | if present, specify the power, the capacity of the system, and whether there is a storage tank; |

**Table A1.** *Cont.*

| **Characteristics of the appliances in the house** | |
|---|---|
| appliances in the kitchen: | hours of daily use of each appliance; |
| type of fridge and volume in litres: | energy class of the appliance; |
| washing clothes (washing machine and/or tumble dryer): | load capacity and number of washes per week performed; |
| | energy class of the appliance; |
| presence of other appliances: | total daily usage in minutes; |
| type of lighting system; | |
| presence and number of TVs: | size of each TV; |
| | energy class of the TV; |
| | total daily hours of use; |
| other electrical devices: | total daily hours of use; |
| presence and number of PCs: | total daily hours of use; |
| type of Internet connection for surfing the Internet; | |
| presence and type of printer: | average number of prints per day; |
| type of personal care appliances: | total daily usage minutes; |
| **Energy consumption and costs** | |
| usual or seasonal residence; | |
| electricity: | type of contract (free or captive market); |
| | actual consumption for various months in kW/h; |
| | actual expenses of the various months in euros; |
| gas: | type of contract (free or captive market); |
| | actual consumption for various months in $Sm^3$; |
| | actual expenses of the various months in EUR; |

## References


1. World Economic Forum. The Global Risks Report 2024—19th Edition—Insight Report. Cologne. 2024. Available online: https://www3.weforum.org/docs/WEF_The_Global_Risks_Report_2024.pdf (accessed on 3 August 2024).
2. United Nations Environment Programme. Emissions Gap Report 2022: The Closing Window—Climate Crisis Calls for Rapid Transformation of Societies. Nairobi. 2022. Available online: https://www.unep.org/emissions-gap-report-2022 (accessed on 3 August 2024).
3. World Green Building Council. Global Policy Principles for a Sustainable Built Environment. London. 2023. Available online: https://worldgbc.org/wp-content/uploads/2023/04/WorldGBC-Global-Policy-Principles_FINAL.pdf (accessed on 3 August 2024).
4. World Economic Forum. *The Building Sector Is Key to the Fight Against Climate Change*; World Economic Forum: Cologny, Switzerland, 2024.
5. Ashrafuzzaman, M.; Furini, G.L. Climate change and human health linkages in the context of globalization: An overview from global to southwestern coastal region of Bangladesh. *Environ. Int.* **2019**, *127*, 402–411. [CrossRef] [PubMed]
6. Zhang, Y.; Wang, C.H.; Davies, H.A. Net-zero energy buildings: The integration of renewable energy sources and efficient design strategies. *Renew. Energy* **2021**, *164*, 1131–1145.
7. Albadry, S.; Tarabieh, K.; Sewilam, H. Achieving Net Zero-Energy Buildings through Retrofitting Existing Residential Buildings Using PV Panels. *Energy Procedia* **2017**, *115*, 195–204. [CrossRef]
8. World Green Building Council. Beyond Buildings. Why An Integrated Approach to Buildings and Infrastructure Is Essential for Climate Action and Sustainable Development. London. 2021. Available online: https://worldgbc.s3.eu-west-2.amazonaws.com/wp-content/uploads/2022/08/10190455/Beyond-Buildings_reduced_compressed.pdf (accessed on 3 August 2024).
9. Romano, G.; Mancini, F. Transformation of a historical building into a Nearly Zero Energy Building (nZEB). *J. Phys. Conf. Ser.* **2022**, *2385*, 012008. [CrossRef]

10. United Nations Environment Programme. Global Status Report for Buildings and Construction: Beyond Foundations: Mainstreaming Sustainable Solutions to Cut Emissions from the Buildings Sector. Nairobi. 2024. Available online: https://www.unep.org/resources/report/global-status-report-buildings-and-construction (accessed on 3 August 2024).
11. Zhang, R.; Yin, L.; Jia, J.; Yin, Y. Application of ATS-GWIFBM Operator Based on Improved Time Entropy in Green Building Projects. *Adv. Civ. Eng.* **2019**, *2019*, 3519195. [CrossRef]
12. REN21. Renewables 2024 Global Status Report. 2024. Available online: https://www.lerenovaveis.org/contents/lerpublication/2024_abr_ren21_renewables-2024--global-status-report.pdf (accessed on 3 August 2024).
13. EC; Communication from the Commission to the European Parliament; the European Council; the Council; the European Economic and Social Committee; the Committee of the Regions. The European Green Deal. COM(2019) 640 Final, Bruxelles. 2019. Available online: https://eur-lex.europa.eu/resource.html?uri=cellar:b828d165-1c22-11ea-8c1f-01aa75ed71a1.0006.02/DOC_1&format=PDF (accessed on 3 August 2024).
14. Haas, R.; Rainer, K.R.; Kecskés, H.R.B. Achieving 2030 targets: The role of national energy and climate plans in EU member states. *Energy Rep.* **2022**, *8*, 109–124.
15. MiSE; MATTM; MIT. Piano Nazionale Integrato per l'Energia e il Clima, Roma. 2020. Available online: https://www.mimit.gov.it/images/stories/documenti/PNIEC_finale_17012020.pdf (accessed on 3 August 2024).
16. Baroni, G.; Manzolini, M.; Gattari, V. Piano Nazionale Integrato per l'Energia e il Clima: Analisi delle misure per la sostenibilità energetica e ambientale in Italia. *Riv. Ital. Energ.* **2020**, *37*, 112–129.
17. Civita, M.; Giacobazzi, L.D.; Bolognini, T. Le Politiche Energetiche Italiane Verso il 2030: Un'analisi del PNIEC e dei Suoi Impatti. 2022. Available online: https://www.fondazionesviluppsostenibile.org/wp-content/uploads/dlm_uploads/PNIEC-analisi-preliminare-Stakeholder-Forum-I4C-20-luglio.pdf (accessed on 3 August 2024).
18. Camera dei Deputati Servizio Studi XVIII Legislatura. Il Piano Nazionale di Ripresa e Resilienza (PNRR), Roma. 2022. Available online: https://www.governo.it/sites/governo.it/files/PNRR.pdf (accessed on 3 August 2024).
19. Sartori, I.; Sechi, P.L.; Castaldi, G. Il Piano Nazionale Integrato per l'Energia e il Clima: Obiettivi e Misure per il 2030 in Italia. 2021. Available online: https://www.efficienzaenergetica.enea.it/glossario-efficienza-energetica/lettera-p/piano-nazionale-integrato-per-l-energia-e-il-clima-2030-pniec.html (accessed on 3 August 2024).
20. López, A.; Pardo, F. National Integrated Energy and Climate Plans: An analysis of European Union member states' strategies. *Renew. Sustain. Energy Rev.* **2020**, *127*, 109847.
21. Moschetti, R.; Brattebø, H. Sustainable Business Models for Deep Energy Retrofitting of Buildings: State-of-the-art and Methodological Approach. *Energy Procedia* **2016**, *96*, 435–445. [CrossRef]
22. Ge, J.; Wang, Y.; Zhou, D.; Gu, Z.; Meng, X. Effects of urban vegetation on microclimate and building energy demand in winter—An evaluation using coupled simulations. *Sustain. Cities Soc.* **2024**, *102*, 105199. [CrossRef]
23. OJEU. Directive (EU) 2018/844 of the European Parliament and of the Council of 30 May 2018 Amending Directive 2010/31/EU on the Energy Performance of Buildings and Directive 2012/27/EU on Energy Efficiency—L 156/75 of 19.6.2018. 2018. Available online: https://eur-lex.europa.eu/legal-content/EN/TXT/PDF/?uri=CELEX:32018L0844 (accessed on 3 August 2024).
24. OJEU. Directive (EU) 2018/2002 of the European Parliament and of the Council of 11 December 2018 Amending Directive 2012/27/EU on Energy Efficiency—L 328/210 of 21.12.2018. 2018. Available online: https://eur-lex.europa.eu/legal-content/EN/TXT/PDF/?uri=CELEX:32018L2002 (accessed on 3 August 2024).
25. IEA. Net Zero by 2050 A Roadmap for the Global Energy Sector. 2021. Available online: https://iea.blob.core.windows.net/assets/deebef5d-0c34-4539-9d0c-10b13d840027/NetZeroby2050-ARoadmapfortheGlobalEnergySector_CORR.pdf (accessed on 3 August 2024).
26. Georgakopoulos, D.; Schuster, H.; Baker, D.; Cichocki, A. Managing escalation of collaboration processes in crisis mitigation situations. In Proceedings of the 16th International Conference on Data Engineering (Cat. No.00CB37073), San Diego, CA, USA, 29 February–3 March 2000; pp. 45–56.
27. Mousazadeh, H.; Sadeghi, M.M.; Ahmed, R.S. Advancements in smart grid technologies and their impact on local renewable energy adoption and emissions reduction. *J. Clean. Prod.* **2022**, *335*, 130261.
28. Sovacool, B.K.; Baum, C.M.; Low, S.; Roberts, C.; Steinhauser, J. Climate policy for a net-zero future: Ten recommendations for Direct Air Capture. *Environ. Res. Lett.* **2022**, *17*, 074014. [CrossRef]
29. Hesslerová, P.; Pokorný, J.; Huryna, H.; Seják, J.; Jirka, V. The impacts of greenery on urban climate and the options for use of thermal data in urban areas. *Prog. Plan.* **2022**, *159*, 100545. [CrossRef]
30. Madushika, U.G.D.; Ramachandra, T.; Karunasena, G.; Udakara, P.A.D.S. Energy retrofitting technologies of buildings: A review-based assessment. *Energies* **2023**, *16*, 4924. [CrossRef]
31. Miller, W.; Thornton, P.T. Comprehensive renovation strategies for existing buildings: From energy efficiency to zero emissions. *J. Build. Perform.* **2021**, *12*, 66–81.
32. Tucci, F.; Altamura, P.; Cecafosso, V.; Turchetti, G.; Giampaoletti, M.; Nava, F.; Pani, M.M.; Romano, G.; Tulelli, V.; Dalsasso, C.; et al. Rapporto "Verso la Neutralità Climatica Delle Green City: Approcci, Indirizzi, Strategie e Azioni", Sapienza Università di Roma, Roma. 2021. Available online: https://www.greencitynetwork.it/wp-content/uploads/stati_generali_green_economy_La_neutralita_climatica_delle_citta.pdf (accessed on 3 August 2024).
33. Ma, K.; Yu, Y.; Yang, B.; Yang, J. Demand-Side Energy Management Considering Price Oscillations for Residential Building Heating and Ventilation Systems. *IEEE Trans. Ind. Inform.* **2019**, *15*, 4742–4752. [CrossRef]

34. Baiani, S.; Altamura, P.; Turchetti, G.; Romano, G. Transizione Energetica e Circolare del Patrimonio Industriale. Il Caso Dell'ex SNIA a Roma/Energy and Circular Transition of the Industrial Heritage. The Ex SNIA case in Rome. *Agathón* **2024**, *15*, 190–203.
35. Cappelletti, F.; Dalla Mora, T.; Peron, F.; Romagnoni, P.; Ruggeri, P. Building renovation: Which kind of guidelines could be proposed for policy makers and professional owners? *Energy Procedia* **2015**, *78*, 2366–2371. [CrossRef]
36. Centro Ricerche Economiche, Sociologiche e di Mercato nell'Edilizia (CRESME). Incentivi e Riduzione del Rischio Sismico in Italia, Cosa Fare, Come Fare. 2018. Available online: https://www.cresme.it/wp-content/uploads/2023/04/Italia_Antisismica_abstract.pdf (accessed on 3 August 2024).
37. OJEU. Communication from the Commission to the European Parliament, The Council, The European Economic and Social Committee and the Committee of the Regions 'Fit for 55': Delivering the EU's 2030 Climate Target on the Way to Climate Neutrality—COM(2021) 550 Final, Bruxelles. 2021. Available online: https://eur-lex.europa.eu/legal-content/EN/TXT/HTML/?uri=CELEX:52021DC0550 (accessed on 3 August 2024).
38. OJEU (2024), Directive (EU) 2024/1275 of the European Parliament and of the Council of 24 April 2024 on the Energy Performance of Buildings (Recast), 2024/1275. Bruxelles. Available online: https://eur-lex.europa.eu/legal-content/EN/TXT/PDF/?uri=OJ:L_202401275 (accessed on 3 August 2024).
39. Bellocchi, S.; De Falco, M.; Gambini, M.; Manno, M.; Stilo, T.; Vellini, M. Opportunities for power-to-Gas and Power-to-liquid in CO2-reduced energy scenarios: The Italian case. *Energy* **2019**, *175*, 847–861. [CrossRef]
40. MITE. Dipartimento Energia—Direzione Generale Infrastrutture E Sicurezza—La situazione energetica nazionale nel 2021, Roma. Available online: https://dgsaie.mise.gov.it/pub/sen/relazioni/relazione_annuale_situazione_energetica%20_nazionale_dati%20_2021.pdf (accessed on 3 August 2024).
41. REN21. Renewables 2022 Global Status Report. REN21 Secretariat c/o UN Environment Programme, Paris. 2022. Available online: https://www.ren21.net/wp-content/uploads/2019/05/GSR2022_Full_Report.pdf (accessed on 3 August 2024).
42. King, R.; Vullnetari, J. From shortage economy to second economy. *J. Rural. Stud.* **2016**, *44*, 198–207. [CrossRef]
43. Gonzalez Benson, O.; Smith, R.; Song, C. Resettlement cities: A mixture model analysis of the dispersion and placement strategy. *Cities* **2022**, *126*, 103698. [CrossRef]
44. Ahmed, S.; Ali, A.; D'Angola, A. A Review of Renewable Energy Communities: Concepts, Scope, Progress, Challenges, and Recommendations. *Sustainability* **2024**, *16*, 1749. [CrossRef]
45. Genave, A.; Blancard, S.; Garabedian, S. An assessment of energy vulnerability in Small Island Developing States. *Ecol. Econ.* **2020**, *171*, 106595. [CrossRef]
46. Elahee, K. Energy Management Challenges in a Small Island Economy. *Energy Environ.* **2010**, *21*, 803–813. [CrossRef]
47. Battistelli, F.; Messina, A.; Tomassetti, L.; Montiroli, C.; Manzo, E.; Torre, M.; Tratzi, P.; Segreto, M.; Chu, C.-Y.; Paolini, V.; et al. Assessment of Energy, Mobility, Waste, and Water Management on Italian Small Islands. *Sustainability* **2023**, *15*, 11490. [CrossRef]
48. Corsini, A.; Delibra, G.; Pizzuti, I.; Tajalli-Ardekani, E. Challenges of renewable energy communities on small Mediterranean islands: A case study on Ponza island. *Renew. Energy* **2023**, *215*, 118986. [CrossRef]
49. Leal Filho, W.; Balogun, A.-L.; Surroop, D.; Salvia, A.L.; Narula, K.; Li, C.; Hunt, J.D.; Gatto, A.; Sharifi, A.; Feng, H.; et al. Realising the Potential of Renewable Energy as a Tool for Energy Security in Small Island Developing States. *Sustainability* **2022**, *14*, 4965. [CrossRef]
50. Romano, G.; Baiani, S.; Mancini, F. Energy Consumption Profile. Profilo di Consumo Energetico. Google Docs. Available online: https://docs.google.com/forms/d/e/1FAIpQLSeNhaE08Z8wVL_k9eWVmwAouT39VQ5qMabEJ_sk0CkUuTDKiQ/viewform (accessed on 3 August 2024).
51. Mancini, F.; Romano, S.; Lo Basso, G.; Cimaglia, J.; De Santoli, L. How the Italian Residential Sector Could Contribute to Load Flexibility in Demand Response Activities: A Methodology for Residential Clustering and Developing a Flexibility Strategy. *Energies* **2020**, *13*, 3359. [CrossRef]
52. Suomalainen, K.; Eyers, D.; Ford, R.; Stephenson, J.; Anderson, B.; Jack, M. Detailed comparison of energy-related time-use diaries and monitored residential electricity demand. *Energy Build.* **2019**, *183*, 418–427. [CrossRef]
53. Soto, A.M.; Jentsch, M.F. Comparison of prediction models for determining energy demand in the residential sector of a country. *Energy Build.* **2016**, *128*, 38–55. [CrossRef]
54. ARERA. Analisi dei Dati dei Clienti Domestici. Milano. 2022. Available online: https://www.arera.it/dati-e-statistiche/dettaglio/analisi-dei-consumi-dei-clienti-domestici (accessed on 3 August 2024).
55. Schweiker, M.; Shukuya, M. Comparison of theoretical and statistical models of air-conditioning-unit usage behaviour in a residential setting under Japanese climatic conditions. *Build. Environ.* **2009**, *44*, 2137–2149. [CrossRef]
56. Ren, X.; Yan, D.; Wang, C. Air-conditioning usage conditional probability model for residential buildings. *Build. Environ.* **2014**, *81*, 172–182. [CrossRef]
57. Cosenza, G. *Gli Spazi Dell'architettura di Procida*, 1st ed.; N. 8698 BIS.; Edizioni Scientifiche Italiane: Napoli, Italy, 1968.
58. Muro, G.; Muro, V.; Ridda, L. *Piano del Colore dei Centri Storici Dell'isola di Procida di Primo Ambito*; Procida: Napoli, Italy, 2002.
59. Fellella, F.; Ambrosino, R.; Perrella, S.; Rossello, C.; Stasi, A.; Squillante, R.; Massarotti, N.; Vanoli, L.; De Luca, G.; Fabozzi, S. Piano D'azione Per L'energia Sostenibile (PAES)—Comune di Procida (NA). 2015. Available online: https://www.paesitalia.it/azioni/?r=4 (accessed on 3 August 2024).
60. Miraglia, F. *Atlante delle Tecniche Costruttive di Terra di Lavoro Murature in Tufo Grigio (XIII-XV)*; Armando Caramanica Editore: Marina di Minturno, Italy, 2018; ISBN 978-88-7425-110-0.

61. GURI. DECRETO LEGISLATIVO 19 Agosto 2005, n. 192 | Attuazione della Direttiva 2002/91/CE Relativa al Rendimento Energetico Nell'edilizia. (GU Serie Generale n.222 del 23-09-2005—Suppl. Ordinario n. 158). 2005. Available online: https://www.gazzettaufficiale.it/eli/id/2005/09/23/005G0219/sg (accessed on 3 August 2024).
62. Regione Campania. Rapporto Annuale Sulle Attestazioni di Prestazione Energetica (APE) | Indice di Prestazione Energetica Globale non Rinnovabile (EPgl,n_ren) Medio. 2023. Available online: https://www.regione.campania.it/regione/it/tematiche/certificazione-energetica-edifici-ape/rapporto-annuale-sulle-attestazioni-di-prestazione-energetica-ape?page=1 (accessed on 3 August 2024).

*sustainability*

MDPI

*Article*

# Thermal Energy Storage in Energy Communities: A Perspective Overview through a Bibliometric Analysis

**Luca Brunelli [1], Emiliano Borri [2], Anna Laura Pisello [3], Andrea Nicolini [4], Carles Mateu [2] and Luisa F. Cabeza [2,*]**

1 CIRIAF-Interuniversity Research Centre, University of Perugia, Via G. Duranti 63, 06125 Perugia, Italy; luca.brunelli@dottorandi.unipg.it
2 GREiA Research Group, Universitat de Lleida, Pere de Cabrera s/n, 25001 Lleida, Spain; emiliano.borri@udl.cat (E.B.); carles.mateu@udl.cat (C.M.)
3 EAPLAB, CIRIAF-Interuniversity Research Centre, University of Perugia, Via G. Duranti 63, 06125 Perugia, Italy; anna.pisello@unipg.it
4 Department of Engineering, UNIPG—University of Perugia, Via G. Duranti 93, 06125 Perugia, Italy; andrea.nicolini@unipg.it
* Correspondence: luisaf.cabeza@udl.cat

**Abstract:** The climate and energy crisis requires immediate countermeasures. Renewable energy communities (RECs) are capable of enhancing the consumption of renewable energy, involving citizens with a leading role in the energy transition process. The main objective of a REC is to maximize the consumption of renewable energy by reducing the mismatch between energy supply and demand. This is possible through the use of strategies and technologies including energy storage systems. Among these, the use of thermal energy storage (TES) is an efficient strategy due to the lower investment required compared to other storage technologies, like electric batteries. This study aims to define the role of TES in RECs, through a bibliometric analysis, in order to highlight research trends and possible gaps. This study shows that the existing literature on TES does not present terms related to RECs, thus presenting a research gap. On the other hand, RESs address the topic of energy storage in the literature, without focusing on TES in particular but considering the general aspect of the topic. Therefore, this leaves open a possibility for the development of research on TES as a possible technology applied to a REC to maximize the renewable energy sharing.

**Keywords:** renewable energy community; thermal energy storage (TES); bibliometric analysis

**Citation:** Brunelli, L.; Borri, E.; Pisello, A.L.; Nicolini, A.; Mateu, C.; Cabeza, L.F. Thermal Energy Storage in Energy Communities: A Perspective Overview through a Bibliometric Analysis. *Sustainability* **2024**, *16*, 5895. https://doi.org/10.3390/su16145895

Academic Editors: Elena Lucchi, Tianyi Chen and Wen Zhang

Received: 17 June 2024
Revised: 3 July 2024
Accepted: 8 July 2024
Published: 10 July 2024



## 1. Introduction

Year after year, the climate crisis represents an increasingly concrete threat against which it is now urgent to take countermeasures. In the stocktaking at the last UN Climate Change Conference (COP 28), the goal of peaking global greenhouse gas emissions by 2025 and reducing them by 43% by 2030 and 60% by 2035 compared to 2019 levels in order to limit global warming to 1.5 °C was confirmed [1].

In the context of climate change mitigation, energy is one of the sectors where emission reductions can have the greatest short-term effects. This is why there is an increasing need for a transition to renewable energy sources and a consequent shift away from fossil fuels.

The European Union manifested this objective in recent years through the "Clean Energy Package" [2] and the "Fit for 55" [3], measures promoted by the European Commission that set a perspective of climate neutrality from fossil fuels from 2050 onwards. The Clean Energy Package focus aims to transform the economy and society according to a new sustainability paradigm, in which citizens should be encouraged to actively participate in the energy market, even in aggregated forms, to foster their empowerment and make them responsible for their consumption. To achieve this objective, consumers should be able to directly manage their energy.

In this context, renewable energy communities (RECs) [4] represent a social innovation capable of advancing more equitable models for managing, consuming, and producing energy, enhancing democratic decision-making and control over renewable energy, and placing it in the hands of communities and individuals. This citizens' empowerment further contributes to increasing social acceptance of renewable energy generation plants thanks to the underlying participatory approaches that provide citizens an active role in the decision-making process [5].

Introduced by the European Renewable Energy Directive (RED II), RECs are autonomous legal entities, in which the members and stakeholders controlling it can participate openly and voluntarily, staying close to renewable energy production plants. This participation must not generate financial gains, but can give economic, social, environmental, and energy benefits. Members of an energy community can be classified into energy producers, energy consumers, and energy prosumers, i.e., producers of energy that self-consume on site. The main objective of a REC is to maximize the sharing of energy produced through renewable energy sources (RESs) in a collective self-consumption scheme [6]. However, the use of RESs entails the usual problem of solving the time mismatch between production and user demand for energy.

In this context, the so-called power-to-X strategies [7] consist of energy carrier conversions in order to efficiently use the available energy storage technologies. Power-to-heat [8] is an example of a strategy for which thermal energy storage (TES) systems play an important role. With the use of a heat pump, it is possible to convert electrical energy into a heat carrier, storing it to meet the heat demand of a utility. In these systems, TES [9] consisting of hot water tanks is a simple and economical solution for storing excess renewable energy [10].

This can be a way of increasing the sharing of renewable energy within a REC. In fact, the use of TES allows for storing the excess electrical energy produced by the plants of the producers that are part of the REC. Thus, the use of energy from renewable sources is guaranteed to the different consumers of the REC, not only in the periods of electrical energy demand peaks, but also in the thermal energy demand peaks.

This solution proves to be more sustainable than the installation of an electric storage battery. This is evidenced by Fambri et al. [11], who in their study compare the use of an electric battery and a heat pump thermal energy storage system in a REC.

However, the use of TES also has additional benefits. For example, it can decrease the thermal energy demand of buildings; counteract the randomness of renewable energy production, both thermal and electrical; and constitute a source of heat recovery [12].

Furthermore, in a larger field than the individual building, e.g., in the case of districts [13–15], more relatable to a REC, they can foster the diffusion, development, and flexibility of district heating and cooling systems powered by renewables, providing thermal energy to all consumers in them.

This study shows that although there is an energy saving from increased self-consumption of renewables through electrical storage, the economic benefit is not such as to justify the investment. On the other hand, the use of TES, despite having lower results in terms of energy savings, proves to be more cost-effective. The use of TES in a REC is also taken up in other works as a strategy to utilize surplus energy [16–18].

The aim of this paper is to give a detailed overview and report on the state of the art in the role and involvement of TES within RECs. Thus, a bibliometric analysis, which is a recognized technique of interest used in order to study scientific advances in a given topic, is carried out. A quantitative analysis is conducted regarding the number of publications, authors, and institutions researching this topic.

The results show that studies about the integration of TES to RECs, two very recent topics that have grown considerably in recent years, are still only marginal, as evidenced by the small number of papers about it.

Furthermore, to the best of the authors' knowledge, there are still no papers in the literature that fill this gap, nor are there any that do so through the bibliometric analysis approach.

The objective of this study is also to provide a guide for researchers wishing to investigate the main networks and institutions working in the field of TES as a tool for sharing renewable energy within RECs, and to use the main research gaps identified to inspire future research on this topic, filling this deficit.

This research can also be for industries, public administrations, and private citizens who intend to become part of a REC, seeking to deepen their understanding of TES as a tool for sharing renewable energy surplus.

## 2. Materials and Methods

This section reports the methodology applied to develop the bibliometric analysis shown in the following sections. For this work, Scopus was preferred as the reference database. Web of Science was not considered due to its low number of documents on technology topics [19], while ResearchGate and Google Scholar were also not considered due to their low reliability in the bibliometric field [20]. In order to identify research trends and gaps in the application of TES in energy communities, bibliometric data were collected using two queries (Table 1) on the database Scopus, with the last access on 12 January 2024. A first query was used to collect bibliometric data related to the studies on TES in buildings including districts [21]. Compared to the work by Borri et al. [17], the query used in this study includes both buildings and districts, whereas previously, these categories were analyzed separately. In addition, the category "roads and bridges" is not considered here. Moreover, in this case, the transport sector was excluded in the query. The analysis of TES was carried out regarding its application to buildings and districts, in order to consider not only the dimension of the individual building but also a broader one, as a community.

**Table 1.** Queries used for the Scopus database.

| Topic | Query |
|---|---|
| TES in buildings and districts | TITLE-ABS-KEY ("thermal energy storage*" AND ("building*" OR "built environment" OR "district*" OR "municipal*" OR "neighbo$rhood") AND NOT ("aircraft" OR "transport* application" OR "transport sector")) |
| Renewable energy communities | "energy communit*" AND "renewable*" AND NOT ("renewable energy forecasting" OR "building energy community" OR "energy corridors" OR " solar renewable energy community" OR "NGH" OR "Cu2Se"OR "SDG 7"OR "computational intelligence (CI) community" OR "MEG" OR "nuclear" OR "materials, chemistry, and renewable energy communities") |

A second, more specific query, was then formulated to draw the state of the art in TES, specifically in energy communities. The concept of "energy community" was combined with the aspect of "renewable" sources. This is because renewable energy communities are often referred to simply as "energy communities". However, the term "renewable" is added, as sometimes the term "energy community" may refer to a broader and diversified concept of community, understood as a group of researchers and scientists in the specific field of energy, which is different from the concept intended in this paper. Therefore, the query excludes the keywords "building energy community", "solar renewable energy community", "computational intelligence (CI) community", "materials, chemistry, and renewable energy communities", and "energy corridors". Similarly, terms such as "nuclear", "MEG", "NGH", "Cu2Se", and "SDG 7" were excluded, as they are related to papers not relevant to the research in question.

In this study, bibliometric data were used to identify the main trends in the number of publications. In order to identify the research gaps, keywords were visually mapped and analyzed through the open-source software VOSviewer (version 1.6.20) [22]. Through the latter, it is possible to graphically visualize and analyze bibliometric networks based on the links between different items of the research publication. These links can be about co-authorship, co-occurrence, citation, bibliographic coupling, or co-citation. The elements available for generating maps from publications include authors, institutions, countries, and keywords. The practicality of these analyses conducted with VOSviewer is that the output data can be easily exported and displayed in other resources [23].

Furthermore, the great advantage of VOSviewer is that through its ease of use and accessibility, it enables accurate bibliometric analyses using clusters even for those without extensive knowledge of clustering or software skills [24].

For this reason, VOSviewer is considered a validated tool for investigating gaps and research opportunities in the scientific literature. In particular, it is interesting to see how this software is already used to support bibliometric analysis within studies concerning energy storage [25–28], and also those concerning prosumers and energy communities [29–32].

In this paper, the findings regarding the co-occurrence of author keywords are presented through a network visualization. Each element (author keyword) is represented by a colored circle, with its size reflecting its weight in terms of occurrences or link strength. The colors denote the clusters to which different elements belong, and the distance between circles indicates the relationship between the items. Finally, the two indicated queries were merged within VOSviewer and analyzed together in order to highlight possible points of contact between the two topics and possible research gaps.

## 3. Results and Discussion

### 3.1. TES in Buildings and Districts

This section reports the results of the bibliometric analysis carried out with the methodology described in Section 2. Figure 1 shows the number of documents published in the field of TES applied to buildings and districts. The trend indicates a rapid increase in the number of publications per year since 2009, with over 500 publications in the year 2023. This demonstrates that TES is gaining momentum in built environment applications, becoming a pivotal component of the energy transition.

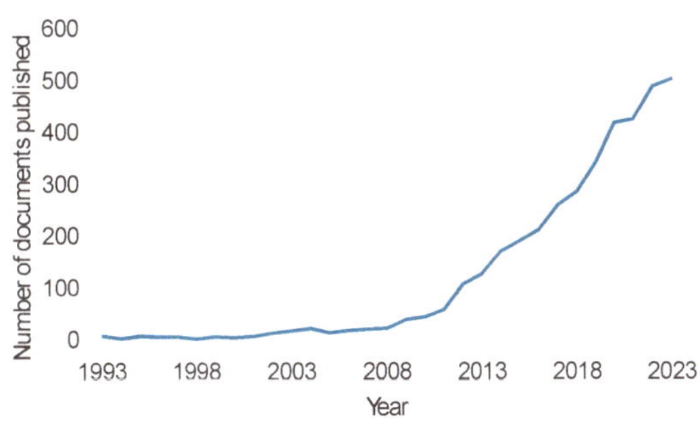


**Figure 1.** Number of documents published per year in thermal energy storage.

Figure 2 illustrates the publication trend over the years for the leading territories that have published papers on TES for building and district applications. As already illustrated by Borri et al. [17], the data reveal that the United States (USA) began publishing documents related to TES before 1990, whereas Europe and China commenced their publications after 2008. In the USA, most cited publications are focused on phase change materials (PCMs) [33–35], optimizing the control of TES [36,37], and the use of TES in district heating and cooling systems [38].

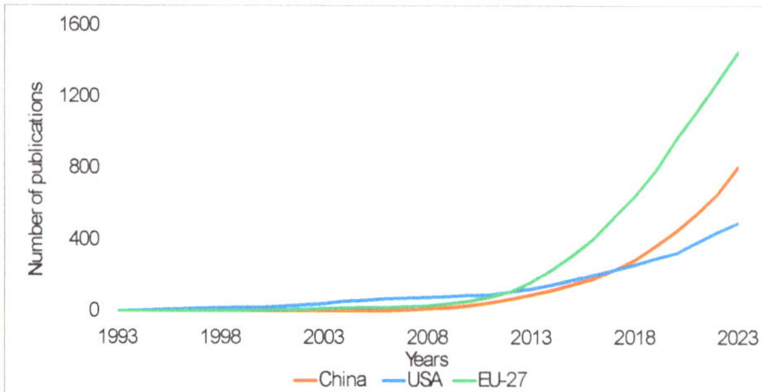


**Figure 2.** Number of publications over the years on TES in buildings and districts for the top 3 territories between 1993 and 2023.

Research on this topic in Europe has increased rapidly as a result of directives about the energy performance of buildings and energy efficiency introduced by the European Union [39,40]. The most cited papers from Europe on Scopus predominantly consist of review articles focusing on PCMs in buildings [41–44].

In addition to these, within the same scope lies the most cited article, published by Cabeza et al. [45] in 2011, which constitutes a review on the use of this technology, classification of materials, materials available, and problems and possible solutions on the application of such materials in buildings. Additional highly cited documents include general reviews on TES and the materials used in their application [9,46].

From Figure 2, it can be observed that China follows a similar trend to that of the European Union, albeit with a lower number of publications. This is because, like Europe, China has implemented a substantial plan to reduce emissions from fossil fuels by 2030 compared to those in 2005 [47].

Among the most cited articles in this field, there are general reviews on TES [48] and the application of latent heat thermal energy storage [49]. Additionally, numerous reviews regarding the study and use of PCMs [50–53] are available on Scopus. The most cited paper concerns precisely this topic and was published in 2012 by Zhou et al. [54]. In Figure 3, the main European countries publishing documents regarding the application of TES in the context of buildings and districts are shown.

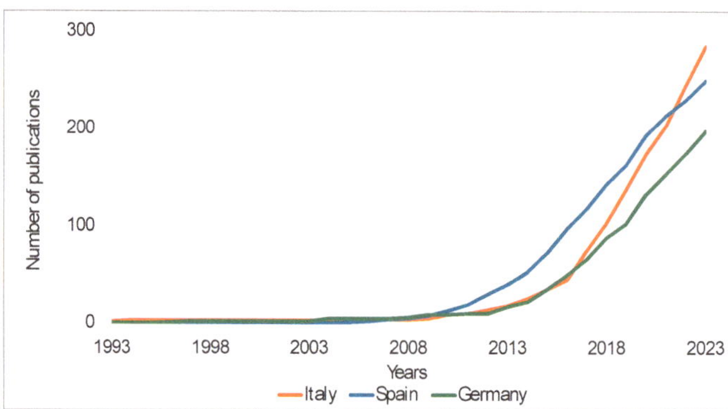


**Figure 3.** Number of publications over the years on TES in buildings and districts for the top EU27 countries between 1993 and 2023.

It is possible to notice that Spain, Italy, and Germany have a similar trend of publications that started to rapidly increase after 2010. As of 2023, Italy is the country with the highest number of publications, among which the most cited ones focus on energy management strategies [55–57], cogeneration [58,59], district heating and cooling systems [60], and the use of thermal energy storage techniques in thermal insulation [61]. The second country with the highest number of publications is Spain, whose studies focus PCMs [62,63] and the application of TES in the domain of buildings [64,65].

Finally, the third country with the highest number of publications is Germany, which covers various topics including the application of TES to buildings to enhance their development [66,67], the study of PCMs [68,69], aquifer TES [70], TES for solar power plants [71,72], and district heating [73].

Figure 4 illustrates the journals with the most publications related to TES applied to buildings and districts. The best is *Energy and Buildings*, with 297 publications as of December 2023. It is interesting that the journal *Energies* appears in Figure 4, with 143 open-access publications. The most cited paper from this journal was published in 2020 by Enescu et al. [74]. The authors and the institutions with most publications on TES applied to buildings and districts are shown in Tables 2 and 3, respectively. The author with the most publications is Luisa F. Cabeza, affiliated with the "Universitat de Lleida", in Spain, with 121 publications as of December 2023. Her most cited article on TES applied to buildings and districts dates back to 2011 and it shows a review of the publications on the use of PCMs in buildings [45]. Some of his other most cited works concern the application of PCMs [75–78] and TES technologies integrated [79,80] to the field of buildings. This makes Universitat de Lleida the affiliation with the highest number of publications on this topic.

The second most published author is Ahmet Sari, affiliated with Karadeniz Technical University, Turkey, which is also among the affiliations with the highest number of publications on TES in buildings and districts. His most cited articles focus on the thermal properties of PCMs [81–84]. As China is the second most published territory, two Chinese affiliations are among those with the most published articles: the Ministry of Education of the People's Republic of China, and Tsinghua University.

Figure 5 shows the co-occurrence of keywords obtained through the software VOSviewer for the first query related to TES in buildings and districts. As stated in the methodology, the keywords belonging to the same topic were grouped to represent the macro-areas of research. In order to obtain the results displayed in the figure, the keywords were filtered with a minimum occurrence of 10. In the right part of the figure (green cluster), it is possible to notice that a lot of studies are related to the development and the improvement of storage materials, in particular PCMs. On the left side of the figure, it is possible to notice that the bigger cluster (red) is related to TES applications, including keywords such as "control", "optimization", "demand side management", "numerical model", and "district heating". The blue cluster is related to the applications of TES as a system in buildings, including keywords such as "heating", "cooling", "buildings", and "energy saving". The other main cluster (yellow) is related to methods to analyze improve the thermal parameters of storage materials.

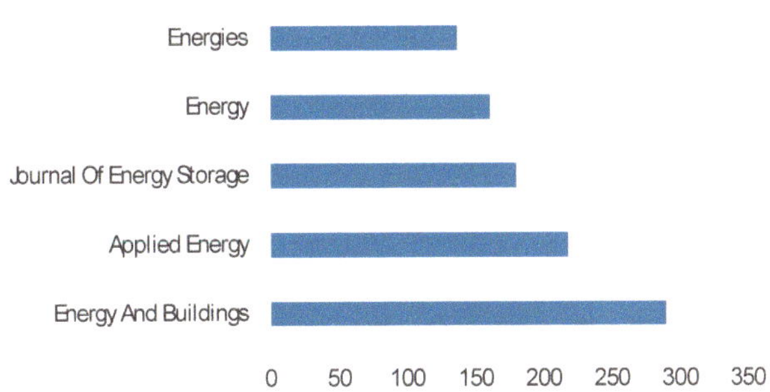


**Figure 4.** Top 5 journals publishing on TES applied to buildings and districts.

**Table 2.** Top 5 institutions with documents published on TES in buildings and districts.

| Affiliation | Publication | Country |
|---|---|---|
| Universitat de Lleida | 123 | Spain |
| Ministry of Education of the People's Republic of China | 108 | China |
| Tsinghua University | 71 | China |
| Karadeniz Technical University | 59 | Turkey |
| King Fahd University of Petroleum and Minerals | 58 | Saudi Arabia |

**Table 3.** Top 10 authors on documents published on TES in buildings and districts.

| Author Name | # Publications | Affiliation | Country |
|---|---|---|---|
| Cabeza, L.F. | 121 | Universitat de Lleida | Spain |
| Sari, A. | 84 | Karadeniz Technical University | Saudi Arabia |
| Barreneche, C. | 40 | Universitat de Barcelona | Spain |
| Tyagi, V.V. | 37 | Shri Mata Vashini Devi University | India |
| Fernández, A.I. | 34 | Universitat de Barcelona | Spain |
| Hekimoğlu, G. | 34 | Karadeniz Technical University | Turkey |
| Kim, S. | 33 | Yonsei University | South Korea |
| de Gracia, A. | 28 | Universitat de Lleida | Spain |
| Gencel, O. | 25 | Bartin Üniversitesi | Turkey |
| Pisello, A.L. | 25 | Università degli Studi di Perugia | Italy |

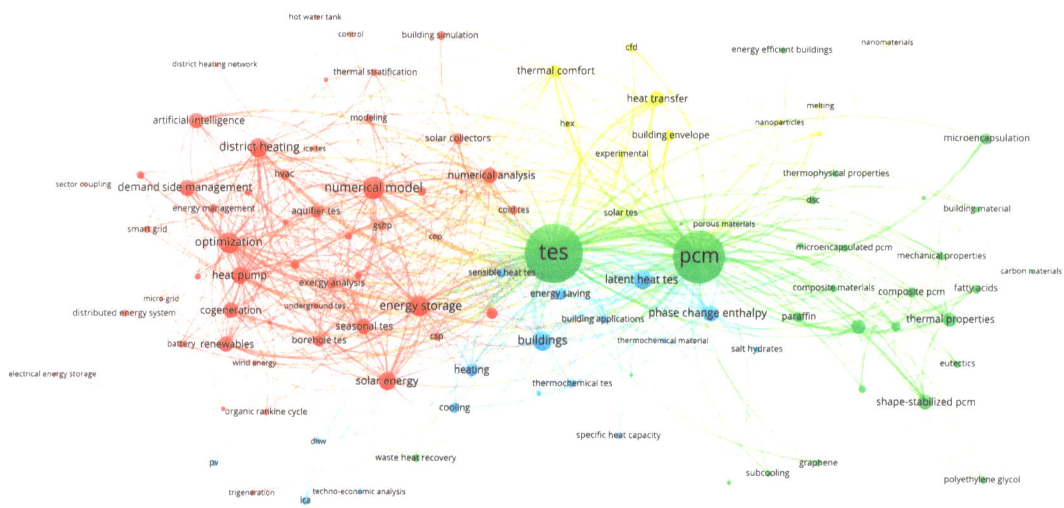


**Figure 5.** Network visualization of author keywords on "TES on buildings and districts" elaborated with VOSviewer.

Table 4 shows the author keywords with the highest number of occurrences on "TES on buildings and districts". While the keyword with the most occurrences is obviously "TES", it can be seen that the second is "PCM (phase change materials)". This is evidenced by the large number of articles that were published on this topic in different territories and countries. A further keyword with a high occurrence is "numerical model", due to the widespread use in the literature of numerical modelling techniques to simulate the behavior of TES, in particular PCMs [35,85–87].

**Table 4.** Highest number of occurrences on "TES on buildings and districts".

| Keyword | Number of Occurrences |
|---|---|
| TES (thermal energy storage) | 1609 |
| PCM (phase change material) | 1257 |
| numerical model | 230 |
| optimization | 204 |
| energy storage | 184 |
| district heating | 168 |
| latent heat TES | 164 |
| solar energy | 155 |
| heat pump | 138 |
| phase change enthalpy | 124 |

The keyword "optimization" mainly refers to optimization studies for the integration of thermal storage in building systems, in order to reduce the mismatch between demand and supply of renewable energy and minimize costs [67,88–91]. The scope of the districts taken into consideration through the query is highlighted by the keyword "district heating". Among the most frequently cited articles with this keyword are applications of TES to combined heat and power (CHP) systems for district heating [15,92,93], review articles [38,60,94], and studies on the thermal inertia of buildings as thermal storage in district heating and cooling systems [95,96].

A research gap emerges from Figure 5. In fact, there are no terms or keywords related to RECs. However, there are keywords such as "distributed energy system", "smart grid", "micro-grid", and "sector coupling", which are related to a broader concept than the single building and can be approached as energy sharing. Indeed, these terms are related to a collective dimension of energy consumption and management, attributable to the concept of RECs. The keyword "distributed energy system" refers to a system based on small-scale installations distributed across the territory, in stark contrast to the traditional model of centrally produced power plants. It is precisely on this basis that the REC is built, aiming to facilitate the spread of renewable energies in the territory, meeting both environmental and community needs [97]. The concept of a "microgrid" is instead linked to energy communities through the themes of energy self-sufficiency and energy sharing. The members of an energy community are no longer seen as mere consumers, but become "prosumers", meaning they are energy producers who consume and are capable of exchanging energy with each other through "peer-to-peer" sharing mechanisms [98,99]. Furthermore, through energy efficiency systems and smart management energy systems, linked to the maximization of renewable energy sharing, "smart grids" become a fundamental tool for RECs [100–102]. Lastly, the possibility of having a highly integrated energy system, capable of maximizing the use of renewable energy produced either through storage [103] or across different consumption sectors [104,105], makes "sector coupling" a fundamental element in optimizing energy communities.

Figures 6 and 7 show the co-occurrence of these keywords and their links. The figures highlight, given the small number of links, the low relevance of these terms in the literature. Figure 6 shows the co-occurrence of the keyword "distributed energy system". The few links present all concern the red cluster, in the context of TES applications. The most cited paper on distributed energy systems applied in TES was published in 2020, from Wirtz et al. [73], and it is about a low-temperature network of buildings equipped with heat pumps, chillers, and heat exchangers for thermal energy storage, connected to minimize annual energy costs and to reduce emissions. This concept is very close to that of RECs and is also approached by other studies on this topic [106–108].

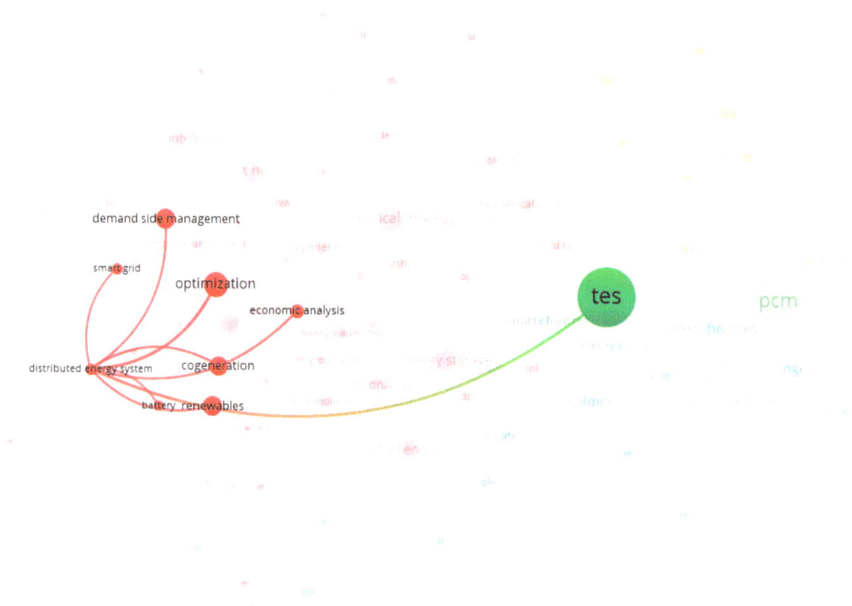


(a)

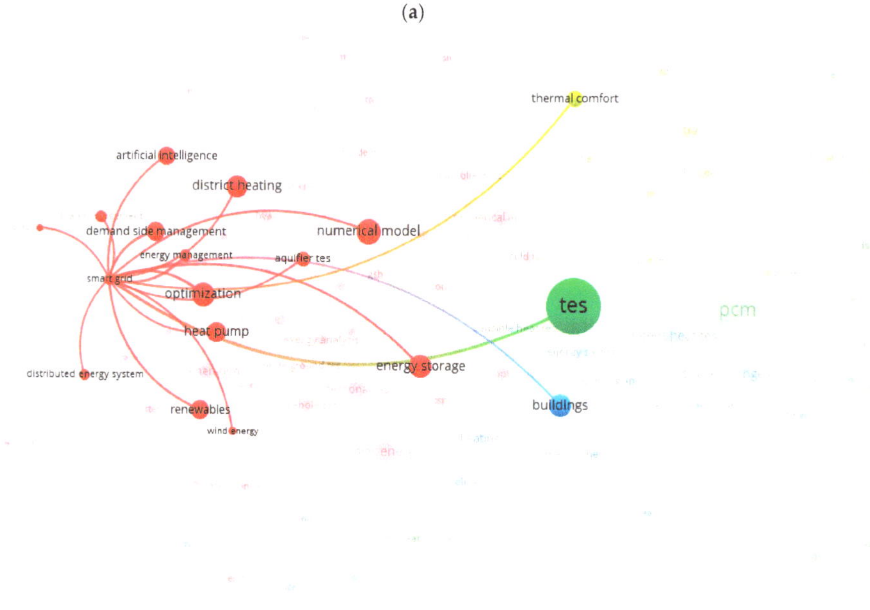


(b)

**Figure 6.** Co-occurrence and links concerning the keywords (**a**) "distributed energy system" and (**b**) "smart grid".

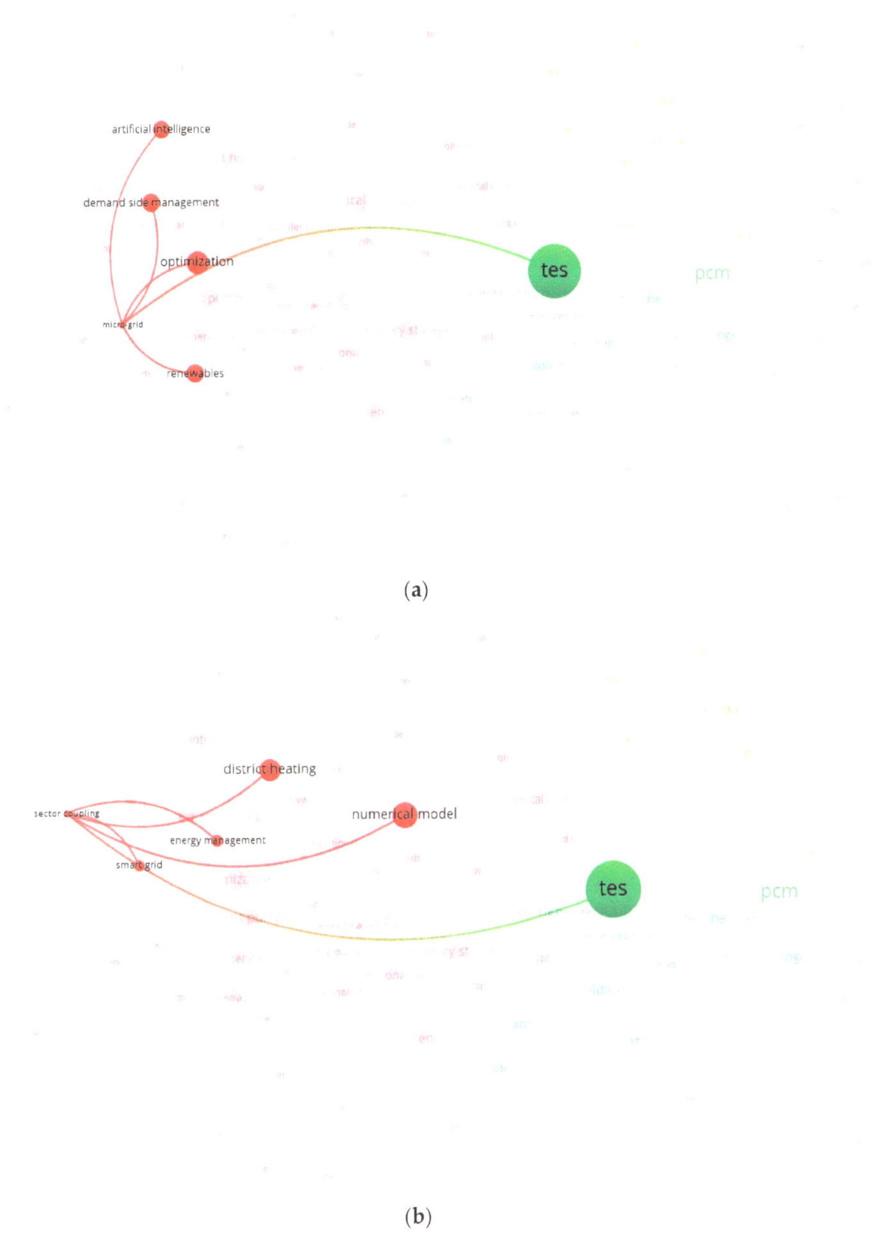


**Figure 7.** Co-occurrence and links regarding the keywords (**a**) "micro-grid", and (**b**) "sector coupling".

Figure 7 shows how microgrids and sector coupling have the lowest occurrences and thus represent research gaps in the applications of TES. In fact, the term microgrid has only four links besides TES: "demand side management", "artificial intelligence", "optimization", and "renewables". It is no coincidence that the main studies of TES on this topic concern optimal energy management models applied to microgrids [109–111]. Also, the term sector coupling has just four links besides TES, and they are "district heating", "energy management", "numerical model", and "smart grid". It is interesting to note that the most cited article related to this keyword in the context of TES, published by Bartolini et al. [112] in 2020, has as its subject the concept of an energy community based on a multi-energy system to which various storage technologies are applied.

The most cited article in the field of microgrids applied to TES dates back to 2021 and was published by Li et al. [113]. The most cited article on sector coupling applied to TES is by Zeyen et al. and it explores strategies to mitigate winter heating demand peaks in the European Union, crucial for cost and emissions reduction goals [114]. In the field of materials development, they are all related to the keyword "TES", while in the field of TES applications, they are related to terms such as "demand side management", "optimization", and "numerical model".

### *3.2. Renewable Energy Communities*

The number of documents published related to renewable energy communities (RECs) is shown in Figure 8. The trend shows a steady and progressive increase in publications since 2018. This shows how the energy community is considered a key tool in the energy transition process.

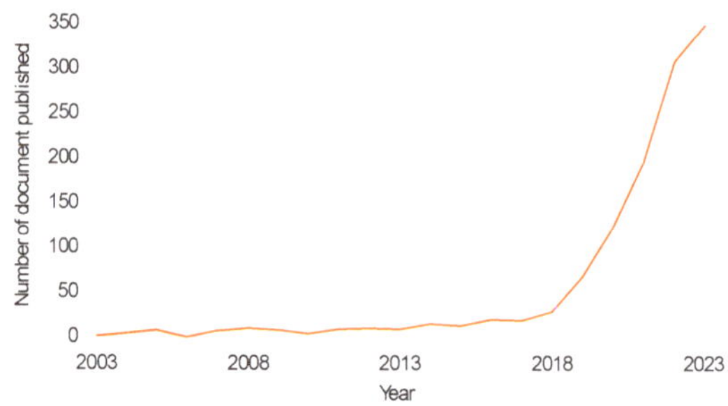


**Figure 8.** Number of publications per year in RECs.

Figure 9 illustrates the trend in publications for the main territories that have published on renewable energy communities (RECs). The trend shows that, compared to the United States and China, the European Union has significantly increased the number of publications on RECs since 2018. This increase is due to the European Parliament approval of Directive 2018/2001/EU (RED II) [115] and Directive 2019/944/EU (IEM) [116]. These directives are part of the Clean Energy Package [2], a collection of regulations adopted by the European Union in 2019 to decarbonize the European energy system. Through the RED II and IEM directives, the concept of an energy community was introduced in Europe as well as the concept of renewable energy sharing.

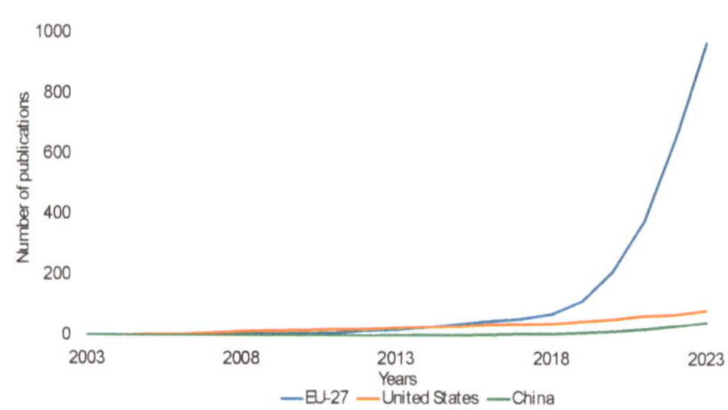


**Figure 9.** Number of publications over the years on RECs for the top 3 territories between 1993 and 2023.

The article with the highest number of citations in Europe was published by Lowitzsch et al. [117] in 2020 and contained a definition of the European regulatory context of the RED II and IEM directives, providing a preliminary practical interpretation while awaiting implementation by the member states of the European Union. Furthermore, the most cited papers focus on the innovations brought about by energy communities as a tool for transitioning towards renewable energies, from regulatory [118], economic [119], as well as environmental and social perspectives [120–123].

Despite significantly fewer publications, United States ranks second in terms of the highest number of citations. In the USA, there is a notable focus on research concerning net zero energy communities [98,124–126], where the primary goal is achieving zero emissions from fossil fuel combustion. Furthermore, several studies address the application in energy communities of optimization strategies [127,128] and of battery storage systems [98,129] for these communities. Regarding this topic, the most cited paper in the USA was published by Barbour et al. in 2018 [130].

Finally, China has started showing interest in RECs quite recently, since 2016, as seen in Figure 9. The most cited article is from 2020, authored by Feng et al. [127], and evaluates a coalitional game model to manage an energy community. The majority of the most cited articles in China also focus on net zero energy communities [131–135].

Figure 10 shows the main European countries by the number of publications between 1993 and 2023. Among these, Italy is the country that has increased its number of publications on this topic the most in recent years. In the figure, there is a noticeable positive deviation in the trend starting from 2018 and a further one starting from 2021. These two dates correspond to the implementation of the RED II and IEM directives in Italian legislation. In 2018, there was an early implementation through Law 8/2020 [136], while in 2021, there was the final implementation through Legislative Decree 199/2021 [137], which defined the characteristics of RECs in terms of energy and economics by establishing a shared energy incentive system [138].

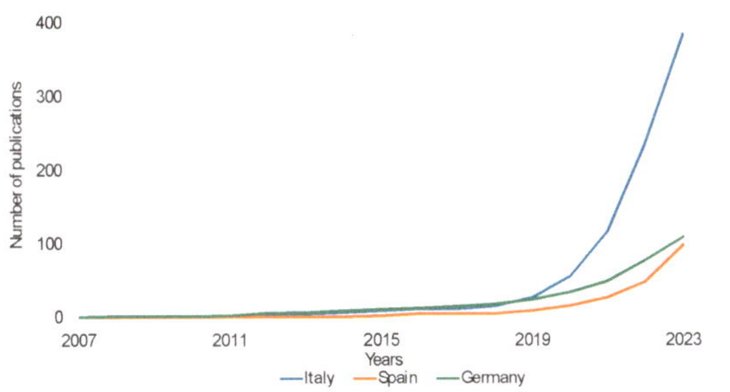


**Figure 10.** Number of publications during the year in RECs for the top 3 EU27 countries between 1993 and 2023.

The implementation of these directives in Italian legislation has provided a boost to research and development in energy communities as a new tool for energy management. The most cited Italian article is by Moroni et al. [139], dating back to 2019, and represents an introduction to energy communities through a taxonomic approach, attempting to explain what they are and their importance in the energy transition. Additional highly cited articles focus on the application of battery energy storage to RECs [112,129], smart energy systems in so-called "smart energy communities" [140,141], and the definition of optimization models for RECs [142,143].

In comparison to Italy, as can also be seen from Figure 10, in Germany, energy communities have represented a minor element of novelty, because they have already been established since the early 2000s in the German energy system, as "energy cooperatives" [144]. Furthermore, already in 2017, the Renewable Energy Sources Act 2017

(EEG) [145] defined self-supply and introduced the definition of an "energy community" into German legislation.

The most cited paper in Germany about RECs is from Lowitzsh et al. [117] and introduces the novelties of the Clean Energy Package in the European energy scenario. Further articles among the most cited focus on the importance of the new European directives [118,146]. In addition, as evidence of the established presence of energy communities in the German system, among the most cited articles is one by Schweizer-Ries et al. [147], dating back to 2008, which investigates the social dimension of the acceptability of renewable energies in the territory.

Spain represents the third European country in terms of the number of citations on the topic of energy communities. In 2019, Royal Decree 244/2019 [148] of April 5 established a legal definition of self-consumption and allowed for shared or collective self-consumption, both in the internal network and in nearby installations over the network. Indeed, Figure 10 shows an increase in the publication trend starting from 2019. The most cited article on energy communities in Spain was published in 2019 by Lezama et al. [119] and deals with the role of energy communities in local markets and how local markets can facilitate energy trading, thereby increasing the tolerable penetration of renewable resources and facilitating the energy transition. Among the most cited articles, there are also definitions of optimal REC models [149–151], as well as an analysis of sustainable energy communities, in a comparative analysis between Spain and Germany [151].

Figure 11 shows the main journals containing publications related to RECs. In this case, the top journal is *Energies*, with a total number of 124 publications updated to December 2023, all in open-access form. The authors and the institutions with the most publications on REC are shown in Tables 5 and 6, respectively. Looking at Table 5, one can see that among the affiliations with the highest number of publications on RECs, the top three positions are occupied by Italian universities. This is mainly due to the recent transposition of European directives and the introduction of the REC as a model in the Italian energy system. For this reason, several studies have been published in recent years to illustrate what a REC is and what its benefits are, introducing reproducible models and case studies in the national energy context [139,142,152–154].

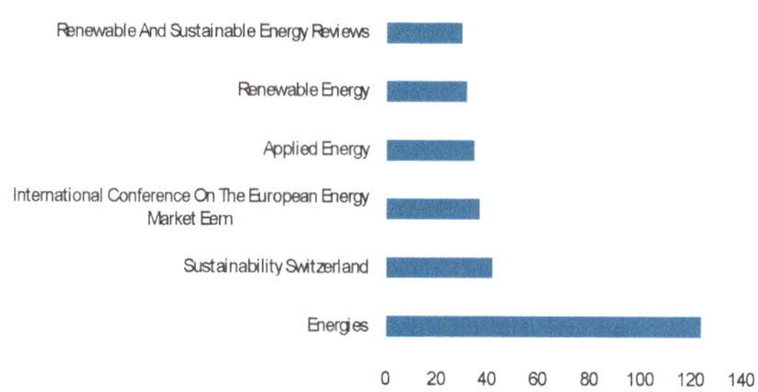


**Figure 11.** Top 5 journals publishing on RECs.

**Table 5.** Top 5 institutions with documents published on RECs.

| Affiliation | Publications | Country |
|---|---|---|
| Politecnico di Torino | 47 | Italy |
| Sapienza Università di Roma | 46 | Italy |
| Politecnico di Milano | 27 | Italy |
| Delft University of Technology | 26 | The Netherlands |
| Technische Universität Wien | 26 | Austria |

**Table 6.** Top 10 authors with documents published on RECs.

| Author Name | # Publications | Affiliation | Country |
|---|---|---|---|
| Vale, Z. | 20 | Instituto Politécnico do Porto | Portugal |
| Ghiani, E. | 16 | Università degli Studi di Cagliari | Italy |
| Martirano, L. | 15 | Sapienza University of Rome | Italy |
| Menniti, D. | 12 | Università della Calabria | Italy |
| Pinnarelli, A. | 12 | Università della Calabria | Italy |

Furthermore, the Politecnico di Torino university supported the research activity that led to the creation of the first energy community in Italy [155,156], in Magliano Alpi, in the province of Cuneo. It is interesting to note that Delft University of Technology is among the affiliations with the highest number of publications. The Netherlands, in contrast to Italy, is a pioneer state on the subject of RECs, having already introduced them into its legislation in 1998 and counting over 500 projects in the country [157]. The most cited article from that university was published in 2018 by Koirala et al. [122] and emphasizes the importance of citizen involvement in order to create a low-carbon community.

Figure 12 shows the co-occurrence of keywords obtained through the software VOSviewer for the second query related to RECs. The results shown in the figure were obtained by filtering the keywords with a minimum occurrence of 7. From the figure, it can be observed that the keyword "energy community" is linked to various types of energy-sharing schemes. Indeed, there are connections with terms such as "renewable energy communities" and "citizen energy community" (green cluster), types introduced in Europe through the RED II and IEM directives of the Clean Energy Package [118,158], as well as local energy communities (yellow cluster), understood as energy communities where citizens are extensively involved in local renewable energy system projects [158]. Table 7 shows the author keywords with the highest number of occurrences on "renewable energy communities".

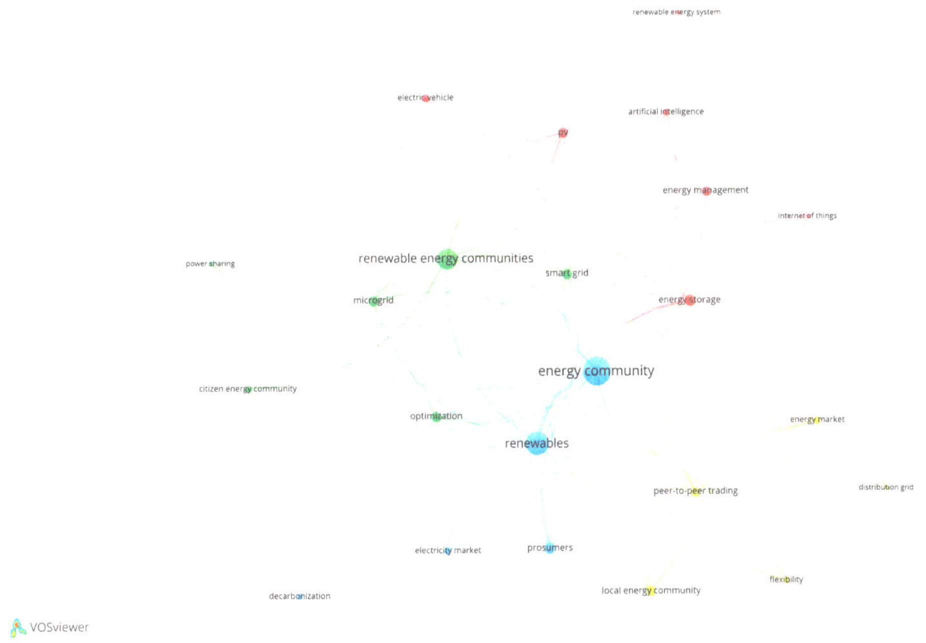


**Figure 12.** Co-occurrence of keywords related to RECs.

**Table 7.** Author keywords with the highest number of occurrences on "renewable energy communities".

| Keyword | Number of Occurrences |
|---|---|
| energy community | 401 |
| renewables | 258 |
| renewable energy communities | 219 |
| energy storage | 63 |
| prosumers | 62 |
| smart grid | 58 |
| microgrid | 57 |
| optimization | 55 |
| energy management | 43 |

Among the various items related to RECs, there is no specific reference to TES, but only the general term "energy storage" is mentioned. The subsequent Figure 13 illustrates the co-occurrence of the keyword "energy storage". Through Figure 13, it is evident that the keyword "energy storage" has a high number of connections both within the general scope of "blue cluster" energy communities and within the more specific realms of renewable and citizen energy communities (green cluster), as well as within the energy market and relationships among consumers of the CER (yellow cluster). Currently, the figure shows that the implementation of energy storage in energy communities is linked with "PV". Furthermore, there are some keywords that describe an optimization of the energy storage system as artificial intelligence and internet of things (red cluster).

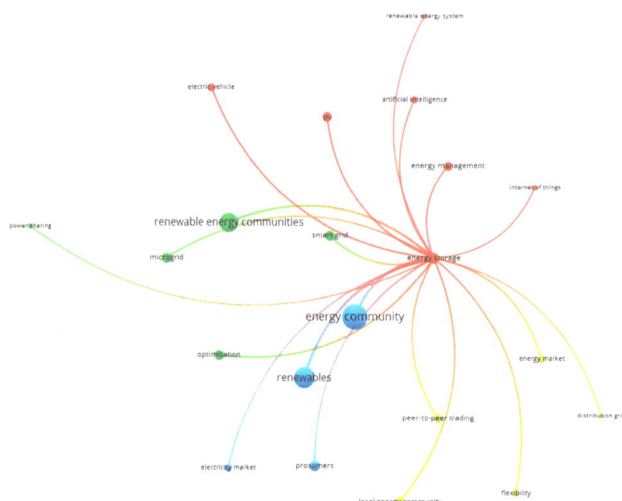


**Figure 13.** Co-occurrences and links with the keyword "energy storage".

The most cited article in the field of energy storage applied to RECs was published in 2018 by Barbour et al. [130] It highlights the necessity for the development of community-level storage technologies. While the literature predominantly features articles on battery energy storage [101,128,158–161], among the most cited articles is a study by Liu et al. [132], which applies a multi-energy storage system, both electrical and thermal, to achieve self-sufficiency in a net zero energy community. Additionally, Bartolini et al. [112] conduct a comparison between battery storage and TES to determine the optimal utilization of excess energy produced by a facility in a REC. The combination of battery and TES has also been taken up by further studies to create a hybrid system capable of maximizing the use of renewable energy [160,162,163].

In addition, Figure 13 indicates that energy storage applied to RECs is linked to the realm of smart grids and microgrids, corroborating what was previously analyzed in the co-occurrence analysis of TES. Finally, analyzing the latest publications on energy storage applied to RECs, it can be observed that solutions involving TES are increasingly prevalent, either in standalone systems or hybrid configurations with battery storage. For this reason, TES systems can represent a feasible possibility of application to RECs as a solution for maximizing renewable energy sharing.

### 3.3. Combined Results from Both Queries

Figure 14 shows the co-occurrence of keywords obtained by combining the queries concerning RECs and TES systems in buildings and districts into the VOSviewer software. The image shows how there is a clear separation between the two topics. It can be seen that the keywords related to the field of RECs are concentrated on the left-hand side, while those related to TES are on the right-hand side.

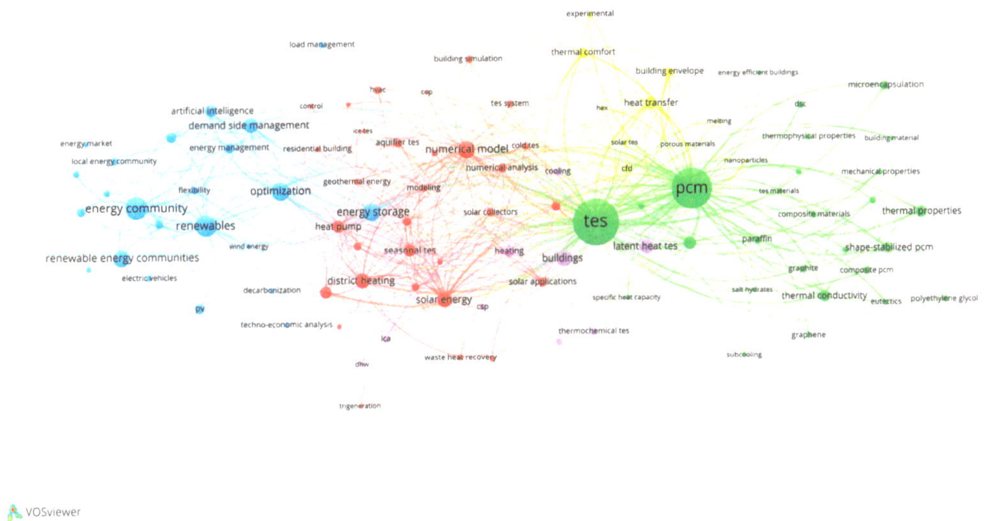


**Figure 14.** Co-occurrence of the keywords related to the merged queries.

This gap of the implementation of TES in RECs is confirmed with Figure 15, highlighting the main links to the keywords "energy community" and "TES". It can be seen that the two keywords are not related to each other. Furthermore, numerous keywords related to the sphere of TES are clearly separated from RECs. A clear example is "PCM", located on the opposite side of the figure from "energy community". However, it is possible to highlight several common points between TES and RECs that can be starting points for future research developments. For example, the term "TES" has links to the words "prosumer", "smart grid", "microgrid", "pv", and "demand side management", which are related to the field of RECs and the collective sharing of renewables.

One of these is the keyword "energy storage", the co-occurrence of which is shown in Figure 16a. As already shown in the previous section, most of the REC publications devoted to energy storage mostly concern electric storage batteries, as also shown by the terms "electric vehicle" and "electricity market", so a further gap is evident. Furthermore, the presence of the term "seasonal tes" indicates that from the perspective of TES systems, the focus of research is more on large systems, in contrast to smaller, distributed technologies that are more suited to the scope of RECs. However, the keywords "heat pump", "cogeneration", "trigeneration", and "district heating", also present in the co-occurrence of another common keyword, "solar energy" (Figure 16b), open up future research possibilities not only on the electrical side, but also on the thermal side. For this reason, the topic of energy storage could represent a bridge between TES and RECs. "Heat pump" is an important common point, which means that electrification of heating systems in energy communities could be the main focus of TES implementation. The keyword "district heating" once again highlights the importance of TES in areas of application greater than the individual building, expanding the possibility of sharing renewable thermal energy between several consumers. Finally, the terms "cogeneration" and "trigeneration" refer to the already mentioned possible benefits of theses in terms of recovering surplus heat.

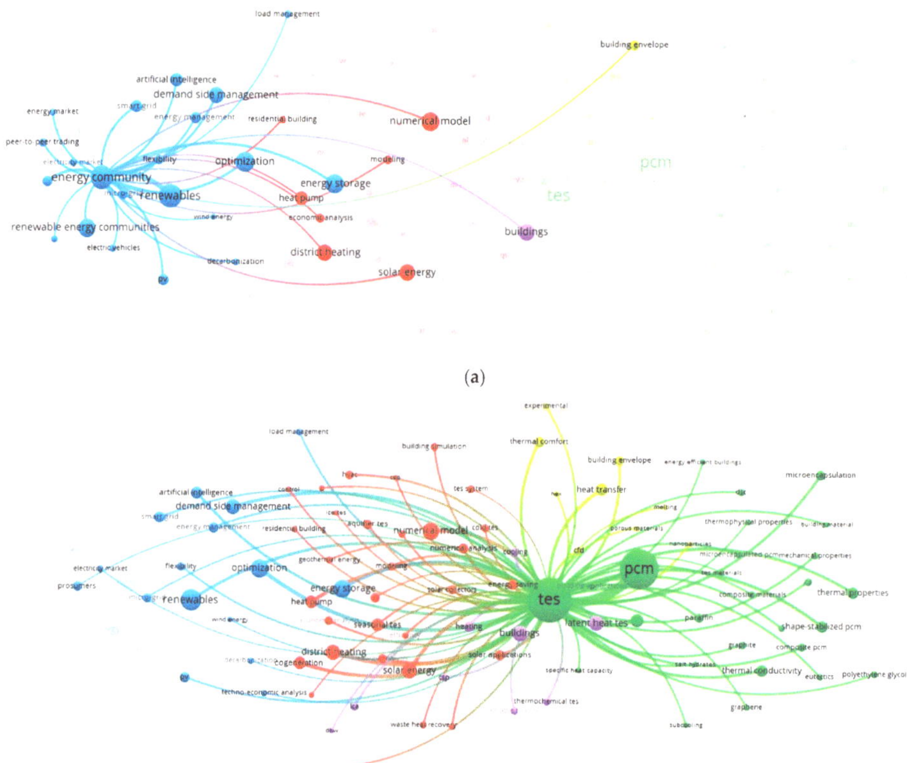


(a)

(b)

**Figure 15.** Co-occurrence of the keywords (**a**) "energy community" and (**b**) "TES".

Figure 17, on the other hand, shows the co-occurrence of the keyword "numerical model", for which there are several links to both RECs and TES. However, this shows a further gap, as it highlights the absence of the experimental field in the application of these two topics.

Finally, it is interesting to note that terms such as "building envelope", "thermal comfort", and "energy efficient buildings", which can be traced back to the residential sector, are repeated in the occurrences of these keywords common to TES and RECs. This could represent a further possibility to research the use of TES as a domestic solution to achieve energy efficiency for consumers and members of a REC.

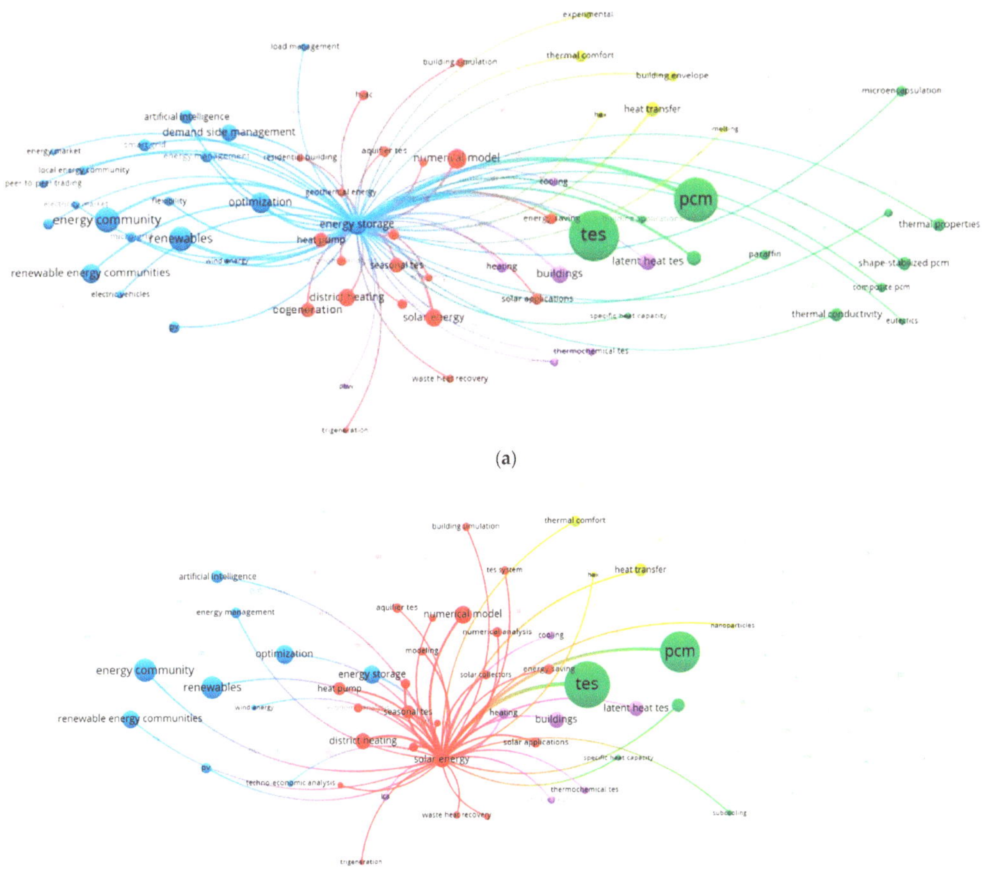


**Figure 16.** Co-occurrence of the keywords (**a**) "energy storage" and (**b**) "solar energy".

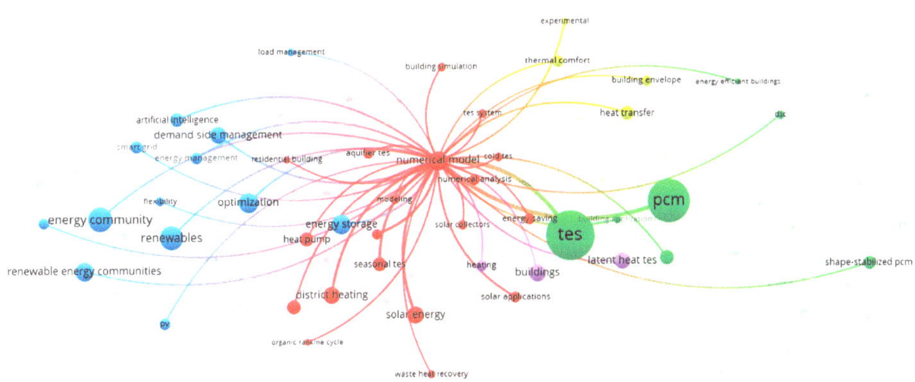


**Figure 17.** Co-occurrence of the keywords "numerical model".

Figure 18 shows the overlay visualization created using the VOSviewer software, allowing for us to observe the trends of different research items. It is evident that both TES and RECs are relatively recent research topics. As we move from TES to RECs from right to left, RECs appear to be an increasingly recent topic. While terms related to materials in TESs, such as "thermal properties", "fatty acids", and "building material" are positioned in 2017 [164–166], the items in the field of RECs, including "prosumer", "peer-to-peer trading", "local energy community", and "citizen energy community", are very recent, dating back to 2021 [167–170]. It can be concluded that the application of TES in the field of RECs is still relatively unexplored at the research level. However, it represents a recently emerging topic, still in its initial phase, with several possibilities for expansion and exploration in the future.

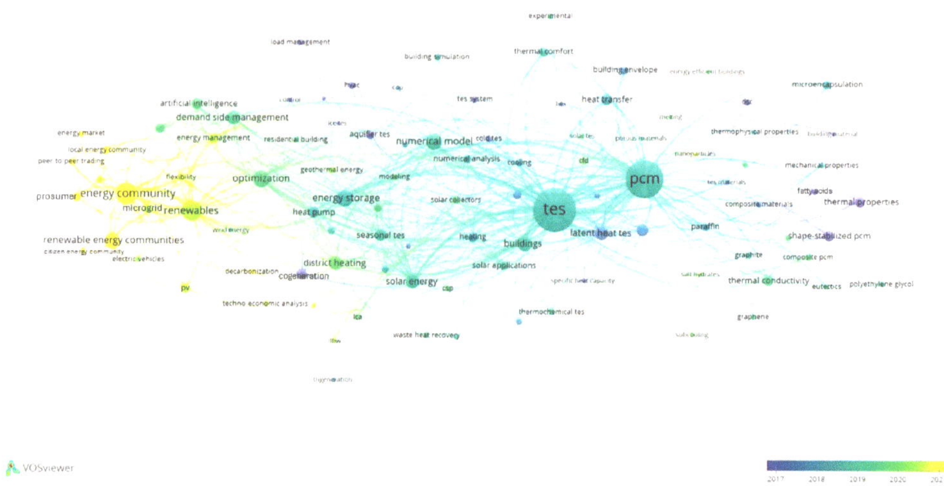


Figure 18. Overlay visualization for merged queries.

## 4. Conclusions

Renewable energy communities (RECs) represent a fundamental tool for the energy transition to renewable energy sources. Thanks to the distributed generation of renewable energy and the involvement of community members, who are no longer mere consumers but active and aware participants, they can effectively constitute a new paradigm for energy production, management, and consumption. This paradigm starkly contrasts with the fossil fuel-dependent model, addressing not only environmental and energy needs but also social ones.

The main goal of a REC is to maximize the sharing of renewable energy among its members. To achieve this, it is necessary to align periods of energy demand and supply. In this regard, there are various strategies, including the use of energy storage systems. Among these, TES represents a solution. Thanks to the power-to-heat strategy, which involves converting excess electrical energy into heat; its versatility; and its lower cost compared to other types of storage such as battery storage, TES presents a tangible opportunity to maximize the utilization of renewable energy in a REC. Furthermore, in a collective dimension of self-consumption such as RECs, TES systems can represent an opportunity for the implementation of district heating and heating systems fueled by renewables, providing flexibility and the possibility of bridging the mismatch between demand and availability of energy.

This study aimed to define the perspective of TES in RECs with a detailed bibliometric analysis. An analysis of TES systems was carried out regarding their application to buildings and districts, in order to consider not only the dimension of the individual building but also a broader one, traceable to a community. An initial analysis of the results shows that these two topics are very recent, with a significant increase in the trend of publications only in recent years. TES research, for example, has seen a considerable increase in publications since 2010, especially in Europe, following the approval of energy efficiency targets. Even more recent is the development of REC studies, which have grown exponentially since 2018, also based on the approval of European directives.

For this reason, most of the publications on these topics are very recent.

This work showed that energy communities constitute a research gap in the TES literature. In the co-occurrence analysis of keywords referring to TES, no terms related to RECs emerged. However, the presence of some terms characteristic of a larger storage dimension than the individual user, like "district heating", "smart grid", and "micro grid", leaves open the possibility of research in the area of collective self-consumption, not only from an electrical point of view, but also from that of thermal energy.

Furthermore, through bibliometric analysis, it became apparent that energy storage is applied in RECs, although it is referred to in the literature, through keywords, by this general definition, without focusing explicitly on thermal storage. However, among the most cited publications were several papers on the comparison between battery storage and TES. Moreover, more recent studies focus more on thermal or hybrid battery–TES systems. This allows for not only reducing the unreasonable investment costs of battery storage, but also for maximizing the sharing of renewable energies, through another energy carrier, towards increasing self-sufficiency.

Analyzing the merged keywords related to TES and RECs underscores the gap between these two topics. However, it also shows how the concepts of "energy storage", "solar energy", and "numerical model", which are relevant to both, could be a starting point for future research. These common points, with the big novelty that characterizes the subject, give room for potential investigation of TES technologies applied into RECs, in particular in Europe, where the research of these two topics is focused.

**Author Contributions:** Conceptualization, L.F.C. and E.B.; methodology, E.B.; formal analysis, C.M. and A.L.P.; investigation, L.B. and E.B.; resources, A.L.P. and L.F.C.; data curation, L.F.C.; writing—original draft preparation, L.B.; writing—review and editing, E.B., A.L.P., A.N., C.M. and L.F.C.; visualization, L.B.; supervision, A.N. and L.F.C.; project administration, A.L.P. and L.F.C.; funding acquisition, A.L.P. and L.F.C. All authors have read and agreed to the published version of the manuscript.

**Funding:** This work was partially funded by the Ministerio de Ciencia e Innovación—Agencia Estatal de Investigación (AEI) (PID2021-123511OB-C31—MCIN/AEI/10.13039/501100011033/FEDER, UE) and by the Ministerio de Ciencia e Innovación—Agencia Estatal de Investigación (AEI) (RED2022-134219-T). L. Brunelli acknowledges Fondazione Perugia for the support given through the "EDUCER" project (21082, 2022.0406).

**Institutional Review Board Statement:** Not applicable.

**Informed Consent Statement:** Not applicable.

**Data Availability Statement:** Data available upon request to the corresponding author.

**Acknowledgments:** The authors at the University of Lleida thank the Generalitat de Catalunya for the quality accreditation granted to the GREiA research group (2021 SGR 01615). GREiA is a TECNIO-certified agent in the category of technological development of the Gobierno de Cataluña. This work is partially supported by ICREA inside the program ICREA Academia. L. Brunelli aknowledgments are due to the PhD school in Energy and Sustainable Development at CIRIAF from the University of Perugia.

**Conflicts of Interest:** The authors declare no conflicts of interest.

## References


1. Smeeth, L.; Haines, A. COP 28: Ambitious climate action is needed to protect health. *BMJ* **2023**, *383*, p2938. [CrossRef] [PubMed]
2. Nouicer, A.; Meeus, L. European University Institute. Robert Schuman Centre for Advanced Studies. The EU Clean Energy Package. Available online: https://energy.ec.europa.eu/topics/energy-strategy/clean-energy-all-europeans-package_en (accessed on 14 March 2024).
3. Schlacke, S.; Wentzien, H.; Thierjung, E.M.; Köster, M. Implementing the EU Climate Law via the 'Fit for 55' package. *Oxf. Open Energy* **2022**, *1*, oiab002. [CrossRef]
4. Walker, G.; Devine-Wright, P. Community renewable energy: What should it mean? *Energy Policy* **2008**, *36*, 497–500. [CrossRef]
5. Ferreira, E.; Sequeira, M.M.; Gouveia, J.P. Sharing Is Caring: Exploring Distributed Solar Photovoltaics and Local Electricity Consumption through a Renewable Energy Community. *Sustainability* **2024**, *16*, 2777. [CrossRef]
6. Giarmanà, E. Managing renewable electricity within collective self-consumption schemes: A systematic private law approach. *Renew. Sustain. Energy Rev.* **2023**, *188*, 113896. [CrossRef]
7. Sternberg, A.; Bardow, A. Power-to-What?-Environmental assessment of energy storage systems. *Energy Environ. Sci.* **2015**, *8*, 389–400. [CrossRef]
8. Pastore, L.M.; Basso, G.L.; Ricciardi, G.; de Santoli, L. Smart energy systems for renewable energy communities: A comparative analysis of power-to-X strategies for improving energy self-consumption. *Energy* **2023**, *280*, 128205. [CrossRef]
9. Sarbu, I.; Sebarchievici, C. A Comprehensive Review of Thermal Energy Storage. *Sustainability* **2018**, *10*, 191. [CrossRef]
10. Pastore, L.M.; Basso, G.L.; Ricciardi, G.; de Santoli, L. Synergies between Power-to-Heat and Power-to-Gas in renewable energy communities. *Renew. Energy* **2022**, *198*, 1383–1397. [CrossRef]
11. Fambri, G.; Marocco, P.; Badami, M.; Tsagkrasoulis, D. The flexibility of virtual energy storage based on the thermal inertia of buildings in renewable energy communities: A techno-economic analysis and comparison with the electric battery solution. *J. Energy Storage* **2023**, *73*, 109083. [CrossRef]
12. Arce, P.; Medrano, M.; Gil, A.; Oró, E.; Cabeza, L.F. Overview of thermal energy storage (TES) potential energy savings and climate change mitigation in Spain and Europe. *Appl. Energy* **2011**, *88*, 2764–2774. [CrossRef]
13. Verrilli, F.; Srinivasan, S.; Gambino, G.; Canelli, M.; Himanka, M.; Del Vecchio, C.; Sasso, M.; Glielmo, L. Model Predictive Control-Based Optimal Operations of District Heating System With Thermal Energy Storage and Flexible Loads. *IEEE Trans. Autom. Sci. Eng.* **2017**, *14*, 547–557. [CrossRef]
14. Dahash, A.; Ochs, F.; Janetti, M.B.; Streicher, W. Advances in seasonal thermal energy storage for solar district heating applications: A critical review on large-scale hot-water tank and pit thermal energy storage systems. *Appl. Energy* **2019**, *239*, 296–315. [CrossRef]
15. Wang, H.; Yin, W.; Abdollahi, E.; Lahdelma, R.; Jiao, W. Modelling and optimization of CHP based district heating system with renewable energy production and energy storage. *Appl. Energy* **2015**, *159*, 401–421. [CrossRef]
16. Fouladvand, J.; Rojas, M.A.; Hoppe, T.; Ghorbani, A. Simulating thermal energy community formation: Institutional enablers outplaying technological choice. *Appl. Energy* **2022**, *306*, 117897. [CrossRef]
17. Pasqui, M.; Vaccaro, G.; Lubello, P.; Milazzo, A.; Carcasci, C. Heat pumps and thermal energy storages centralised management in a Renewable Energy Community. *Int. J. Sustain. Energy Plan. Manag.* **2023**, *38*, 65–82. [CrossRef]
18. Mohiti, M.; Mazidi, M.; Oggioni, N.; Steen, D.; Tuan, L.A. An IGDT-Based Energy Management System for Local Energy Communities Considering Phase-Change Thermal Energy Storage. *IEEE Trans. Ind. Appl.* **2024**, *60*, 4470–4481. [CrossRef]
19. Cabeza, L.F.; Chàfer, M.; Mata, É. Comparative analysis of web of science and scopus on the energy efficiency and climate impact of buildings. *Energies* **2020**, *13*, 409. [CrossRef]
20. Falagas, M.E.; Pitsouni, E.I.; Malietzis, G.A.; Pappas, G. Comparison of PubMed, Scopus, web of science, and Google scholar: Strengths and weaknesses. *FASEB J.* **2008**, *22*, 338–342. [CrossRef]
21. Borri, E.; Zsembinszki, G.; Cabeza, L.F. Recent developments of thermal energy storage applications in the built environment: A bibliometric analysis and systematic review. *Appl. Therm. Eng.* **2021**, *189*, 116666. [CrossRef]
22. van Eck, N.J.; Waltman, L. Software survey: VOSviewer, a computer program for bibliometric mapping. *Scientometrics* **2010**, *84*, 523–538. [CrossRef] [PubMed]
23. Moral-Muñoz, J.A.; Herrera-Viedma, E.; Santisteban-Espejo, A.; Cobo, M.J. Software tools for conducting bibliometric analysis in science: An up-to-date review. *Prof. Inf.* **2020**, *29*, 1–20. [CrossRef]
24. van Eck, N.J.; Waltman, L. Citation-based clustering of publications using CitNetExplorer and VOSviewer. *Scientometrics* **2017**, *111*, 1053–1070. [CrossRef] [PubMed]
25. Ogarek, P.; Wojtoń, M.; Słyś, D. Hydrogen as a Renewable Energy Carrier in a Hybrid Configuration of Distributed Energy Systems: Bibliometric Mapping of Current Knowledge and Strategies. *Energies* **2023**, *16*, 5495. [CrossRef]
26. Paliwal, M.K.; Jakhar, S.; Sharma, V. Nano-enhanced phase change materials for energy storage in photovoltaic thermal management systems: A bibliometric and thematic analysis. *Int. J. Thermofluids* **2023**, *17*, 100310. [CrossRef]
27. Liu, J.; Li, J.; Wang, J. In-depth analysis on thermal hazards related research trends about lithium-ion batteries: A bibliometric study. *J. Energy Storage* **2021**, *35*, 102253. [CrossRef]
28. Omrany, H.; Chang, R.; Soebarto, V.; Zhang, Y.; Ghaffarianhoseini, A.; Zuo, J. A bibliometric review of net zero energy building research 1995–2022. *Energy Build.* **2022**, *262*, 111996. [CrossRef]
29. Parra-Domínguez, J.; Sánchez, E.; Ordóñez, Á. The Prosumer: A Systematic Review of the New Paradigm in Energy and Sustainable Development. *Sustainability* **2023**, *15*, 10552. [CrossRef]

30. Gough, M.; Santos, S.F.; Javadi, M.; Castro, R.; Catalão, J.P.S. Prosumer Flexibility: A Comprehensive State-of-the-Art Review and Scientometric Analysis. *Energies* **2020**, *13*, 2710. [CrossRef]
31. Papatsounis, A.G.; Botsaris, P.N.; Katsavounis, S. Thermal/Cooling Energy on Local Energy Communities: A Critical Review. *Energies* **2022**, *15*, 1117. [CrossRef]
32. Shah, S.H.H.; Lei, S.; Ali, M.; Doronin, D.; Hussain, S.T. Prosumption: Bibliometric analysis using HistCite and VOSviewer. *Kybernetes* 2019, *ahead-of-print*. [CrossRef]
33. Ji, H.; Sellan, D.P.; Pettes, M.T.; Kong, X.; Ji, J.; Shi, L.; Ruoff, R.S. Enhanced thermal conductivity of phase change materials with ultrathin-graphite foams for thermal energy storage. *Energy Environ. Sci.* **2014**, *7*, 1185–1192. [CrossRef]
34. Song, M.; Niu, F.; Mao, N.; Hu, Y.; Deng, S. Review on building energy performance improvement using phase change materials. *Energy Build.* **2018**, *158*, 776–793. [CrossRef]
35. Al-Saadi, S.N.; Zhai, Z. Modeling phase change materials embedded in building enclosure: A review. *Renew. Sustain. Energy Rev.* **2013**, *21*, 659–673. [CrossRef]
36. Braun, J.E. Reducing energy costs and peak electrical demand through optimal control of building thermal storage. *ASHRAE Trans.* **1990**, *96*, 876–888.
37. Ma, Y.; Borrelli, F.; Hencey, B.; Coffey, B.; Bengea, S.; Haves, P. Model Predictive Control for the Operation of Building Cooling Systems. *IEEE Trans. Control Syst. Technol.* **2012**, *20*, 796–803. [CrossRef]
38. Lake, A.; Rezaie, B.; Beyerlein, S. Review of district heating and cooling systems for a sustainable future. *Renew. Sustain. Energy Rev.* **2017**, *67*, 417–425. [CrossRef]
39. European Commission. Energy Performance of Buildings Directive. 2018. Available online: https://ec.europa.eu/energy/topics/energy-efficiency/energy-efficient-buildings/energy-performance-buildings-directive_en (accessed on 14 March 2024).
40. Horizon. The EU Framework Programme for Research and Innovation. 2020. Available online: https://ec.europa.eu/programmes/horizon2020/en/ (accessed on 12 February 2019).
41. Cabeza, L.F.; Castello, C.; Nogués, M.; Medrano, M.; Leppers, R.; Zubillaga, O. Use of microencapsulated PCM in concrete walls for energy savings. *Energy Build.* **2007**, *39*, 113–119. [CrossRef]
42. Kuznik, F.; David, D.; Johannes, K.; Roux, J.-J. A review on phase change materials integrated in building walls. *Renew. Sustain. Energy Rev.* **2011**, *15*, 379–391. [CrossRef]
43. Soares, N.; Costa, J.J.; Gaspar, A.R.; Santos, P. Review of passive PCM latent heat thermal energy storage systems towards buildings' energy efficiency. *Energy Build.* **2013**, *59*, 82–103. [CrossRef]
44. de Gracia, A.; Cabeza, L.F. Phase change materials and thermal energy storage for buildings. *Energy Build.* **2015**, *103*, 414–419. [CrossRef]
45. Cabeza, L.F.; Castell, A.; Barreneche, C.; De Gracia, A.; Fernández, A.I. Materials used as PCM in thermal energy storage in buildings: A review. *Renew. Sustain. Energy Rev.* **2011**, *15*, 1675–1695. [CrossRef]
46. Tatsidjodoung, P.; Le Pierrès, N.; Luo, L. A review of potential materials for thermal energy storage in building applications. *Renew. Sustain. Energy Rev.* **2013**, *18*, 327–349. [CrossRef]
47. Li, Y.; Wei, Y.; Dong, Z. Will China Achieve Its Ambitious Goal?—Forecasting the $CO_2$ Emission Intensity of China towards 2030. *Energies* **2020**, *13*, 2924. [CrossRef]
48. Alva, G.; Lin, Y.; Fang, G. An overview of thermal energy storage systems. *Energy* **2018**, *144*, 341–378. [CrossRef]
49. Zhang, Y.; Zhou, G.; Lin, K.; Zhang, Q.; Di, H. Application of latent heat thermal energy storage in buildings: State-of-the-art and outlook. *Build. Environ.* **2007**, *42*, 2197–2209. [CrossRef]
50. Lin, Y.; Jia, Y.; Alva, G.; Fang, G. Review on thermal conductivity enhancement, thermal properties and applications of phase change materials in thermal energy storage. *Renew. Sustain. Energy Rev.* **2018**, *82*, 2730–2742. [CrossRef]
51. Huang, X.; Chen, X.; Li, A.; Atinafu, D.; Gao, H.; Dong, W.; Wang, G. Shape-stabilized phase change materials based on porous supports for thermal energy storage applications. *Chem. Eng. J.* **2019**, *356*, 641–661. [CrossRef]
52. Zhao, C.Y.; Zhang, G.H. Review on microencapsulated phase change materials (MEPCMs): Fabrication, characterization and applications. *Renew. Sustain. Energy Rev.* **2011**, *15*, 3813–3832. [CrossRef]
53. Pomianowski, M.; Heiselberg, P.; Zhang, Y. Review of thermal energy storage technologies based on PCM application in buildings. *Energy Build.* **2013**, *67*, 56–69. [CrossRef]
54. Zhou, D.; Zhao, C.-Y.; Tian, Y. Review on thermal energy storage with phase change materials (PCMs) in building applications. *Appl. Energy* **2012**, *92*, 593–605. [CrossRef]
55. Arteconi, A.; Hewitt, N.J.; Polonara, F. State of the art of thermal storage for demand-side management. *Appl. Energy* **2012**, *93*, 371–389. [CrossRef]
56. Arteconi, A.; Hewitt, N.J.; Polonara, F. Domestic demand-side management (DSM): Role of heat pumps and thermal energy storage (TES) systems. *Appl. Therm. Eng.* **2013**, *51*, 155–165. [CrossRef]
57. Comodi, G.; Giantomassi, A.; Severini, M.; Squartini, S.; Ferracuti, F.; Fonti, A.; Cesarini, D.N.; Morodo, M.; Polonara, F. Multi-apartment residential microgrid with electrical and thermal storage devices: Experimental analysis and simulation of energy management strategies. *Appl. Energy* **2015**, *137*, 854–866. [CrossRef]
58. Barbieri, E.S.; Melino, F.; Morini, M. Influence of the thermal energy storage on the profitability of micro-CHP systems for residential building applications. *Appl. Energy* **2012**, *97*, 714–722. [CrossRef]

59. Barbieri, E.S.; Spina, P.R.; Venturini, M. Analysis of innovative micro-CHP systems to meet household energy demands. *Appl. Energy* **2012**, *97*, 723–733. [CrossRef]
60. Guelpa, E.; Verda, V. Thermal energy storage in district heating and cooling systems: A review. *Appl. Energy* **2019**, *252*, 113474. [CrossRef]
61. Pisello, A.L. State of the art on the development of cool coatings for buildings and cities. *Sol. Energy* **2017**, *144*, 660–680. [CrossRef]
62. Cao, V.D.; Pilehvar, S.; Salas-Bringas, C.; Szczotok, A.M.; Rodriguez, J.F.; Carmona, M.; Al-Manasir, N.; Kjøniksen, A.L. Microencapsulated phase change materials for enhancing the thermal performance of Portland cement concrete and geopolymer concrete for passive building applications. *Energy Convers. Manag.* **2017**, *133*, 56–66. [CrossRef]
63. Saffari, M.; de Gracia, A.; Fernández, C.; Cabeza, L.F. Simulation-based optimization of PCM melting temperature to improve the energy performance in buildings. *Appl. Energy* **2017**, *202*, 420–434. [CrossRef]
64. Lizana, J.; Chacartegui, R.; Barrios-Padura, A.; Valverde, J.M. Advances in thermal energy storage materials and their applications towards zero energy buildings: A critical review. *Appl. Energy* **2017**, *203*, 219–239. [CrossRef]
65. Navarro, L.; de Gracia, A.; Colclough, S.; Browne, M.; McCormack, S.J.; Griffiths, P.; Cabeza, L.F. Thermal energy storage in building integrated thermal systems: A review. Part 1. Active storage systems. *Renew. Energy* **2016**, *88*, 526–547. [CrossRef]
66. Stinner, S.; Huchtemann, K.; Müller, D. Quantifying the operational flexibility of building energy systems with thermal energy storages. *Appl. Energy* **2016**, *181*, 140–154. [CrossRef]
67. Henze, G.P.; Felsmann, C.; Knabe, G. Evaluation of optimal control for active and passive building thermal storage. *Int. J. Therm. Sci.* **2004**, *43*, 173–183. [CrossRef]
68. Lazaro, A.; Peñalosa, C.; Solé, A.; Diarce, G.; Haussmann, T.; Fois, M.; Zalba, B.; Gshwander, S.; Cabeza, L.F. Intercomparative tests on phase change materials characterisation with differential scanning calorimeter. *Appl. Energy* **2013**, *109*, 415–420. [CrossRef]
69. Youssef, Z.; Delahaye, A.; Huang, L.; Trinquet, F.; Fournaison, L.; Pollerberg, C.; Doetsch, C. State of the art on phase change material slurries. *Energy Convers. Manag.* **2013**, *65*, 120–132. [CrossRef]
70. Fleuchaus, P.; Godschalk, B.; Stober, I.; Blum, P. Worldwide application of aquifer thermal energy storage—A review. *Renew. Sustain. Energy Rev.* **2018**, *94*, 861–876. [CrossRef]
71. Stein, W.H.; Buck, R. Advanced power cycles for concentrated solar power. *Sol. Energy* **2017**, *152*, 91–105. [CrossRef]
72. Laing, D.; Lehmann, D.; Fi, M.; Bahl, C. Test results of concrete thermal energy storage for parabolic trough power plants, Journal of Solar Energy Engineering. *Trans. ASME* **2009**, *131*, 0410071–0410076. [CrossRef]
73. Wirtz, M.; Kivilip, L.; Remmen, P.; Müller, D. 5th Generation District Heating: A novel design approach based on mathematical optimization. *Appl. Energy* **2020**, *260*, 114158. [CrossRef]
74. Enescu, D.; Chicco, G.; Porumb, R.; Seritan, G. Thermal Energy Storage for Grid Applications: Current Status and Emerging Trends. *Energies* **2020**, *13*, 340. [CrossRef]
75. Marin, P.; Saffari, M.; de Gracia, A.; Zhu, X.; Farid, M.M.; Cabeza, L.F.; Ushak, S. Energy savings due to the use of PCM for relocatable lightweight buildings passive heating and cooling in different weather conditions. *Energy Build.* **2016**, *129*, 274–283. [CrossRef]
76. Moreno, P.; Solé, C.; Castell, A.; Cabeza, L.F. The use of phase change materials in domestic heat pump and air-conditioning systems for short term storage: A review. *Renew. Sustain. Energy Rev.* **2014**, *39*, 1–13. [CrossRef]
77. Ibanez, M.; Lázaro, A.; Zalba, B.; Cabeza, L.F. An approach to the simulation of PCMs in building applications using TRNSYS. *Appl. Therm. Eng.* **2005**, *25*, 1796–1807. [CrossRef]
78. Ferrer, G.; Solé, A.; Barreneche, C.; Martorell, I.; Cabeza, L.F. Corrosion of metal containers for use in PCM energy storage. *Renew. Energy* **2015**, *76*, 465–469. [CrossRef]
79. Navarro, L.; de Gracia, A.; Niall, D.; Castell, A.; Browne, M.; McCormack, S.J.; Griffiths, P.; Cabeza, L.F. Thermal energy storage in building integrated thermal systems: A review. Part 2. Integration as passive system. *Renew. Energy* **2016**, *85*, 1334–1356. [CrossRef]
80. Solé, A.; Martorell, I.; Cabeza, L.F. State of the art on gas–solid thermochemical energy storage systems and reactors for building applications. *Renew. Sustain. Energy Rev.* **2015**, *47*, 386–398. [CrossRef]
81. Karaipekli, A.; Sarı, A. Capric–myristic acid/vermiculite composite as form-stable phase change material for thermal energy storage. *Sol. Energy* **2009**, *83*, 323–332. [CrossRef]
82. Sarı, A.; Karaipekli, A. Preparation, thermal properties and thermal reliability of palmitic acid/expanded graphite composite as form-stable PCM for thermal energy storage. *Sol. Energy Mater. Sol. Cells* **2009**, *93*, 571–576. [CrossRef]
83. Sarı, A. Thermal energy storage characteristics of bentonite-based composite PCMs with enhanced thermal conductivity as novel thermal storage building materials. *Energy Convers. Manag.* **2016**, *117*, 132–141. [CrossRef]
84. Karaipekli, A.; Sarı, A. Capric–myristic acid/expanded perlite composite as form-stable phase change material for latent heat thermal energy storage. *Renew. Energy* **2008**, *33*, 2599–2605. [CrossRef]
85. Arkar, C.; Medved, S. Free cooling of a building using PCM heat storage integrated into the ventilation system. *Sol. Energy* **2007**, *81*, 1078–1087. [CrossRef]
86. Guarino, F.; Athienitis, A.; Cellura, M.; Bastien, D. PCM thermal storage design in buildings: Experimental studies and applications to solaria in cold climates. *Appl. Energy* **2017**, *185*, 95–106. [CrossRef]
87. Kuznik, F.; Virgone, J. Experimental investigation of wallboard containing phase change material: Data for validation of numerical modeling. *Energy Build.* **2009**, *41*, 561–570. [CrossRef]

88. Wang, Z.; Hong, T.; Piette, M.A. Building thermal load prediction through shallow machine learning and deep learning. *Appl. Energy* **2020**, *263*, 114683. [CrossRef]
89. Finck, C.; Li, R.; Kramer, R.; Zeiler, W. Quantifying demand flexibility of power-to-heat and thermal energy storage in the control of building heating systems. *Appl. Energy* **2018**, *209*, 409–425. [CrossRef]
90. Good, N.; Karangelos, E.; Navarro-Espinosa, A.; Mancarella, P. Optimization under Uncertainty of Thermal Storage-Based Flexible Demand Response with Quantification of Residential Users' Discomfort. *IEEE Trans. Smart Grid* **2015**, *6*, 2333–2342. [CrossRef]
91. Lu, Y.; Wang, S.; Sun, Y.; Yan, C. Optimal scheduling of buildings with energy generation and thermal energy storage under dynamic electricity pricing using mixed-integer nonlinear programming. *Appl. Energy* **2015**, *147*, 49–58. [CrossRef]
92. Dai, Y.; Chen, L.; Min, Y.; Chen, Q.; Hao, J.; Hu, K.; Xu, F. Dispatch Model for CHP with Pipeline and Building Thermal Energy Storage Considering Heat Transfer Process. *IEEE Trans. Sustain. Energy* **2019**, *10*, 192–203. [CrossRef]
93. Nuytten, T.; Claessens, B.; Paredis, K.; Van Bael, J.; Six, D. Flexibility of a combined heat and power system with thermal energy storage for district heating. *Appl. Energy* **2013**, *104*, 583–591. [CrossRef]
94. Vandermeulen, A.; van der Heijde, B.; Helsen, L. Controlling district heating and cooling networks to unlock flexibility: A review. *Energy* **2018**, *151*, 103–115. [CrossRef]
95. Kensby, J.; Trüschel, A.; Dalenbäck, J.-O. Potential of residential buildings as thermal energy storage in district heating systems—Results from a pilot test. *Appl. Energy* **2015**, *137*, 773–781. [CrossRef]
96. Romanchenko, D.; Kensby, J.; Odenberger, M.; Johnsson, F. Thermal energy storage in district heating: Centralised storage vs. storage in thermal inertia of buildings. *Energy Convers. Manag.* **2018**, *162*, 26–38. [CrossRef]
97. Musolino, M.; Maggio, G.; D'Aleo, E.; Nicita, A. Three case studies to explore relevant features of emerging renewable energy communities in Italy. *Renew. Energy* **2023**, *210*, 540–555. [CrossRef]
98. Novoa, L.; Flores, R.; Brouwer, J. Optimal renewable generation and battery storage sizing and siting considering local transformer limits. *Appl. Energy* **2019**, *256*, 113926. [CrossRef]
99. Spertino, F.; Ciocia, A.; Di Leo, P.; Fichera, S.; Malgaroli, G.; Ratclif, A. Toward the Complete Self-Sufficiency of an nZEBs Microgrid by Photovoltaic Generators and Heat Pumps: Methods and Applications. *IEEE Trans. Ind. Appl.* **2019**, *55*, 7028–7040. [CrossRef]
100. Kloppenburg, S.; Boekelo, M. Digital platforms and the future of energy provisioning: Promises and perils for the next phase of the energy transition. *Energy Res. Soc. Sci.* **2019**, *49*, 68–73. [CrossRef]
101. Liu, J.; Yang, H.; Zhou, Y. Peer-to-peer trading optimizations on net-zero energy communities with energy storage of hydrogen and battery vehicles. *Appl. Energy* **2021**, *302*, 117578. [CrossRef]
102. Ghiani, E.; Giordano, A.; Nieddu, A.; Rosetti, L.; Pilo, F. Planning of a Smart Local Energy Community: The Case of Berchidda Municipality (Italy). *Energies* **2019**, *12*, 4629. [CrossRef]
103. Radl, J.; Fleischhacker, A.; Revheim, F.H.; Lettner, G.; Auer, H. Comparison of Profitability of PV Electricity Sharing in Renewable Energy Communities in Selected European Countries. *Energies* **2020**, *13*, 5007. [CrossRef]
104. Pastore, L.M. Combining Power-to-Heat and Power-to-Vehicle strategies to provide system flexibility in smart urban energy districts. *Sustain. Cities Soc.* **2023**, *94*, 104548. [CrossRef]
105. Backe, S.; Korpås, M.; Tomasgard, A. Heat and electric vehicle flexibility in the European power system: A case study of Norwegian energy communities. *Int. J. Electr. Power Energy Syst.* **2021**, *125*, 106479. [CrossRef]
106. Zhang, J.; Cho, H.; Mago, P.J.; Zhang, H.; Yang, F. Multi-Objective Particle Swarm Optimization (MOPSO) for a Distributed Energy System Integrated with Energy Storage. *J. Therm. Sci.* **2019**, *28*, 1221–1235. [CrossRef]
107. Li, Z.; Zhi, X.; Wu, Z.; Qian, G.; Jiang, R.; Wang, B.; Huang, R.; Yu, X. Role of different energy storage methods in decarbonizing urban distributed energy systems: A case study of thermal and electricity storage. *J. Energy Storage* **2023**, *73*, 108931. [CrossRef]
108. Powell, K.M.; Kim, J.S.; Cole, W.J.; Kapoor, K.; Mojica, J.L.; Hedengren, J.D.; Edgar, T.F. Thermal energy storage to minimize cost and improve efficiency of a polygeneration district energy system in a real-time electricity market. *Energy* **2016**, *113*, 52–63. [CrossRef]
109. Hussain, A.; Bui, V.-H.; Kim, H.-M.; Im, Y.-H.; Lee, J.-Y. Optimal Energy Management of Combined Cooling, Heat and Power in Different Demand Type Buildings Considering Seasonal Demand Variations. *Energies* **2017**, *10*, 789. [CrossRef]
110. Tooryan, F.; HassanzadehFard, H.; Dargahi, V.; Jin, S. A cost-effective approach for optimal energy management of a hybrid CCHP microgrid with different hydrogen production considering load growth analysis. *Int. J. Hydrog. Energy* **2022**, *47*, 6569–6585. [CrossRef]
111. Martirano, L.; Habib, E.; Parise, G.; Greco, G.; Manganelli, M.; Massarella, F.; Parise, L. Demand Side Management in Microgrids for Load Control in Nearly Zero Energy Buildings. *IEEE Trans. Ind. Appl.* **2017**, *53*, 1769–1779. [CrossRef]
112. Bartolini, A.; Carducci, F.; Muñoz, C.B.; Comodi, G. Energy storage and multi energy systems in local energy communities with high renewable energy penetration. *Renew. Energy* **2020**, *159*, 595–609. [CrossRef]
113. Li, Z.; Xu, Y.; Feng, X.; Wu, Q. Optimal Stochastic Deployment of Heterogeneous Energy Storage in a Residential Multienergy Microgrid with Demand-Side Management. *IEEE Trans. Ind. Inf.* **2021**, *17*, 991–1004. [CrossRef]
114. Zeyen, E.; Hagenmeyer, V.; Brown, T. Mitigating heat demand peaks in buildings in a highly renewable European energy system. *Energy* **2021**, *231*, 120784. [CrossRef]

115. Directive (EU) 2018/2001. Directive (EU) 2018/2001 of the European Parliament and of the Council of 11 December 2018 on the Promotion of the Use of Energy from Renewable Sources (Recast). Available online: https://joint-research-centre.ec.europa.eu/welcome-jec-website/reference-regulatory-framework/renewable-energy-recast-2030-red-ii_en (accessed on 3 April 2024).
116. Directive (EU) 2019/944. Directive (EU) 2019/944 of the European Parliament and of the Council of 5 June 2019 on Common Rules for the Internal Market for Electricity and Amending Directive 2012/27/EU. Available online: https://eur-lex.europa.eu/legal-content/IT/LSU/?uri=CELEX:32019L0944 (accessed on 14 November 2023).
117. Lowitzsch, J.; Hoicka, C.E.; van Tulder, F.J. Renewable energy communities under the 2019 European Clean Energy Package—Governance model for the energy clusters of the future? *Renew. Sustain. Energy Rev.* **2020**, *122*, 109489. [CrossRef]
118. Inês, C.; Guilherme, P.L.; Esther, M.-G.; Swantje, G.; Stephen, H.; Lars, H. Regulatory challenges and opportunities for collective renewable energy prosumers in the EU. *Energy Policy* **2020**, *138*, 111212. [CrossRef]
119. Lezama, F.; Soares, J.; Hernandez-Leal, P.; Kaisers, M.; Pinto, T.; Vale, Z. Local Energy Markets: Paving the Path toward Fully Transactive Energy Systems. *IEEE Trans. Power Syst.* **2019**, *34*, 4081–4088. [CrossRef]
120. Dóci, G.; Vasileiadou, E.; Petersen, A.C. Exploring the transition potential of renewable energy communities. *Futures* **2015**, *66*, 85–95. [CrossRef]
121. Dóci, G.; Vasileiadou, E. "Let's do it ourselves" Individual motivations for investing in renewables at community level. *Renew. Sustain. Energy Rev.* **2015**, *49*, 41–50. [CrossRef]
122. Koirala, B.P.; Araghi, Y.; Kroesen, M.; Ghorbani, A.; Hakvoort, R.A.; Herder, P.M. Trust, awareness, and independence: Insights from a socio-psychological factor analysis of citizen knowledge and participation in community energy systems. *Energy Res. Soc. Sci.* **2018**, *38*, 33–40. [CrossRef]
123. Gjorgievski, V.Z.; Cundeva, S.; Georghiou, G.E. Social arrangements, technical designs and impacts of energy communities: A review. *Renew. Energy* **2021**, *169*, 1138–1156. [CrossRef]
124. Mittal, A.; Krejci, C.C.; Dorneich, M.C.; Fickes, D. An agent-based approach to modeling zero energy communities. *Sol. Energy* **2019**, *191*, 193–204. [CrossRef]
125. Gaiser, K.; Stroeve, P. The impact of scheduling appliances and rate structure on bill savings for net-zero energy communities: Application to West Village. *Appl. Energy* **2014**, *113*, 1586–1595. [CrossRef]
126. Lopes, R.A.; Martins, J.; Aelenei, D.; Lima, C.P. A cooperative net zero energy community to improve load matching. *Renew. Energy* **2016**, *93*, 1–13. [CrossRef]
127. Feng, C.; Wen, F.; You, S.; Li, Z.; Shahnia, F.; Shahidehpour, M. Coalitional Game-Based Transactive Energy Management in Local Energy Communities. *IEEE Trans. Power Syst.* **2020**, *35*, 1729–1740. [CrossRef]
128. Cosic, A.; Stadler, M.; Mansoor, M.; Zellinger, M. Mixed-integer linear programming based optimization strategies for renewable energy communities. *Energy* **2021**, *237*, 121559. [CrossRef]
129. Campana, P.E.; Cioccolanti, L.; François, B.; Jurasz, J.; Zhang, Y.; Varini, M.; Stridh, B.; Yan, J. Li-ion batteries for peak shaving, price arbitrage, and photovoltaic self-consumption in commercial buildings: A Monte Carlo Analysis. *Energy Convers. Manag.* **2021**, *234*, 113889. [CrossRef]
130. Barbour, E.; Parra, D.; Awwad, Z.; González, M.C. Community energy storage: A smart choice for the smart grid? *Appl. Energy* **2018**, *212*, 489–497. [CrossRef]
131. Liu, Z.; Fan, G.; Sun, D.; Wu, D.; Guo, J.; Zhang, S.; Yang, X.; Lin, X.; Ai, L. A novel distributed energy system combining hybrid energy storage and a multi-objective optimization method for nearly zero-energy communities and buildings. *Energy* **2022**, *239*, 122577. [CrossRef]
132. Liu, Z.; Guo, J.; Wu, D.; Fan, G.; Zhang, S.; Yang, X.; Ge, H. Two-phase collaborative optimization and operation strategy for a new distributed energy system that combines multi-energy storage for a nearly zero energy community. *Energy Convers. Manag.* **2021**, *230*, 113800. [CrossRef]
133. Fan, G.; Liu, Z.; Liu, X.; Shi, Y.; Wu, D.; Guo, J.; Zhang, S.; Yang, X.; Zhang, Y. Energy management strategies and multi-objective optimization of a near-zero energy community energy supply system combined with hybrid energy storage. *Sustain. Cities Soc.* **2022**, *83*, 103970. [CrossRef]
134. Liu, J.; Zhou, Y.; Yang, H.; Wu, H. Uncertainty energy planning of net-zero energy communities with peer-to-peer energy trading and green vehicle storage considering climate changes by 2050 with machine learning methods. *Appl. Energy* **2022**, *321*, 119394. [CrossRef]
135. Alabi, T.M.; Lu, L.; Yang, Z.; Zhou, Y. A novel optimal configuration model for a zero-carbon multi-energy system (ZC-MES) integrated with financial constraints. *Sustain. Energy Grids Netw.* **2020**, *23*, 100381. [CrossRef]
136. Legge del 28 Febbraio 2020, n. 8. Available online: https://www.gazzettaufficiale.it/eli/id/2020/02/29/20G00021/sg (accessed on 9 April 2024).
137. Decreto Legislativo 199/2021. DECRETO-LEGISLATIVO-8-novembre-2021-n.-199. 2021. Available online: https://www.gazzettaufficiale.it/atto/serie_generale/caricaDettaglioAtto/originario?atto.dataPubblicazioneGazzetta=2021-11-30&atto.codiceRedazionale=21G00214&elenco30giorni=true (accessed on 14 November 2023).
138. Tatti, A.; Ferroni, S.; Ferrando, M.; Motta, M.; Causone, F. The Emerging Trends of Renewable Energy Communities' Development in Italy. *Sustainability* **2023**, *15*, 6792. [CrossRef]
139. Moroni, S.; Alberti, V.; Antoniucci, V.; Bisello, A. Energy communities in the transition to a low-carbon future: A taxonomical approach and some policy dilemmas. *J. Environ. Manag.* **2019**, *236*, 45–53. [CrossRef] [PubMed]

140. Ceglia, F.; Esposito, P.; Marrasso, E.; Sasso, M. From smart energy community to smart energy municipalities: Literature review, agendas and pathways. *J. Clean. Prod.* **2020**, *254*, 120118. [CrossRef]
141. Moreno-Munoz, A.; Bellido-Outeirino, F.J.; Siano, P.; Gomez-Nieto, M.A. Mobile social media for smart grids customer engagement: Emerging trends and challenges. *Renew. Sustain. Energy Rev.* **2016**, *53*, 1611–1616. [CrossRef]
142. Martirano, L.; Rotondo, S.; Kermani, M.; Massarella, F.; Gravina, R. Power Sharing Model for Energy Communities of Buildings. *IEEE Trans. Ind. Appl.* **2021**, *57*, 170–178. [CrossRef]
143. Fioriti, D.; Frangioni, A.; Poli, D. Optimal sizing of energy communities with fair revenue sharing and exit clauses: Value, role and business model of aggregators and users. *Appl. Energy* **2021**, *299*, 117328. [CrossRef]
144. Krug, M.; Di Nucci, M.R.; Caldera, M.; De Luca, E. Mainstreaming Community Energy: Is the Renewable Energy Directive a Driver for Renewable Energy Communities in Germany and Italy? *Sustainability* **2022**, *14*, 7181. [CrossRef]
145. The 2017 Renewable Energy Sources Act. Available online: https://www.bmwk.de/Redaktion/EN/Artikel/Energy/res-2017.html (accessed on 9 April 2024).
146. Hanke, F.; Lowitzsch, J. Empowering Vulnerable Consumers to Join Renewable Energy Communities—Towards an Inclusive Design of the Clean Energy Package. *Energies* **2020**, *13*, 1615. [CrossRef]
147. Schweizer-Ries, P. Energy sustainable communities: Environmental psychological investigations. *Energy Policy* **2008**, *36*, 4126–4135. [CrossRef]
148. Real Decreto 244/2019, de 5 de Abril, por el que se Regulan las Condiciones Administrativas, Técnicas y Económicas del Autoconsumo de Energía Eléctrica. Available online: https://www.boe.es/buscar/doc.php?id=BOE-A-2019-5089 (accessed on 9 April 2024).
149. Faia, R.; Soares, J.; Pinto, T.; Lezama, F.; Vale, Z.; Corchado, J.M. Optimal Model for Local Energy Community Scheduling Considering Peer to Peer Electricity Transactions. *IEEE Access* **2021**, *9*, 12420–12430. [CrossRef]
150. Tostado-Véliz, M.; Kamel, S.; Hasanien, H.M.; Turky, R.A.; Jurado, F. Optimal energy management of cooperative energy communities considering flexible demand, storage and vehicle-to-grid under uncertainties. *Sustain. Cities Soc.* **2022**, *84*, 104019. [CrossRef]
151. Romero-Rubio, C.; de Andrés Díaz, J.R. Sustainable energy communities: A study contrasting Spain and Germany. *Energy Policy* **2015**, *85*, 397–409. [CrossRef]
152. Minuto, F.D.; Lanzini, A. Energy-sharing mechanisms for energy community members under different asset ownership schemes and user demand profiles. *Renew. Sustain. Energy Rev.* **2022**, *168*, 112859. [CrossRef]
153. D'Adamo, I.; Mammetti, M.; Ottaviani, D.; Ozturk, I. Photovoltaic systems and sustainable communities: New social models for ecological transition. The impact of incentive policies in profitability analyses. *Renew. Energy* **2023**, *202*, 1291–1304. [CrossRef]
154. Moncecchi, M.; Meneghello, S.; Merlo, M. A Game Theoretic Approach for Energy Sharing in the Italian Renewable Energy Communities. *Appl. Sci.* **2020**, *10*, 8166. [CrossRef]
155. Ghiani, E.; Trevisan, R.; Rosetti, G.L.; Olivero, S.; Barbero, L. Energetic and Economic Performances of the Energy Community of Magliano Alpi after One Year of Piloting. *Energies* **2022**, *15*, 7439. [CrossRef]
156. Olivero, S.; Ghiani, E.; Rosetti, G.L. The first Italian Renewable Energy Community of Magliano Alpi. In Proceedings of the 2021 IEEE 15th International Conference on Compatibility, Power Electronics and Power Engineering (CPE-POWERENG), Florence, Italy, 14–16 July 2021; IEEE: Piscataway, NJ, USA, 2021; pp. 1–6. [CrossRef]
157. Tarpani, E.; Piselli, C.; Fabiani, C.; Pigliautile, I.; Kingma, E.J.; Pioppi, B.; Pisello, A.L. Energy Communities Implementation in the European Union: Case Studies from Pioneer and Laggard Countries. *Sustainability* **2022**, *14*, 12528. [CrossRef]
158. Azarova, V.; Cohen, J.; Friedl, C.; Reichl, J. Designing local renewable energy communities to increase social acceptance: Evidence from a choice experiment in Austria, Germany, Italy, and Switzerland. *Energy Policy* **2019**, *132*, 1176–1183. [CrossRef]
159. Hahnel, U.J.J.; Herberz, M.; Pena-Bello, A.; Parra, D.; Brosch, T. Becoming prosumer: Revealing trading preferences and decision-making strategies in peer-to-peer energy communities. *Energy Policy* **2020**, *137*, 111098. [CrossRef]
160. Kim, M.-H.; Kim, D.; Heo, J.; Lee, D.-W. Techno-economic analysis of hybrid renewable energy system with solar district heating for net zero energy community. *Energy* **2019**, *187*, 115916. [CrossRef]
161. Cielo, A.; Margiaria, P.; Lazzeroni, P.; Mariuzzo, I.; Repetto, M. Renewable Energy Communities business models under the 2020 Italian regulation. *J. Clean. Prod.* **2021**, *316*, 128217. [CrossRef]
162. Zheng, S.; Jin, X.; Huang, G.; Lai, A.C.K. Coordination of commercial prosumers with distributed demand-side flexibility in energy sharing and management system. *Energy* **2022**, *248*, 123634. [CrossRef]
163. Ding, X.; Guo, Q.; Qiannan, T.; Jermsittiparsert, K. Economic and environmental assessment of multi-energy microgrids under a hybrid optimization technique. *Sustain. Cities Soc.* **2021**, *65*, 102630. [CrossRef]
164. Ke, H. Phase diagrams, eutectic mass ratios and thermal energy storage properties of multiple fatty acid eutectics as novel solid-liquid phase change materials for storage and retrieval of thermal energy. *Appl. Therm. Eng.* **2017**, *113*, 1319–1331. [CrossRef]
165. Alva, G.; Lin, Y.; Liu, L.; Fang, G. Synthesis, characterization and applications of microencapsulated phase change materials in thermal energy storage: A review. *Energy Build.* **2017**, *144*, 276–294. [CrossRef]
166. Wahid, M.A.; Hosseini, S.E.; Hussen, H.M.; Akeiber, H.J.; Saud, S.N.; Mohammad, A.T. An overview of phase change materials for construction architecture thermal management in hot and dry climate region. *Appl. Therm. Eng.* **2017**, *112*, 1240–1259. [CrossRef]
167. Horstink, L.; Wittmayer, J.M.; Ng, K. Pluralising the European energy landscape: Collective renewable energy prosumers and the EU's clean energy vision. *Energy Policy* **2021**, *153*, 112262. [CrossRef]

168. Hashemipour, N.; del Granado, P.C.; Aghaei, J. Dynamic allocation of peer-to-peer clusters in virtual local electricity markets: A marketplace for EV flexibility. *Energy* **2021**, *236*, 121428. [CrossRef]
169. Jasiński, J.; Kozakiewicz, M.; Sołtysik, M. Determinants of Energy Cooperatives' Development in Rural Areas—Evidence from Poland. *Energies* **2021**, *14*, 319. [CrossRef]
170. Frieden, D.; Tuerk, A.; Antunes, A.R.; Athanasios, V.; Chronis, A.-G.; d'Herbemont, S.; Kirac, M.; Marouço, R.; Neumann, C.; Catalayud, E.P.; et al. Are We on the Right Track? Collective Self-Consumption and Energy Communities in the European Union. *Sustainability* **2021**, *13*, 12494. [CrossRef]

*sustainability*

MDPI

*Article*

# Harnessing Curtailed Wind-Generated Electricity via Electrical Water Heating Aggregation to Alleviate Energy Poverty: A Use Case in Ireland

**Ciara Ahern [1,2,*], Ronan Oliver [1] and Brian Norton [1,2]**

1 Dublin Energy Lab & Built Environment Research and Innovation Centre, Technological University Dublin, D07 EWV4 Dublin, Ireland; ronan.oliver@tudublin.ie (R.O.); brian.norton@tudublin.ie (B.N.)
2 MaREI—SFI Research Centre for Energy, Climate and Marine Research, P43 C573 Cork, Ireland
* Correspondence: ciara.ahern@tudublin.ie

**Abstract:** Ireland experiences high energy poverty rates alongside surplus wind energy resources. With 77% of Irish households equipped with electrical immersion heaters for domestic hot water (DHW) generation, this study proposes an Electrical Water Heating Aggregation (EWHA) scheme. The scheme allocates surplus wind-generated electricity to provide DHW to fuel-poor households, thereby alleviating energy poverty through harnessing curtailed wind energy. Through a developed wind-generated electricity allocation model and half-hourly data analysis for a weather year, this research assesses the feasibility and economic viability of the EWHA scheme, focusing on the householder as the primary benefactor from the scheme (as opposed to ancillary grid service provision). The results suggest an optimal aggregation size where maximum curtailment and carbon offset coincide with maximum benefits for participants. The findings indicate that fuel-poor households in Ireland could receive a full DHW tank every three weeks using surplus wind energy, harnessing 89% of overnight curtailed wind energy and offsetting 33 Mkg$CO_2$ annually. Moreover, the scheme could potentially save the Irish state approximately EUR 4 million by 2030, increasing to EUR 11 million by 2050, in carbon costs. Overall, this research demonstrates the potential of EWHA schemes to alleviate energy poverty, optimise wind energy utilisation, and contribute significantly to carbon emission reduction targets.

**Keywords:** demand-side management; wind energy; curtailment; electrical water heating aggregation; fuel poverty; renewables energies in built environment

**Citation:** Ahern, C.; Oliver, R.; Norton, B. Harnessing Curtailed Wind-Generated Electricity via Electrical Water Heating Aggregation to Alleviate Energy Poverty: A Use Case in Ireland. *Sustainability* **2024**, *16*, 4470. https://doi.org/10.3390/su16114470

Academic Editors: Elena Lucchi, Tianyi Chen and Wen Zhang

Received: 25 March 2024
Revised: 9 May 2024
Accepted: 20 May 2024
Published: 24 May 2024



## 1. Introduction

Increased penetration of wind and solar energy into power systems is being sought to offset fossil fuels and decarbonise electricity generation [1]. Driven by policies, incentives and declining costs, wind and solar electricity generation has increased substantially [1,2] and is set to continue as follows:

- The share of renewable energy in the European Union (EU) is targeted to double from 17.5% in 2017 [3] to 42.5% in 2030 [4];
- The United States is targeting a 50% to 52% reduction in greenhouse gas (GHG) emissions from 2005 levels by 2030 [5].
- China has pledged that 25% of its primary energy will be derived from non-fossil fuels by 2030 [6].

Global electricity generation capacity from wind has multiplied over five-fold from 200 GW in 2010 to 651 GW in 2019 [7] and 1021 GW in 2024 [8], providing 7.33% of the world's electricity demand in 2022 [9]. Wind met a significant amount of the national electricity demand in 2019 in Denmark (47.2%), Ireland (30.7%), Portugal (26.3%), Germany (21.2%) and Spain (20.8%) [10].

Increases in variable, less predictable, non-synchronous wind- and solar-generated electricity result in electrical power systems encountering transmission and/or operational constraints to maintain system stability [1,11]. To ensure that there is sufficient dispatchable synchronous generation on the system to maintain the security of supply and system stability, system operators sometimes accept less wind- or solar-generated electricity than is potentially available [12]. System Non-Synchronous Penetration (SNSP) [12], also referred to as a variable renewable energy (VRE) limit [13], is a real-time measure of the percentage of generation from non-synchronous sources relative to the system demand given by Equation (1):

$$SNSP = \frac{Non-synchronous\ power}{Synchronous\ and\ non-synchronous\ on\ the\ system} = \frac{Wind + Solar + High\ Voltage\ Direct\ Current}{Electricity\ Demand + Exports} \quad (1)$$

Transmission system operators (TSOs) limit SNSP when the following applies [12]:

1. Wind power production exceeds the SNSP limit, addressed by dispatching down wind generators across the entire national grid, referred to as curtailment.
2. Line or cable capacity cannot transmit the electricity produced to serve demand because of one of the following [14]:
    - More wind generation than the localised carrying capacity of the network;
    - An outage for maintenance, upgrade works or faults,

   referred to as constraint.
3. Generation exceeds demand, referred to as energy balancing.

Increasing renewable generated electricity energy requires TSOs to increase SNSP limits [15]. For example, since 2010, the SNSP limit in Ireland has increased from 50% in 2015 to 75% in 2022, with strategies in place to meet a 95% target by 2030 [16]. Ireland has a relatively isolated electrical system, so it experiences high amounts of unharnessed wind-generated electricity [12,14,17] while energy-balancing requirements are forecasted to increase significantly with greater penetration of wind energy [11,12,17,18].

Otherwise, unused VRE has monetary value; based on annual wind curtailment figures [15,19–21] matched to the average yearly retail price of electricity [22], the total retail value of unused VRE in Ireland between 2017 and 2021 is estimated at EUR 1.105 billion, shown in Table 1 to range from EUR 85M in 2017 to EUR 441M in 2020.

**Table 1.** Average market retail value of unused VRE electricity from 2017 to 2021 [15,19–22].

| | 2017 | 2018 | 2019 | 2020 | 2021 | Total |
|---|---|---|---|---|---|---|
| GWh | 386 | 707 | 1008 | 1909 | 752 | 4762 |
| €/kWh | 0.22053 | 0.22607 | 0.23213 | 0.2312 | 0.2456 | |
| € Total M | € 85 | € 160 | € 234 | € 441 | € 185 | € 1105 |

Curtailment, constraint and energy balancing have traditionally been viewed as wasted energy [23] or as a limitation on the penetration of VRE [24]. However, in the last two years, studies have shown the advantages of obtaining more:

- Electrical services, through electric vehicle charging [25,26], heating [27], cooling [28] and energy harvesting through battery storage [21,23].
- Financial services, through profit generation from the provision of energy to support energy users, such as bitcoin mining [29].
- Social services, through provisioning potentially left-behind groups with otherwise unused energy [27,30].
- Grid services, through the provision of demand services supporting an increase in VRE/SNSP limits [31] that can mitigate localised constraint issues [30] if provision is localised to a constraint (enabled by GIS mapping).

Unused VRE electricity can *inter alia* be stored for later use in batteries, pumped hydro, compressed air, and electro-thermal storage. Demand-side management/response

(DSM/DSR) can also provide system stability and flexibility [32]. DSM comprises a set of load management strategies, incorporating planning, integration, and monitoring of pre-assigned routine activities based on the consumer's usage pattern [33] whereas DSR can be defined [34] as "changes in electric use by demand-side resources from their normal consumption patterns in response to changes in the price of electricity, or to incentive payments". DSM/DSR systems aim to efficiently use available energy without installing new transmission infrastructure. Article 32 of the EU 2019 Clean Energy Package [35] requires TSOs to consider flexibility in grid planning as an alternative to system expansion [30].

The residential sector accounts for 25% of the overall electrical load in Ireland [32], so it correspondingly impacts overall power balance, stability, and efficient power management. Article 32 further requires market access for residential consumers. Due to regulatory barriers as well as a lack of market products suitable for small end users, the demand-side flexibility potential in the residential sector in Ireland is underutilised [36,37]. Residential customers in Ireland can only participate in DSM/DSR through tariff-based schemes where they are encouraged to move their usage to cheaper off-peak night-time hours.

Agbonaye et al. [38] explored the value of demand flexibility for managing wind energy constraint and curtailment in Northern Ireland, finding that the optimum aggregation size and household savings varied based on whether constraint or curtailment was the primary control parameter, further making the point that there could be other competing uses for unused wind such as the production of green hydrogen for industrial use, grid-scale storage, and district heating schemes. Agbonaye et al. [27], in a different study, highlighted that households at risk of fuel poverty do not have the capital to invest in heat pumps, photovoltaics, batteries, etc., while the electrification of heat and transport is likely to result in increased network costs, disproportionately affecting vulnerable households; they concluded that it makes sense to prioritise the use of flexibility for vulnerable consumers so that they are not left behind in the transition to clean energy. Agbonaye et al. [30] went on to develop a methodology for identifying vulnerable neighbourhoods along with a flexibility prioritisation framework that ensured a fair distribution of flexibility opportunities.

Figure 1 illustrates that hot water tanks make up around a quarter of the capacity used in residential demand-side flexibility applications [36], due to the ready use of stored hot water. Smart controls can provide customers with control over how much hot water they need at particular times [39].

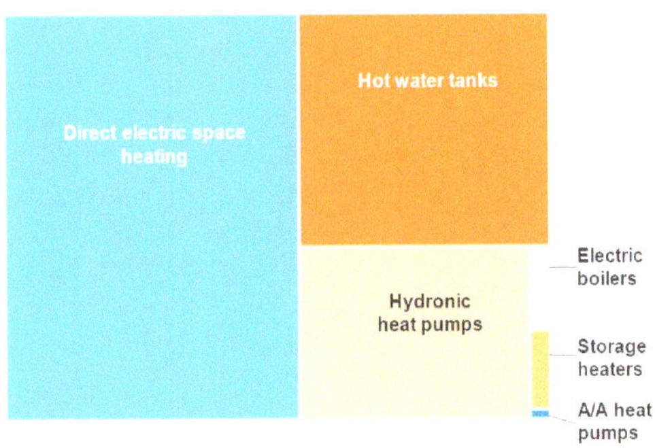


**Figure 1.** Share of flexible electric heating capacity currently used in demand response, by asset type (total capacity ~1.4 GW) [33].

Across Europe, water heating accounted for 17% of energy consumption in households in 2012 [40]. Due to their energy storage characteristics, EWHs are considered ideal candidates for demand response [41], and it is proven that EWHA schemes are useful for frequency regulation and power balancing [42–44]. France leads the market with installed electrical heating and cooling technology assets, estimated to be over 200 MW [36]. To create a more constant day/night electrical base load suiting nuclear energy generation, Électricité de France (EDF) has used Electrical Water Heating Aggregation (EWHA), along with tariffs that incentivise consumers to use Electrical Water Heaters (EWHs) during off-peak hours [45]. In 2017, more than 13 million EWHs were installed in French households, representing around 50% of residential hot water systems (the rest are supplied by gas and oil heating systems), accounting for an annual consumption of 20 TWh and a peak demand of 8 GW [45]. EWHA schemes have also been implemented in the US; Florida Power & Light have been installing controllable electric EWHs in hundreds of thousands of homes since the 1970s [46], and since the early 1990s, Great River Energy (Minnesota) have implemented an EWHA system in over 110,000 homes [46,47].

Mabina et al., 2021 [48], reviewed control methods concerned with the application of demand response to control EWHA schemes, focusing particularly on studies that sought to provide ancillary grid-balancing services and supporting an increase in VRE in power systems. Most studies [48] focused on controlling a single parameter, such as minimising peak load [49–56] or frequency control [57–60], with one study focused on increasing the integration of wind [61]. Studies that sought to control multiple (two to three) parameters sought to control (i) load shifting and peak load reduction [53], (ii) voltage and load reduction [62], as well as (iii) frequency regulation, load shifting, and peak load reduction [63]. Interestingly to this study, none of the EWHA studies reviewed sought to maximise the benefit to the householder as a control parameter, so this study is novel in its approach.

"Real Value" [64] and "EU-SysFlex" [32] are two large DSM studies with the Irish TSO as a participant. "RealValue" [64] ran from 2015 to 2018 and deployed Smart Electric Thermal Storage (SETS) in 1250 homes in Germany, Latvia, and Ireland (volumes per country are unspecified), demonstrating scalability through modelling and virtual simulation [65,66] and finding that SETS can meet householders' space and water heating needs in a low-cost and energy-efficient manner whilst enabling residential storage heaters to connect on a cloud aggregation platform to charge based on grid constraints and market price signals. "EU-SysFlex" [32], carried out between 2018 and 2020, trialled "Residential Service Provision" through the "Power Off and Save project" [32,67], in which over 1400 households in Ireland were engaged to reduce their electricity consumption at peak times. This trial showed that switching off the standard 3 KW electrical hot water immersions at peak times was a cost-effective way to control domestic electrical load [67] without impacting occupant comfort.

Due to a historically high price of electricity in Ireland, compared to other European countries, most homes in Ireland heat and generate hot water through fossil-fuelled hydronic radiator systems and indirect domestic hot water cylinders and thus do not have electric space heating installed [68]. Most (77%) Irish homes have a secondary electric hot water immersion fitted to domestic hot water tanks that acts as a backup or as an alternative to the boiler. Ireland, whose predominant fuel source is oil, has seen the average fuel price for households (EUR cent/kWh) [18] rise for electricity (22%), gas (32%), and oil (70%), with energy poverty in Ireland being at its highest recorded rate in 2022 [69]. It is known that to ration energy use, some fuel-poor householders routinely go without heat or electricity, with many allowing themselves very limited hot water [69].

*Research Aim and Objectives*

Unharnessed curtailed wind, while a growing resource, is a finite variable resource. Using Ireland as a case study, this research assesses how unharnessed curtailed wind energy in Ireland could be effectively redeployed to DHW in households, with a focus on maximising benefits for fuel-poor households (as opposed to the provision of ancillary grid services). This research aim will be addressed through the following objectives:

1. Characterise curtailed wind energy in Ireland for a representative weather year.
2. Review the literature to establish DHW consumption profiles to characterise DHW loads for Irish households, especially for fuel-poor households, where explicit data may be lacking.

3. Develop a wind-generated electricity allocation model utilising half-hourly wind data to assess the feasibility and economics of reallocating surplus wind energy to DHW.
4. Establish model parameters to ensure equitable energy allocation within an EWHA scheme, considering various aggregation sizes.
5. Evaluate the benefit to the householder of participating in an EWHA scheme as a function of aggregation size.

## 2. Materials and Methods

The overarching methodological approach is to redeploy unharnessed variable wind-generated electricity to heat hot water by matching or balancing hot water demand to meet the available supply. It was necessary to characterise curtailed wind energy in Ireland (Section 2.1) as well as typical DHW loads in Irish households (Section 2.2). Model parameters were set up (Section 2.3) to ensure that each household aggregated within an EWHA scheme received an equitable allocation of energy, while the annualised benefit to householders from participating in an EWHA scheme was assessed for various aggregation sizes. DHW profiles for Irish households are not explicitly available and no profiles for fuel-poor households in Ireland exist; therefore, a review of hot water consumption profiles was carried out in Section 2.2. Equitable distribution across the aggregation was validated in Section 2.5.

### *2.1. Curtailed Wind Profile*

Curtailment, constraint, and energy balancing each pose different challenges to network operators. Curtailment is managed at a nationwide level, whereas constraint is caused by a local and often transitory issue, while energy balancing is required at very high (>70%) penetrations of variable renewable energy. This study only considers curtailed wind-generated electricity. As wind curtailment data for 2020 are considered unrepresentative due to the SARS-CoV-2 pandemic, 2019 curtailment data for Ireland, totalling 319 GWh, were used. As shown in Figure 2, 73% of curtailments, totalling 233 GWh, occur at night between 10 p.m. and 7 a.m. coinciding with low electricity demand. As hot water is generally consumed during the day [70], overnight 2019 curtailed wind data were used as a case study year.

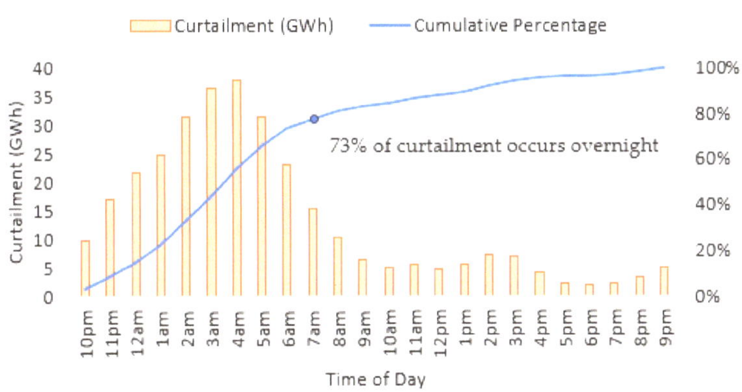


**Figure 2.** Average daily curtailment distribution for Ireland in 2019.

Further to Figure 2, the number of nights that overnight wind curtailment was present is examined in Figure 3a. It is seen that wind curtailment was not present during any half-hour between 10 p.m. and 7 a.m. for most (61%) or 222 nights in 2019. Notwithstanding when curtailed wind was observed in the other (39%) 143 nights of the year, it was present for a significant duration, as 85% (122/143) of nights saw wind curtailment for at least 6 half-hours—the time required to heat an assumed "cold" hot water tank fully to 60 °C. Figure 3b shows the distribution of the 222 "curtailment nights" over a weekly timeframe;

it is seen that although overnight wind curtailment was not observed in 6 calendar weeks, a night of curtailment was observed for the remaining 46 (88%) calendar weeks of the year.

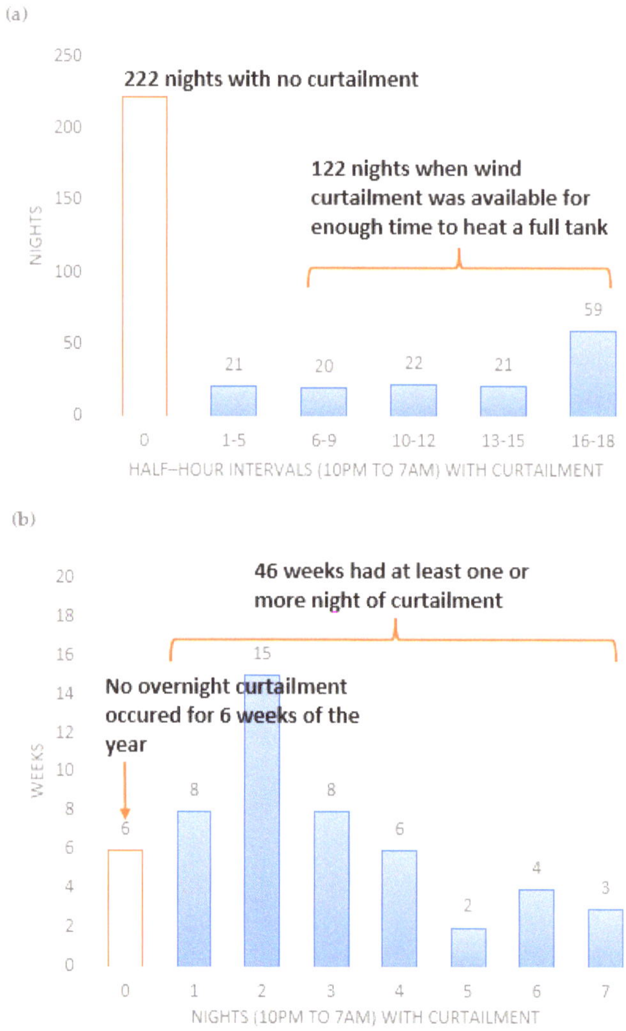


**Figure 3.** (**a**,**b**) Nightly and weekly instances of curtailment (10 p.m. to 7 a.m.) for Ireland in 2019.

### 2.2. DHW Consumption Profiles

DHW loads vary by region and climate; daily DHW consumption in non-fuel-poor dwellings varies from a low of 43 litres/day in Finland to a high of 256 litres/day in Florida, while the global average is 202 litres/day [71]. The day of the week, season, outdoor temperature, and precipitation levels also have distinct effects on water demand [72–74]. DHW consumption is found to increase on hot days in some regions (New York) but reduce in others (Florida) [72–74]. Conversely, in German-speaking regions in Switzerland [75], no significant decrease in consumption during weekends was found, and neither was a correlation with weather conditions established. Water consumption patterns also vary by occupant profiles [76,77]; older people use toilets more often and so use more water and young children take a bath more often, while teenagers take long showers [77]. Behaviour is also influenced by environmental factors and socioeconomic status [71,78].

Notwithstanding the variances listed, hot water usage is largely simultaneous [71,79–82] with two day-time peaks, one in the morning and one in the evening, with the balance spread over the day. Peak characteristics vary by country; for instance, in Finland, the peak is higher in the evening and lower in the morning, whereas the behaviour is the opposite in Germany [75] and in Ireland [83].

DHW consumption profiles have been studied for single-family dwellings in the US [60], UK [70], Netherlands [77], Germany and Switzerland [84,85], Finland [86], and Sweden [87]. DHW consumption profiles of apartments have been characterised in Greece [88]. No metered DHW study has been carried out in the Irish context (the aforementioned study in Ireland [83] characterised standard profiles for building archetype models based on studies conducted across European countries). In the absence of an Irish study, the closest region climatically and geographically is the UK. A hot water consumption study [70] of significance, used in many DHW modelling studies [40,61,89], was carried out by the UK Energy Saving Trust in 2006. Domestic hot water consumption in 112 dwellings was recorded for one year at ten-minute intervals. When water run-off was detected, the sampling rate increased to five seconds. The data was then resampled at constant intervals of 1, 2, and 3 h. This study found the following [70]:

- The mean water consumption rate per UK household was 122 litres/day with a 95% confidence interval of ±18 litres/day.
- Hot water heating time was 2.6 h/day, estimated with a 95% confidence interval of ±0.35 h/day, finding that some households heated water as and when it was required, and the remainder generally heated water between 8:00 a.m. and 10:00 a.m., and again between 6:00 p.m. and 11:00 p.m.
- Storage temperatures were significantly below the widely assumed value of 60 °C, with a mean value of 51.9 °C estimated with a 95% confidence interval of ±1.3 °C.
- A key factor influencing consumption is the number of occupants.

The typical hot water consumption profile for UK dwellings published in the study [70] did not account for days of the week, seasonal variations, or occupancy.

Fuel-poor and non-fuel-poor households consume hot water differently. The Building Research Establishment (BRE) [90], in a 2018 UK study, found that fuel-poor households were significantly (60%) more likely than non-fuel-poor households (45%) to be taking less than one shower and/or bath per person per day. Households in fuel poverty were less likely (21%) than those not in fuel poverty (31%) to take a shower and/or bath once per person per day. Fuel-poor households were less likely (84%) to heat their water with a central heating system, compared with non-fuel-poor households (92%), and fuel-poor households were almost twice as likely (13%) to heat their water with an electric immersion heater than non-fuel-poor households (7%). Fuel-poor households were more likely (47%) to use an electric shower than non-fuel-poor households (35%) and were less likely to use a shower that was pumped from the main hot water system (10% compared with 16%). Regrettably, the UK BRE study [90] did not publish hourly water consumption patterns for the fuel-poor households studied.

Electrical DHW immersion heaters in Ireland are typically rated at 3 kilowatts (3 kW), supplied with a 230-volt single-phase supply limited to 13 amperes. A standard 3 kW electrical immersion will consume 9 kWh of electricity in 3 h to heat 155 litres of 60 °C hot water from 10 °C. Typical domestic hot water cylinders in Ireland store an average hot water requirement of 9.63 kWh/day [91]. At a boiler efficiency of 85% and an average hot water heating time of 2.6 h per household, this equates to 129 L/day heated to 51.9 °C or 108 litres heated to 60 °C and so correlates with the findings of the UK study [70].

This research hypothesis assumes DHW tanks are heated overnight (before 7 a.m.), with hot water available for morning draw-off. This shifts DHW loads from day to night, coinciding with a lower grid carbon intensity, and simplifies the modelling approach. To meet the DHW load in each household, six 1.5 kW half-hour time periods are allocated overnight to each household within the EWHA scheme. Each household receives 9 kWh to almost meet the national daily average hot water consumption figure of 9.63 kWh [91] at 60 °C. A minimum activation period of 30 min and multiples thereof is applied to limit excessive relay switching and to coincide with the available data and electricity market trading intervals.

### 2.3. Aggregation Size

The minimum aggregation was set at 1000 households. Maximum aggregation size is a function of available curtailed wind energy harmonised to DHW demand, quantified in the Results section of this work. Aggregation sizes between the minimum and maximum were tested and results were presented. To assess the benefit to the fuel-poor households, it was necessary to quantify the fuel-poor aggregation size.

An in-depth bottom-up analysis of energy poverty in Ireland [92] found that 1 in 4 (28%) or up to 461,000 households were potentially in fuel poverty in 2015 [93]. At the time of modelling (2021) and while more recent data on fuel poverty, published by the ESRI and available through the Central Statistics Office [94], were available, the calculations are based on the expenditure method. The expenditure method considers a household to be in energy poverty if said household spends more than 10% of overall household income on energy services. This metric can "give an incomplete picture of energy poverty" as households living in deprivation may not be able to afford to spend 10% of their income on energy, thus living in inadequately heated homes and not being captured by the expenditure method; moreover, more affluent households may spend over 10% of their income on energy costs [95]. The objective method for calculating fuel poverty models the level of fuel expenditure required by a typical household to keep their home heated to the levels recommended by the World Health Organisation and compares this to household income in an attempt to calculate exposure to energy poverty, considered in the Government of Ireland's "Spending Review 2020—*Social Impact Assessment—SEAI Programmes Targeting Energy Poverty*" [95] to be a more robust assessment of energy poverty and thus the fuel-poor aggregation figure (461,000) used in this research hypothesis.

### 2.4. Boundary Parameters

Allocation of curtailed wind energy is based on the following boundary parameters:

(1) Only wind curtailed between 10 p.m. and 7 a.m. overnight is applied; the rationale for targeting overnight curtailed wind initially is as follows:

   a. Over 70% of curtailment happens during night-time hours (see Figure 2).
   b. Night-time use-of-system charges are $\frac{1}{4}$ of day-time use-of-system charges.
      I. This means any rebate applied to account for the household meter, turning with the application of redeployed electrical energy, relating to SNSP limits and synchronous plant base load costs will be minimised.
      II. For the same reason, if an aggregator focused on fuel-poor householders seeks contributions from energy generators and suppliers, contributions will go further during night-time hours as the electricity price is lower, meaning the aggregator will be able to provide a higher subsidy against night-time tariff rates—this renders the delivered energy free or at a significantly lower than typical retail cost.
      III. Heating water at night shifts DHW load from day to night, thereby flattening the day/night curve, in line with the existing EWHA schemes reviewed.
   c. In favour of day-to-night load shifting, market participants and Ireland's Commission for Regulation of Utilities (CRU) will be more supportive of schemes targeted for night-time hours initially.
   d. Water consumption profiles are simplified as little draw-off is universally reported across all hot water consumption studies across regions during night-time hours.

(2) A minimum activation period of 30 min or multiples thereof is applied at each 3 kW immersion to limit excessive relay switching and to coincide with the available data and electricity market trading intervals.
(3) Each immersion remains energised across multiples of 30 min so that, if possible, only one relay activation and deactivation signal is necessary from the aggregator, while individual households receive as much useful hot water as possible when allocated energy.
(4) Each household is allocated up to 6 half-hour time periods receiving 9 kWh to almost meet the national daily average hot water consumption figure of 9.63 kWh [91] at 60 °C. The model allocates 1.5 kWh of energy to as many household immersions as possible in a 30 min period, hence prioritising households who have already received an allocation to provide a full tank and hence a useful amount of hot water periodically.
(5) Step 4 ensures that the tank is heated to 60 °C to reduce the risk of legionella growth.

### 2.5. Wind-Generated Electricity Model Validation

Outputs from the model are articulated in Table 2 for a randomly selected night of the 12th of January, as shown in Table 3. Table 2 articulates the model outputs described in Table 3. It is seen from Table 3 that more than 200,000 smart hot water cylinders would need to be available to harness the curtailed night-time wind energy available on that night. Conversely, as shown in Table 4, on the night of the 3rd of January 2019, when there was relatively low curtailment, only 24,590 homes received excess wind allocations, with none receiving 6 allocations or 9 kWh equating to a full tank of hot water.

**Table 2.** Articulation of Wind Allocation Methodology (refers to Table 3).

| Time Period | Result of Model Application |
|---|---|
| 10:00 p.m. to 10:30 p.m. | Wind curtailments of 92 MWh occurred, which, divided by 1.5 kWh per immersion, equates to sufficient energy for 61,322 household DHW immersions (92,000 kWh/1.5 kWh). |
| 10:30 p.m. to 11 p.m. | Available curtailment is sufficient to deliver 1.5 kWh (1 allocation) of energy to 5488 homes. |
| 11:00 p.m. to 11:30 p.m. | No curtailment occurs and thus no energy is allocated. |
| 11:30 p.m. to 12:00 a.m. | There is sufficient energy for 77,033 homes; therefore, 55,834 households receive a second allocation, 5489 receive a third allocation, and 15,710 (77,033–61,322) receive an allocation of wind energy for the first time. In Table 3, the following is noted:<br>○ Homes identified from 1 to 5488 have used power for three 30 min periods;<br>○ Homes identified from 5489 to 61,322 have used power for two 30 min periods;<br>○ Homes identified from 61,323 to 77,033 have been supplied with energy for the first time. |
| 12:00 a.m. to 1:30 a.m. | Available energy is allocated in a similar manner to previous time periods. |
| 1:30 a.m. to 2:00 a.m. | Households identified from 1 to 5488, having received the maximum of 6 allocations, no longer receive energy as noted by the allocation count at the base of Table 3. It is worth noting that even though there is sufficient energy at this time for 212,286 homes, households identified from 5489 to 200,000 (or up to the total households within the identified aggregation) received energy. |
| 2:00 a.m. to 4:00 a.m.: | Available energy is allocated to the remaining homes until each is provided with 9 kWh or 6 allocations of 1.5 kWh of renewable electricity. |
| 4:30 a.m. to 6:30 a.m. | Even though significant energy is available, this is left unused as the 200,000 homes in the identified aggregation have been provided with their maximum allocation. |

**Table 3.** Potential allocation of curtailment to 200,000 homes overnight on the 12th January 2019.

| Time | Curtailment [MWh] | Potential Households [-] | Household Allocations and Identifiers | | | | | | | | | | | | | |
|---|---|---|---|---|---|---|---|---|---|---|---|---|---|---|---|---|
| 22:00 | 92.0 | 61,322 | 1 | 5488 | 5489 | **61,322** | | | | | | | | | | |
| 22:30 | 8.2 | 5488 | 1 | **5488** | | | | | | | | | | | | |
| 23:00 | 0 | 0 | | | | | | | | | | | | | | |
| 23:30 | 115.6 | 77,033 | 1 | 5488 | 5489 | 61,322 | 61,323 | **77,033** | | | | | | | | |
| 00:00 | 210.5 | 140,321 | 1 | 5488 | 5489 | 61,322 | 61,323 | 77,033 | 77,034 | **140,321** | | | | | | |
| 00:30 | 259.2 | 172,791 | 1 | 5488 | 5489 | 61,322 | 61,323 | 77,033 | 77,034 | 140,321 | 140,322 | **172,791** | | | | |
| 01:00 | 275.6 | 183,700 | 1 | 5488 | 5489 | 61,322 | 61,323 | 77,033 | 77,034 | 140,321 | 140,322 | 172,791 | 172,792 | **183,700** | | |
| 01:30 | 318.4 | 212,286 | | | **5489** | 61,322 | 61,323 | 77,033 | 77,034 | 140,321 | 140,322 | 172,791 | 172,792 | 183,700 | 183,701 | **200,000** |
| 02:00 | 323.0 | 215,357 | | | | | **61,323** | 77,033 | 77,034 | 140,321 | 140,322 | 172,791 | 172,792 | 183,700 | 183,701 | **200,000** |
| 02:30 | 346.6 | 231,058 | | | | | | | **77,034** | 140,321 | 140,322 | 172,791 | 172,792 | 183,700 | 183,701 | **200,000** |
| 03:00 | 397.6 | 265,079 | | | | | | | | | **140,322** | 172,791 | 172,792 | 183,700 | 183,701 | **200,000** |
| 03:30 | 427.3 | 284,835 | | | | | | | | | | | **172,792** | 183,700 | 183,701 | **200,000** |
| 04:00 | 462.1 | 308,044 | | | | | | | | | | | | | **183,701** | **200,000** |
| 04:30 | 455.1 | 303,414 | | | | | | | | | | | | | | |
| 05:00 | 447.6 | 298,430 | | | | | | | | | | | | | | |
| 05:30 | 449.3 | 299,564 | | | | | | | | | | | | | | |
| 06:00 | 440.3 | 293,555 | | | | | | | | | | | | | | |
| 06:30 | 460.0 | 306,668 | | | | | | | | | | | | | | |
| | | Allocation Count | 6 | 6 | 6 | 6 | 6 | 6 | 6 | 6 | 6 | 6 | 6 | 6 | 6 | 6 |

**Table 4.** Potential allocation of curtailment to 200,000 homes overnight on the 3rd January 2019.

| Time | Curtailment [MWh] | Potential House-holds [-] | Household Allocations and Identifiers | | | | | |
|---|---|---|---|---|---|---|---|---|
| 22:00 | 0 | 0 | | | | | | |
| 22:30 | 0 | 0 | | | | | | |
| 23:00 | 0 | 0 | | | | | | |
| 23:30 | 0 | 0 | | | | | | |
| 00:00 | 0 | 0 | | | | | | |
| 00:30 | 0 | 0 | | | | | | |
| 01:00 | 0 | 0 | | | | | | |
| 01:30 | 0 | 0 | | | | | | |
| 02:00 | 29.3 | 19,524 | **1** | 4064 | 4065 | **19,524** | | |
| 02:30 | 36.9 | 24,590 | **1** | 4064 | 4065 | 19,524 | 19,525 | **24,590** |
| 03:00 | 6.1 | 4064 | **1** | **4064** | | | | |
| 03:30 | 0 | 0 | | | | | | |
| 04:00 | 0 | 0 | | | | | | |
| 04:30 | 0 | 0 | | | | | | |
| 05:00 | 0 | 0 | | | | | | |
| 05:30 | 0 | 0 | | | | | | |
| 06:00 | 0 | 0 | | | | | | |
| 06:30 | 0 | 0 | | | | | | |
| | | Allocation Count | 3 | 3 | 2 | 2 | 1 | 1 |

## 3. Results

To establish the capacity of night-time curtailed wind energy to heat DHW tanks, the redeployment of night-time curtailed wind energy was modelled for various EWHA sizes until all available night-time curtailed wind energy was harnessed. Referring to Figure 4, 901,188 households, representing the maximum aggregation size, would have been required to use 100% of the 233 GWh night-time curtailed energy in 2019, meeting 7% of participating households' annual hot water load. If the minimum number of 1000 households were in an EWHA scheme, 0.5% of night-time curtailed wind energy would have been harnessed, while 32% of the participating households' annual DHW load would have been met. This phenomenon arises because curtailed wind is a finite resource, meaning the more households aggregated, the fewer allocations each household receives. If normally curtailed wind is diverted to the fuel-poor aggregation of 416,000 households, 89% of unused wind-generated electricity would be utilised, while 13% of a fuel-poor household's water demand would be met.

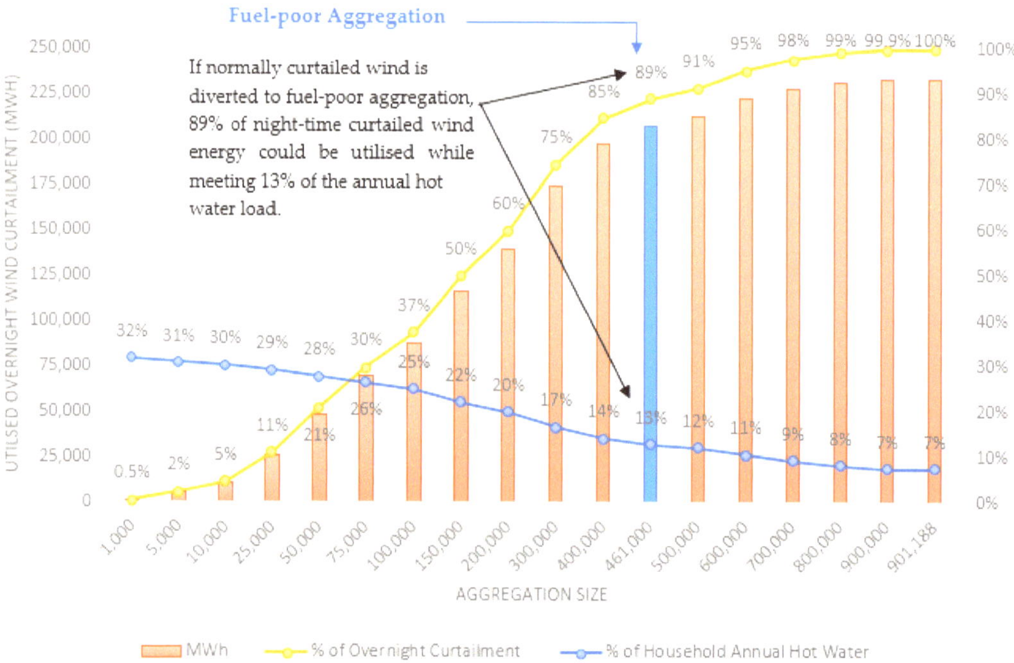


**Figure 4.** Utilised overnight wind curtailment for various "smart" hot water cylinder installations.

## *3.1. Nightly Allocation of Curtailed Wind Energy*

Figure 5 illustrates how many nights each household in each EWHA aggregation set would receive a full allocation of 9 kWh, finding the following:

- Between 1 and 1000 households received a full allocation on 122 nights;
- Between 1001 and 5000 households received a full allocation on 108 nights;
- Between 5001 and 10,000 households received a full allocation on 104 nights.

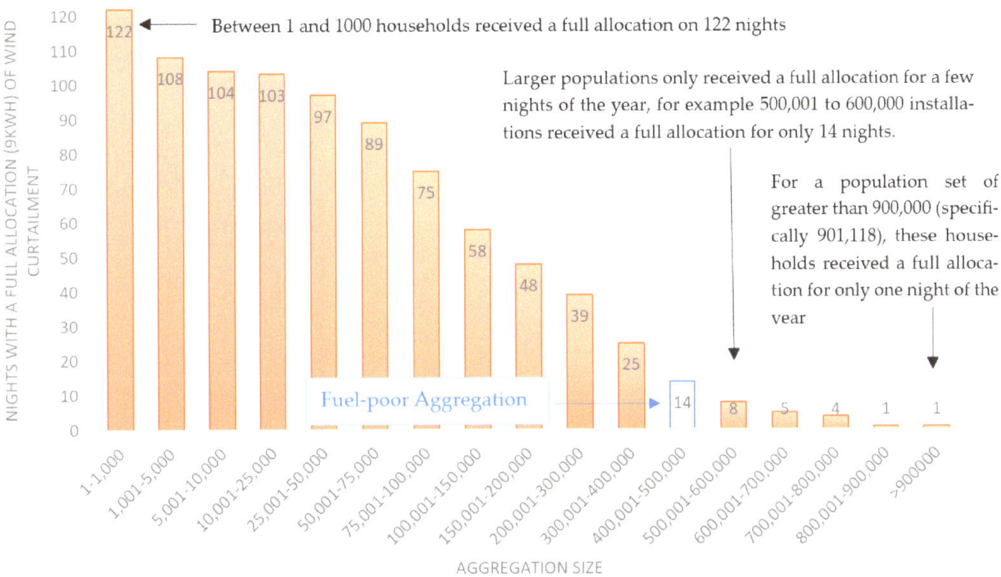


**Figure 5.** Nights with a full allocation (9 kWh) of energy for various population sets.

This continues until it is apparent that the larger populations only received a full allocation for a few nights of the year; for example, between 500,001 and 600,000 installations received a full allocation for only 14 nights or 11% of the nights in which wind curtailment was present, while for a population set greater than 900,000 (specifically 901,118), these households received a full allocation for only one night of the year—coinciding with the Christmas holiday period occurring on Sunday, 29 December 2019, when presumably strong winds and less industrial activity led to significant curtailment.

Following from Figure 5, Figure 6a,b illustrate the half-hourly allocation of curtailed wind energy by aggregation size for the nights wherein participating households received an allocation from the scheme. Figure 6a illustrates the results for the maximum aggregation of 901,188, while Figure 6b illustrates the results for a minimum aggregation of 1000 dwellings. When 901,188 are aggregated, households are provided with a full allocation of six half-hours or 9 kWh of electricity for 82% of the nights they are allocated power under the scheme, whereas if 1000 households are aggregated, the households receive a full allocation for 90% of the nights they are allocated power. It can be concluded, therefore, that under an EWHA scheme distributing normally curtailed wind energy *as a service*, when individual households are allocated energy, they will typically ($\geq$82%) be provided with a full allocation of 9 kWh or a tank of hot water irrespective of the number of households aggregated. This can be attributed to the fact that the model parameters ensure, where possible, that immersions in an individual household remain energised until a full allocation has been received along with the phenomenon that when instances of wind curtailment occur, they typically occur for much of the night (windy nights).

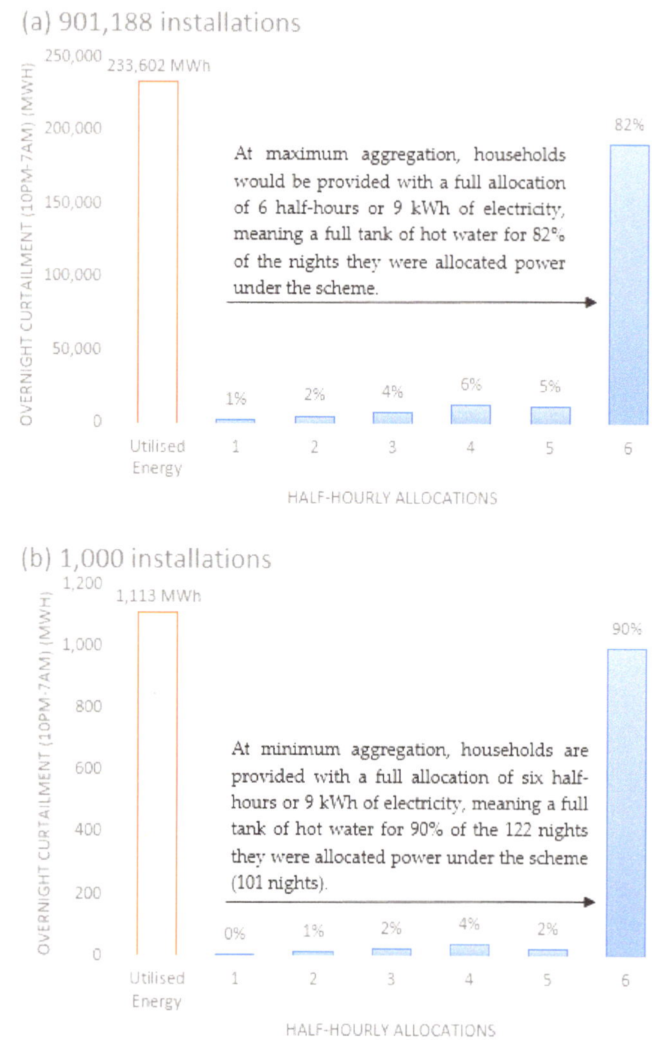


**Figure 6.** (**a**,**b**) Utilised energy and distribution of half-hourly allocations by aggregation size.

### 3.2. *Weekly Allocation of Curtailed Wind Energy*

As this research has a focus on alleviating fuel poverty, the analysis progressed to a pragmatic assessment of the extent to which a full tank of hot water can be provided for at least one night of the week so that important cleaning, bathing, and hygiene tasks within each home can be catered for on a weekly basis. The benefit of this might be much more apparent to fuel-poor households who are known to limit hot water use [96]. The results of the Wind Curtailment Allocation Model are assessed to observe the frequency at which households receive at least one full tank of hot water for the 46 weeks of the year when curtailment was observed (Figure 3b). As shown in Figure 7, it is seen that the smallest (N = 1000) population set received a full allocation for 45/46 (97.8%) weeks, whereas the largest population set (N = 901,188) received such an allocation for 6/46 (13%) weeks that curtailment was available. This latter observation should not be confused with Figure 5, which showed only one *night* of the year when a full allocation could be delivered to the largest population set, because within a *weekly* timeframe, there are seven nights in which to provide households with a full allocation of energy on at least one night of the week. For example, in the week beginning the 3rd of February 2019, there was curtailment on six of seven nights with the potential to provide energy to over one million cylinders.

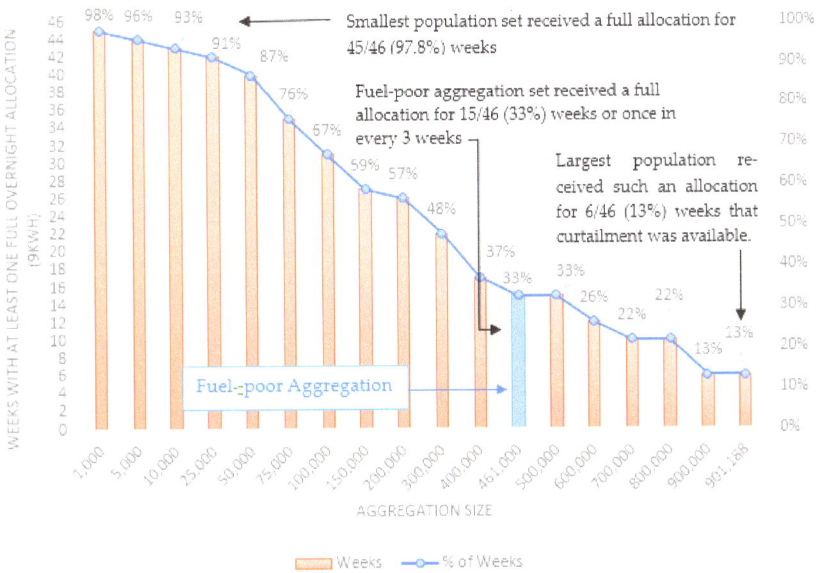


**Figure 7.** Weeks with at least one full overnight allocation of wind curtailment.

Thus, if a roll-out of an EWHA scheme was considered for larger population sets, from say 500,000 households and beyond, not only would they receive a small proportion of their annual hot water demand (≤12% in Figure 4), but these households would only receive at least one night with a full allocation of energy for less than a third of the 46 weeks in which curtailment occurred (≤32.6% in Figure 7). However, if fewer households, say 5000, were aggregated, they would receive almost a third of the annual hot water demand (31% in Figure 4) and at least one night with a full allocation of energy for 44/46 weeks (95.7% in Figure 7) in which curtailment occurred. Therefore, there is a greater observable effect on hot water provision for individual householders if fewer households are aggregated.

### *3.3. Economic and Environmental Analysis*

The Irish householder is estimated to consume 3515 kWh of energy annually to heat hot water [91]. Referencing Figure 8, this costs an average of EUR 284/household/annum, emitting an average of 702 kg/$CO_2$ depending on the heating system and fuel choice. There were 1.86 M occupied dwellings in Ireland in 2022 [95]; therefore, the total energy cost borne by the citizens of Ireland amounts to EUR 529 M, while the carbon emissions associated amount to 1.3 Mt$CO_2$/annum. Figure 8 details, as a function of aggregation size and hence the percentage of hot water load met (cross-referencing Figure 4), DHW fuel costs and resultant carbon emissions per household at the prevailing SNSP of 75%. Referring to Figure 8, it is seen that household cost and $CO_2$ emission savings (from participating in an EWHA scheme) are modest, ranging from 6% or EUR 16/38 kg$CO_2$/annum at the maximum aggregation size of 901,188 dwellings to 20% or EUR 57/136 kg$CO_2$/annum at an aggregation of 100,000 dwellings. This lowers average household DHW bills from EUR 284/annum to EUR 278/annum, at the maximum aggregation, and EUR 284/annum to EUR 227/annum at an aggregation of 100,000 households.

Noting that the model assumes *a standardised water load*, assuming householders can afford to heat hot water in the first place, it is known that in fuel-poor homes, householders limit the generation of hot water to limit financial outlay; therefore, there might be no savings realised as water is not being heated in the first instance, which is a limitation of this work that will be refined in future research.

While the fuel cost is typically paid for by the householder, the "cost" of associated carbon emissions is borne by the state and the citizen. The shadow price of carbon acknowledges the cost to society that carbon emissions create in the form of climate change, air pollution, and other adverse effects sometimes called "externalities". The shadow price of carbon is a theoretical or assumed cost per metric tonne of carbon emissions used to better understand the potential impact of $CO_2$ emissions by monetising (costing) the negative impact of environmental emissions [97,98]. Values for the shadow price of carbon are outlined in a public spending code for Ireland [99]. According to this code, the shadow price of carbon should be based on the estimated future price of $CO_2$ equivalent derived from the EU Emission Trading Scheme (EU ETS) [100]; values for produced carbon of EUR 10 per tonne by 2020, EUR 35 per tonne by 2030, and EUR 100 by 2050 were recommended [98]. However, a 2019 review of carbon pricing carried out by the Irish Government found that the EU ETS failed to price carbon optimally from its perspective and consequently proposed an increase in the shadow price to EUR 32 per tonne by 2020, EUR 100 per tonne by 2030, and EUR 265 by 2050 [98]; it is these shadow prices that are used to estimate the "cost" of carbon emissions associated with average annual domestic hot water production in the 1.86 M occupied dwellings in Ireland, which was EUR 42 M in 2022, rising to EUR 131 M in 2030 and EUR 346 M in 2050 (see Figure 8). At an aggregation level of the fuel-poor (461,000), and referring to data in Figure 8, a potential carbon cost saving to the state of EUR 1 M in 2022 [0.033 $MtCO_2$ × EUR 32/tonne at 75% SNSP], rising to EUR 4 M in 2030 [0.4 $MtCO_2$ × EUR 100/tonne at 95% SNSP] and EUR 10.6 M in 2050 [0.4 $MtCO_2$ × EUR 265/tonne at 95% SNSP], might be realised through redeploying unused wind-generated electricity to heat water in vulnerable Irish households.

The economics of using wind curtailment to displace energy and carbon emissions from conventionally fuelled hot water systems, assessed in Figure 8, are depicted in Figure 9 for various sample population sizes at current (75%, 2022) and future (95%, 2030) SNSPs, noting that synchronous power in Ireland is typically met by gas-fuelled generators with an associated emission factor of 0.2 $kgCO_2/kWh$, at an estimated cost of EUR 7 c/kWh (based on wholesale electricity prices in Ireland in 2019 [22]).

It is again seen that as the percentage of hot water load met decreases with increasing aggregation size (due to excess wind being a finite resource), the aggregated household cost and $CO_2$ savings offset through redeploying unused wind-generated electricity to heat hot water tails off at higher to maximum aggregations. For instance, and referring to Figures 8 and 9, at the prevailing (2023) SNSP of 75%, the displaced fuel cost to the householder at an aggregation size of 100,000 households is EUR 5 M, while the carbon offset is estimated at 14 $MkgCO_2$. If the aggregation size is increased circa 5-fold to 461,000, representing the fuel-poor population in Ireland, the displaced fuel cost rises by 280% from EUR 5 M to EUR 14 M, while the amount of carbon displaced rises by 236%, from 14 $MkgCO_2$ to 33 $MkgCO_2$. If the fuel-poor aggregation size is roughly doubled, increasing to 901,188 dwellings, representing an additional 440,188 dwellings to the scheme and matching the amount of wind dispatched down in 2019, the displaced fuel cost would remain at EUR 14 M, while displaced carbon would rise by only 3%, from 33 $MkgCO_2$ to 34 $MkgCO_2$. The explanation for this is that while a larger number of tanks are being heated, they are heated less often, again because unused wind-generated electricity, as a resource, is finite. Therefore, when sustainability matchmaking, there is an optimum aggregation size wherein near-maximum levels of dispatch-down and carbon are offset concurrent with scheme participants receiving the maximum benefit, in this case hot water, from participating in the scheme. This is the key finding of this research.

Cost and $CO_2$ emissions saving realised from redeployment of excess wind to heat hot water

| Fuel | Fuel cost and $CO_2$ emissions: % | c/kWh | $gCO_2$/kWh | Household Cost and $CO_2$ emissions: € | $kgCO_2$ | Shadow Price Carbon/1000kg $CO_2$: 2020 (€32) | 2030 (€100) | 2050 (€265) | 100% SNSP 75%: € | $kgCO_2$ | 100% SNSP 95%: € | $kgCO_2$ | 25% SNSP 75%: € | $kgCO_2$ | 25% SNSP 95%: € | $kgCO_2$ |
|---|---|---|---|---|---|---|---|---|---|---|---|---|---|---|---|---|
| Electricity – Day | 9% | 22 | 324.5 | €773 | 1141 | €37 | €114 | €302 | €712 | 961 | €761 | 1105 | €178 | 241 | €190 | 277 |
| Electricity – Night | 2% | 10 | 324.5 | €352 | 1141 | €37 | €114 | €302 | €290 | 961 | €339 | 1105 | €73 | 241 | €85 | 277 |
| Natural Gas | 25% | 7.02 | 204.7 | €247 | 720 | €23 | €72 | €191 | €185 | 540 | €234 | 654 | €46 | 135 | €59 | 171 |
| Solid Fuel – Coal | 4% | 5.98 | 340.6 | €210 | 1197 | €38 | €120 | €317 | €149 | 1017 | €198 | 1161 | €37 | 254 | €49 | 290 |
| Solid Fuel – Peat | 5% | 7.09 | 355.9 | €249 | 1251 | €40 | €125 | €332 | €188 | 1071 | €237 | 1215 | €47 | 268 | €59 | 304 |
| Oil | 21% | 5.98 | 257 | €210 | 903 | €29 | €90 | €239 | €149 | 723 | €198 | 867 | €37 | 181 | €49 | 217 |
| Renewables – Biomass | 13% | 5.96 | 25 | €209 | 88 | €3 | €9 | €23 | €148 | | €197 | 52 | €37 | | €49 | 13 |
| Renewables – Solar | 6% | | | | | ## | € - | € - | | | | | | | | |
| Renewables – Geo | 15% | 10 | 162.25 | €352 | 570 | €15 | €57 | €151 | €290 | 390 | €339 | 534 | €73 | 98 | €85 | 134 |
| per household | | | 1 | €284 | 703 | €7 | €22 | €180 | €226 | 545 | €273 | 669 | €57 | 136 | €68 | 167 |
| Energy Cost and associated $CO_2$ emissions – Total (in millions) for various aggregrations | | | 100,000 | €28 | 70 | €2 | €7 | €19 | €23 | 55 | €27 | 67 | €6 | 14 | €7 | 17 |
| | | | 200,000 | €57 | 141 | €4 | €14 | €37 | €45 | 109 | €55 | 134 | | | | |
| | | | 461,000 | €131 | 324 | €10 | €32 | €86 | €104 | 251 | €126 | 308 | | | | |
| | | | 700,000 | €199 | 492 | €16 | €49 | €130 | €159 | 382 | €191 | 468 | | | | |
| | | | 901,188 | €256 | 633 | €20 | €63 | €168 | €204 | 491 | €246 | 603 | | | | |
| | | | 1,860,000 | €529 | 1307 | €42 | €131 | €346 | €421 | 1014 | €507 | 1244 | | | | |

| Fuel | 20% SNSP 75%: € | $kgCO_2$ | 20% SNSP 95%: € | $kgCO_2$ | 13% SNSP 75%: € | $kgCO_2$ | 13% SNSP 95%: € | $kgCO_2$ | 9% SNSP 75%: € | $kgCO_2$ | 9% SNSP 95%: € | $kgCO_2$ | 7% SNSP 75%: € | $kgCO_2$ | 7% SNSP 95%: € | $kgCO_2$ |
|---|---|---|---|---|---|---|---|---|---|---|---|---|---|---|---|---|
| Electricity – Day | €142 | 192 | €152 | 221 | €93 | 125 | €99 | 144 | €64 | 87 | €68 | 100 | €50 | 68 | €53 | 78 |
| Electricity – Night | €58 | 192 | €68 | 221 | €38 | 125 | €44 | 144 | €26 | 87 | €30 | 100 | €20 | 68 | €24 | 78 |
| Natural Gas | €37 | 108 | €47 | 137 | €24 | 70 | €30 | 89 | €17 | 49 | €21 | 61 | €13 | 38 | €16 | 45 |
| Solid Fuel – Coal | €30 | 203 | €40 | 232 | €19 | 132 | €26 | 151 | €13 | 91 | €18 | 104 | €10 | 71 | €14 | 51 |
| Solid Fuel – Peat | €38 | 214 | €47 | 243 | €24 | 139 | €31 | 158 | €17 | 96 | €21 | 109 | €13 | 75 | €17 | 85 |
| Oil | €30 | 145 | €40 | 174 | €19 | 94 | €26 | 113 | €13 | 65 | €18 | 78 | €10 | 51 | €14 | 61 |
| Renewables – Biomass | €30 | | €39 | 10 | €19 | | €26 | 7 | €13 | | €18 | 5 | €10 | | €14 | 9 |
| Renewables – Solar | | | | | | | | | | | | | | | | |
| Renewables – Geo | €58 | 78 | €68 | 107 | €38 | 51 | €44 | 69 | €26 | 35 | €30 | 45 | €20 | 27 | €24 | 37 |
| per household | €45 | 109 | €55 | 134 | €29 | 71 | €35 | 87 | €20 | 49 | €25 | 60 | €16 | 38 | €19 | 45 |
| 100,000 | | | | | | | | | | | | | | | | |
| 200,000 | 9 | 22 | 11 | 27 | | | | | | | | | | | | |
| 461,000 | | | | | €14 | 33 | €16 | 40 | | | | | | | | |
| 700,000 | | | | | | | | | €14 | 34 | €17 | 42 | | | | |
| 901,188 | | | | | | | | | | | | | €14 | 34 | €17 | 43 |
| 1,860,000 | | | | | | | | | | | | | | | | |

Average DHW fuel costs and $CO_2$ emissions per household

Household cost and $CO_2$ emission savings are modest, ranging from 6% or EUR 16/38kg$CO_2$/annum at the maximum aggregation size of 901,188 dwellings to 20% or EUR 57/136kg$CO_2$/annum at an aggregation of 100,000 dwellings

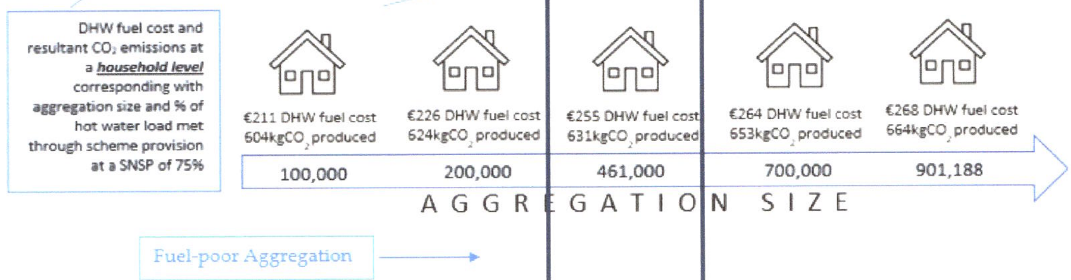


**Figure 8.** Energy cost and $CO_2$ emissions associated with delivered and deployed domestic hot water as a function of aggregation size (cross-referencing Figure 7).

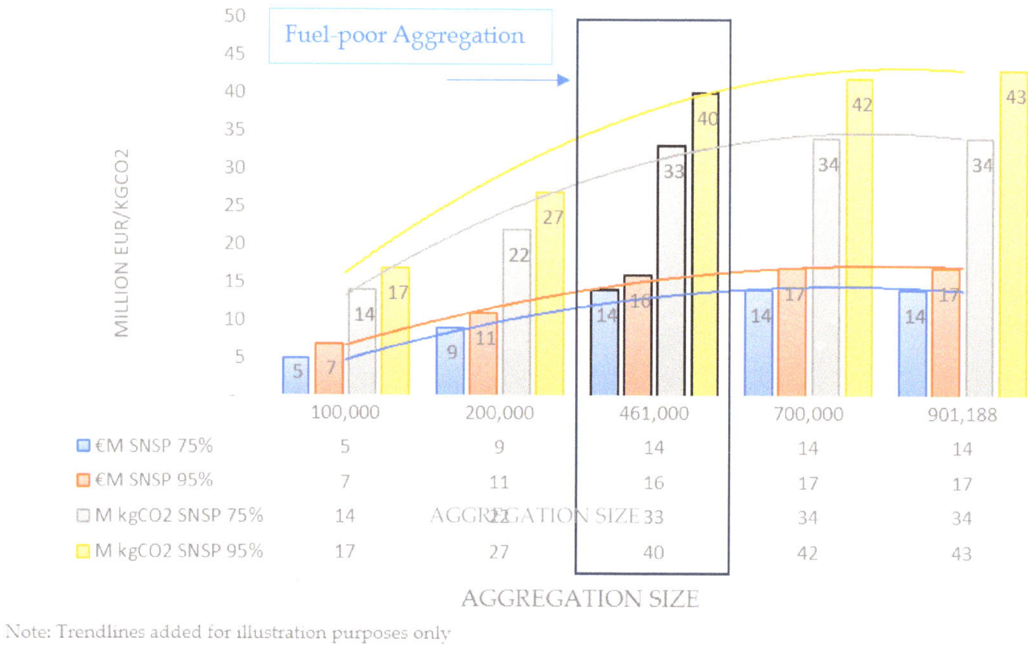


**Figure 9.** Potential displaced cost (to Irish householder) and $CO_2$ emissions (in millions) from dispatch-down powered EWHs for various aggregations.

Optimum aggregation size will alter depending on wind capacity, SNSP limits, and penetration of renewables into the grid. Referring to Table 1, the dispatch-down figure reduced from 1909 GWh in 2020 to 752 GWh in 2021 because 2021 was a less windy year than 2020, while SNSP limits transitioned from 65% to 75% in the same year. Notwithstanding year-to-year variations, it is important to note that dispatch-down due to energy balancing is forecasted to rise significantly in line with greater penetration of wind energy targeted by Ireland's ambitious climate policy [12,101].

Finally, considering payback on the technology to enable household participation in an EWHA scheme, the cost of installation for the largest aggregation possible in 2019 (901,188), estimated at EUR 222/device, would cost EUR 200 M with a simple payback of 14 years. If the fuel-poor population were aggregated (461,000), the theoretical simple payback would reduce to 6.4 years.

## 4. Discussion

It was found that while the use of domestic hot water as an energy store at scale could harness all unused wind-generated electricity in 2019, the value proposition in the form of "free" hot water to the householder diminishes as more tanks are aggregated. Notwithstanding the relatively small saving to the individual householder, it is understood that those in fuel poverty limit their use of hot water and, therefore, the receipt of a free tank of hot water weekly may be a significant boon to those who cannot afford to heat water in the first instance. Fuel poverty affects lone parents, women, people of colour, and the elderly the most and can therefore be seen as a consequence of, and contributor to, injustices linked to gender, ethnicity, and age [102]. Lone parents, overwhelmingly women, and their children are at a higher risk of fuel poverty than all other cohorts of society. It is hypothesised that the receipt of a weekly tank of hot water could become "bath night" for these households. It is also hypothesised that there may be meta benefits arising from a provision of "free" hot water such as positive feelings arising from inclusion in green initiatives/aggregation schemes or through relief of having one less challenge to overcome. Furthermore, it may be possible that these citizens become active actors in the energy system; this could be facilitated through push notifications sent to householders when there is a curtailment event. Such notifications could result in a change in behaviour. For example, the householder might wait to receive an alert to have a "bath night", thereby changing their behaviour to match the supply of renewable energy and, thus, allowing for greater accommodation of VRE and enhanced generation adequacy while creating potential for deferred or avoided network investment. Such outcomes are in line with the ambitions of the EU 2019 Clean Energy Package of using flexibility in grid planning as an alternative to system expansion, simultaneously allowing the effective and non-discriminatory participation of residential customers. Thus, in summary, while, at larger aggregations, the monetary savings from a household perspective are not significant, cumulatively, and as shown in Figure 9, the cost and resultant $CO_2$ savings to and from the residential sector at large are very significant. Therefore, a balance must therefore be struck between alleviating fuel poverty to the greatest extent possible (relatively small aggregation) and the greater carbon benefit realisable by society at large arising for a larger aggregation. Carbon cost savings realised by the state could be rediverted into energy infrastructure. To summarise, as shown in Figure 10, the value proposition of EWH aggregation as a flexible system asset for:

- **The householder or citizen**, who must be balanced against the value to;
- **The state**, in support of meeting the requirements of Article 32 of the EU Clean Energy Package in enabling householders to become actors in the energy system, while creating a citizen-owned energy system that alleviates fuel poverty, reduces reliance on imported fossil fuels, and lowers costs associated with carbon production, to the;
- **The climate**, through facilitating greater penetration of VRE, reducing waste through harnessing unused electricity, and reducing $CO_2$ emissions, and ultimately to;
- **The grid**, by offering *inter alia* frequency control during high-wind conditions along with the creation of new markets while facilitating a higher penetration of VRE through demand response.

Further research is required to establish, as shown in Figure 10, the nexus between a valuable level of demand and frequency control from, say, the perspective of Ireland's TSO, to facilitate a greater penetration of VRE and an increase in SNSP limits, along with the resultant carbon that could be potentially offset at grid level, which will likely favour a larger aggregation as more disaggregated demand load is available versus significantly reducing the impact of fuel poverty at a household level, which this study shows to favour a smaller aggregation.

From the perspective of the "state", favouring a "grid" over a "fuel-poor citizen-first" approach does not, on the face of it, support a just transition as much as taking a "fuel-poor citizen-first" approach might, notwithstanding that the state must balance supporting a just transition and reach the furthest-behind first while, as required by Article 32 of the EU 2019 clean energy package, allowing market access for all, not just fuel-poor, householders to create new markets. The state also seeks to reduce dependence on imported fossil fuel, as well as using demand flexibility to reduce investments required in energy infrastructure. It is evident, thus, that more research is required to establish an optimum solution/aggregation in an Irish but also social context.

Optimum aggregation size will also vary by country; for instance, in South Africa, EWHs contribute up to 40% of a household's total energy consumption [103], and as the households in South Africa rely almost exclusively on electricity for energy, EWHs constitute 30% to 50% of grid load at peak times, accounting for 7% of the country's daily energy requirements [82]. Since 88% of the country's electricity is generated from burning coal, the environmental water heating footprint is substantial [82]. Thus, each country needs to balance a different set of parameters to determine an optimum aggregation size.

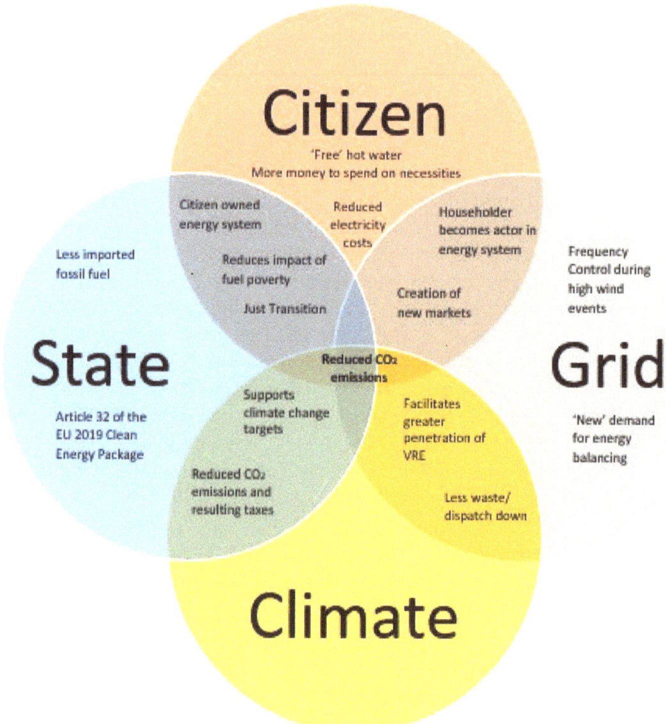


**Figure 10.** Value nexus of a shared, low-cost, flexible system asset capable of eliminating wind wastage.

*Limitations, Recommendations, and Future Research*

No metered DHW study has been carried out in the Irish context, it is recommended that a DHW be carried out to characterise DHW consumption patterns for "standard" and fuel-poor households in Ireland. This study should account for daily, weekly, seasonal, and annual patterns as well as occupant behavioural profiles.

This research assumes *a standardised water load*, assuming householders can afford to heat hot water in the first place; it is known that in fuel-poor homes, householders limit the generation of hot water to limit financial outlay, so there might be no savings realised as water is not being heated in the first instance; this is a limitation of this work that will be refined in future research.

Notwithstanding the simplified modelling approach adopted and understanding that wind- and weather-driven hot water consumption patterns vary year on year, this research provides a theoretical proof of concept that can be validated through a demonstrator EWHA aggregation scheme. As 77% of Irish households already have a DHW immersion, a low-cost retrofit of temperature sensor(s) enabling the accurate measurement of available heat capacity within the cylinder at any time, along with telecommunications to transmit a signal to an aggregator that the cylinder can accept heat input, would be all that would be required to enable a detailed technical and economic feasibility study. This generates an understanding of household (occupancy and system) variations and long-term EWHA scheme viability. A demonstrator project would also enable a comprehensive socio-technical impact analysis of an EWHA initiative, considering inter alia environment and economic benefits as well as the real impact and meta benefits for fuel-poor consumers as well as society at large.

## 5. Conclusions

Dispatch-down has traditionally been viewed as a "bad" thing, in terms of wasted energy; this research presents a case as to how advantages can be taken of unused wind-generated electricity, an increasing resource, to provide more services benefitting society.

This study finds that *all* wind energy curtailed currently can be redeployed into electro-thermal hot water storage, concluding that domestic hot water cylinders offer a currently underutilised large-scale, dispersed, and ubiquitous micro-capacitance which can readily provide some of the needed flexibility to electrical grids, facilitating higher levels of VRE, along with ready market access for residential consumers.

It is shown that redeploying all unused wind-generated electricity to heat hot water in homes, through the addition of low-cost controls on pre-existing infrastructure, has the potential to realise an average EUR 256 M saving to the Irish householder at large to offset 633 $tCO_2$ in 2019, realising a potential shadow carbon cost saving to the state of EUR 6M in 2022, rising to EUR 20 M in 2030 and EUR 162 M in 2050.

While the use of domestic hot water as an energy store at scale was capable of utilising all overnight curtailed wind energy (233 GWh) in 2019, the value proposition in the form of "free" hot water to the householder diminishes as more tanks are aggregated; indeed, at the maximum scale of aggregation, households would only realise a 7% saving on their water fuel bill.

A key finding of this research is that when sustainability matchmaking, there is an optimum aggregation size wherein near-maximum levels of dispatch-down and carbon are offset, concurrent with scheme participants receiving the maximum benefit, in this case hot water, from participating in the scheme. Optimum aggregation size will alter depending on wind capacity, SNSP limits, and penetration of renewables into the grid and will therefore vary from country to country.

As this research has a focus on alleviating fuel poverty, the extent to which a full tank of hot water can be provided for at least one night of the week so that important cleaning, bathing, and hygiene tasks within each home can be catered for on a weekly basis, also known as "bath night", was analysed. It was found that fuel-poor households in Ireland could be provided with a "free" full tank of hot water for 1 in every 3 weeks, capturing almost 90% of overnight curtailed wind energy in 2019, saving a theoretical EUR 29 (10% of their bill) and EUR 14 M across the fuel-poor aggregation, realising a potential carbon cost saving to the state of EUR 327 K in 2022, rising to EUR 1.3 M in 2030 and EUR 10.3 M in 2050, simultaneously alleviating while empowering such households to become electricity market participants.

Considering the primary infrastructure is already in place in most Irish households, secondary enabling infrastructure is a low-cost readily deployable retrofit solution. It is recommended that national policies be put in place to mandate the use of unused wind-generated electricity to provide new services and, in the interest of a just transition, that fuel-poor households be targeted to benefit initially. Due to privacy laws, it is not known publicly which houses are at risk of fuel poverty; therefore, such an initiative would need to be coordinated at the government level to include the Commission for Energy Regulation, the Sustainable Energy Authority of Ireland, the transmission system operator, the distribution system operator, and social housing providers as well as energy suppliers and wind generators.

**Author Contributions:** Conceptualisation, C.A. and R.O.; methodology, R.O. and C.A.; validation, R.O. and C.A.; formal analysis, R.O.; investigation, R.O.; resources, C.A.; data curation, R.O.; writing—original draft preparation, C.A.; writing—review and editing, C.A., R.O., and B.N.; visualisation, R.O. and C.A.; supervision, C.A.; project administration, C.A.; funding acquisition, C.A. All authors have read and agreed to the published version of the manuscript.

**Funding:** This research was partially supported by MaREI, the SFI Research Centre for Energy, Climate and Marine [Grant No: 12/RC/2303-P2].

**Informed Consent Statement:** Desktop study, no humans were involved.

**Data Availability Statement:** Data are contained within the article.

**Conflicts of Interest:** The authors declare no conflicts of interest.

## References

1. Bird, L.; Lew, D.; Milligan, M.; Carlini, E.M.; Estanqueiro, A.; Flynn, D.; Gomez-Lazaro, E.; Holttinen, H.; Menemenlis, N.; Orths, A.; et al. Wind and solar energy curtailment: A review of international experience. *Renew. Sustain. Energy Rev.* **2016**, *65*, 577–586. [CrossRef]
2. Permien, F.H.; Enevoldsen, P. Socio-technical constraints in German wind power planning: An example of the failed interdisciplinary challenge for academia. *Energy Res. Soc. Sci.* **2019**, *55*, 122–133. [CrossRef]
3. Eurostat. Renewable Energy in the EU. Eurostat Newsrelease. 2019. Available online: https://ec.europa.eu/eurostat/documents/2995521/9571695/8-12022019-AP-EN.pdf/b7d237c1-ccea-4adc-a0ba-45e13602b428 (accessed on 19 May 2024).
4. EEA. Share of Energy Consumption from Renewable Sources in Europe. 2024. Available online: https://www.eea.europa.eu/en/analysis/indicators/share-of-energy-consumption-from?activeAccordion=546a7c35-9188-4d23-94ee-005d97c26f2b (accessed on 19 May 2024).
5. US Government. The White House, FACT SHEET: President Biden Sets 2030 Greenhouse Gas Pollution Reduction Target Aimed at Creating Good-Paying Union Jobs and Securing U.S. Leadership on Clean Energy Technologies. 2020. Available online: https://www.whitehouse.gov/briefing-room/statements-releases/2021/04/22/fact-sheet-president-biden-sets-2030-greenhouse-gas-pollution-reduction-target-aimed-at-creating-good-paying-union-jobs-and-securing-u-s-leadership-on-clean-energy-technologies/ (accessed on 19 May 2024).
6. Yang, G.; Zhang, G.; Cao, D.; Zha, D.; Su, B. China's ambitious low-carbon goals require fostering city-level renewable energy transitions. *iScience* **2023**, *26*, 106263. [CrossRef] [PubMed]
7. GWEC. Global Wind Report. 2019. Available online: https://gwec.net/global-wind-report-2019/ (accessed on 4 May 2024).
8. Fernández, L. Global Cumulative Installed Wind Power Capacity 2001–2023, Statista. 2023. Available online: https://www.statista.com/statistics/268363/installed-wind-power-capacity-worldwide/#:~:text=The%20cumulative%20capacity%20of%20installed,about%20946%20gigawatts%20that%20year (accessed on 3 May 2024).
9. Fernández, L. Share of Electricity Generation from Wind Energy Sources Worldwide from 2010 to 2022, Statista. 2022. Available online: https://www.statista.com/statistics/1302053/global-wind-energy-share-electricity-mix/#:~:text=Wind%20energy%20sources%20accounted%20for,percent%20share%20a%20year%20earlier (accessed on 15 May 2024).
10. IEA. 'IEA Wind TCP Annual Report', Technology Collaboration Programme. 2019. Available online: https://iea-wind.org/wp-content/uploads/2022/11/Annual-Report-2019.pdf (accessed on 3 May 2024).
11. Hurtado, M.; Kerci, T.; Tweed, S.; Kennedy, E.; Kamaluddin, N.; Milano, F. Analysis of Wind Energy Curtailment in the Ireland and Northern Ireland Power Systems. In Proceedings of the IEEE Power & Energy Society General Meeting (PESGM), Orlando, FL, USA, 16–20 July 2023.
12. IWEA. *Saving Power, 70 by 30 Implementation Plan*; Irish Wind Energy Association: Dublin Ireland, 2020. Available online: https://www.energystorageireland.com/wp-content/uploads/2020/08/IWEA-ESI-Saving-Power-Report.pdf (accessed on 15 May 2024).
13. Xu, T.; Gao, W.; Qian, F.; Li, Y. The implementation limitation of variable renewable energies and its impacts on the public power grid. *Energy* **2022**, *239*, 121992. [CrossRef]
14. WindEurope. WindEurope Views on Curtailment of Wind Power and Its Links to Priority Dispatch. 2016. Available online: https://windeurope.org/wp-content/uploads/files/policy/position-papers/WindEurope-Priority-Dispatch-and-Curtailment.pdf (accessed on 5 February 2021).
15. Eirgrid. Annual Renewable Constraint and Curtailment Report 2019 in, Dublin, Ireland. 2019. Available online: https://www.eirgridgroup.com/site-files/library/EirGrid/Annual-Renewable-Constraint-and-Curtailment-Report-2019-V1.2.pdf (accessed on 15 May 2024).
16. Eirgrid. Shaping our Electricity Future, in, Dublin, Ireland. 2021. Available online: https://cms.eirgrid.ie/sites/default/files/publications/Shaping-Our-Electricity-Future-Technical-Report-Executive-Summary.pdf (accessed on 15 May 2024).
17. Newbery, D. National Energy and Climate Plans for the island of Ireland: Wind curtailment, interconnectors and storage. *Energy Policy* **2021**, *158*, 112513. [CrossRef]
18. Bashi, M.H.; De Tommasi, L.; Lyons, P. Electricity market integration of utility-scale battery energy storage units in Ireland, the status and future regulatory frameworks. *J. Energy Storage* **2022**, *55*, 105442. [CrossRef]
19. Eirgrid. Annual Renewable Energy Constraint and Curtailment Report 2021, in, Dublin, Ireland. 2021. Available online: https://www.eirgrid.ie/site-files/library/EirGrid/Annual-Renewable-Constraint-and-Curtailment-Report-2021-V1.0.pdf (accessed on 5 May 2024).
20. Eirgrid. Annual Renewable Energy Constraint and Curtailment Report 2018, in, Dublin, Ireland. 2018. Available online: https://www.eirgridgroup.com/site-files/library/EirGrid/Annual-Renewable-Constraint-and-Curtailment-Report-2018-V1.0.pdf (accessed on 5 May 2024).
21. Eirgrid. Annual Renewable Constraint and Curtailment Report 2022. Available online: https://cms-prd.eirgrid.dept.ie/annual-renewable-constraint-and-curtailment-report-2022 (accessed on 15 May 2024).

22. SEAI. Fuel Price Comparision, SEAI. 2023. Available online: https://www.seai.ie/data-and-insights/seai-statistics/key-statistics/prices/ (accessed on 5 May 2024).
23. Chen, H.; Chen, J.; Han, G.; Cui, Q. Winding down the wind power curtailment in China: What made the difference? *Renew. Sustain. Energy Rev.* **2022**, *167*, 112725. [CrossRef]
24. Yasuda, Y.; Bird, L.; Carlini, E.M.; Eriksen, P.B.; Estanqueiro, A.; Flynn, D.; Fraile, D.; Lázaro, E.G.; Martín-Martínez, S.; Hayashi, D.; et al. C-E (curtailment-Energy share) map: An objective and quantitative measure to evaluate wind and solar curtailment. *Renew. Sustain. Energy Rev.* **2022**, *160*, 112212. [CrossRef]
25. Alkhalidi, A.; Alqarra, K.; Abdelkareem, M.A.; Olabi, A.G. Renewable energy curtailment practices in Jordan and proposed solutions. *Int. J. Thermofluids* **2022**, *16*, 100196. [CrossRef]
26. Dumlao, S.M.G.; Ishihara, K.N. Impact assessment of electric vehicles as curtailment mitigating mobile storage in high PV penetration grid. *Energy Rep.* **2022**, *8*, 736–744. [CrossRef]
27. Agbonaye, O.; Keatley, P.; Huang, Y.; Bani-Mustafa, M.; Ademulegun, O.O.; Hewitt, N. Value of demand flexibility for providing ancillary services: A case for social housing in the Irish DS3 market. *Util. Policy* **2020**, *67*, 101130. [CrossRef]
28. Osunmuyiwa, O.O.; Peacock, A.D. What matters? Unlocking householders' flexibility towards cooling automation in India. *Environ. Innov. Soc. Transit.* **2023**, *47*, 100704. [CrossRef]
29. Niaz, H.; Liu, J.J.; You, F. Can Texas mitigate wind and solar curtailments by leveraging bitcoin mining? *J. Clean. Prod.* **2022**, *364*, 132700. [CrossRef]
30. Agbonaye, O.; Keatley, P.; Huang, Y.; Ademulegun, O.O.; Hewitt, N. Mapping demand flexibility: A spatio-temporal assessment of flexibility needs, opportunities and response potential. *Appl. Energy* **2021**, *295*, 117015. [CrossRef]
31. Nikolakakis, T.E.; Gutierrez, T.; Fernandez, L.; Kiviluoma, J.; Rissanen, S.; Lindross, T.J. Power System Flexibility for the Energy Transition: Part 2, IRENA FlexTool Methodology. 2018. Available online: https://www.irena.org/-/media/Files/IRENA/Agency/Publication/2018/Nov/IRENA_Power_system_flexibility_2_2018.pdf?la=en&hash=B7028E2E169CF239269EC9695D53276E084A29AE (accessed on 12 October 2022).
32. Dixon, D. *EU-SysFlex Qualification Trial Process for Technology Integration and Trialling of System Services D4.4*; Eirgrid: Dublin, Ireland, 2018; Available online: https://eu-sysflex.com/wp-content/uploads/2019/03/D4.4-Qualification-Trial-Process-for-technology-Integration-and-trialling-of-System-Services.pdf (accessed on 10 May 2024).
33. Ding, K.; Zhi, J. Chapter 6—Wind Power Peak-Valley Regulation and Frequency Control Technology. In *Large-Scale Wind Power Grid Integration*; Wang, N., Kang, C., Ren, D., Eds.; Academic Press: Oxford, UK, 2016; pp. 211–232.
34. Lee, M.P. Assessment of Demand Response and Advanced Metering, Office of Energy Policy and Innovation, Federal Energy Regulatory Commission. 2012. Available online: https://www.madrionline.org/wp-content/uploads/2013/09/Lee.pdf (accessed on 12 October 2022).
35. ANON. Market Design for Demand Side Reponse, in, Brussels, Belgium, 2015, European Network of Transviewed May 2024. Available online: https://eepublicdownloads.entsoe.eu/clean-documents/Publications/Position%20papers%20and%20reports/entsoe_pp_dsr_web.pdf (accessed on 5 May 2024).
36. Bloom, R.; Sugden, L. *Status and Opportunities for Flexible Residential E-HVAC*; DELTA-EE: Edinburgh, Scotland, 2021.
37. D'Ettorre, F.; Banaei, M.; Ebrahimy, R.; Pourmousavi, S.A.; Blomgren, E.M.V.; Kowalski, J.; Bohdanowicz, Z.; Łopaciuk-Gonczaryk, B.; Biele, C.; Madsen, H. Exploiting demand-side flexibility: State-of-the-art, open issues and social perspective. *Renew. Sustain. Energy Rev.* **2022**, *165*, 112605. [CrossRef]
38. Agbonaye, O.; Keatley, P.; Huang, Y.; Odiase, F.O.; Hewitt, N. Value of demand flexibility for managing wind energy constraint and curtailment. *Renew. Energy* **2022**, *190*, 487–500. [CrossRef]
39. Slowe, J. *Hot Water Tanks Get Smarter, Edinburgh, Scotland*; DELTA-EE: Edinburgh, Scotland, 2020.
40. Kepplinger, P.; Huber, G.; Preißinger, M.; Petrasch, J. State estimation of resistive domestic hot water heaters in arbitrary operation modes for demand side management. *Therm. Sci. Eng. Prog.* **2019**, *9*, 94–109. [CrossRef]
41. Arteconi, A.; Hewitt, N.J.; Polonara, F. State of the art of thermal storage for demand-side management. *Appl. Energy* **2012**, *93*, 371–389. [CrossRef]
42. Elamari, K.I. Using Electric Water Heaters (EWHs) for Power Balancing and Frequency Control in PV-Diesel Hybrid Mini-Grids. Master's Thesis, Concordia University, Montreal, QC, Canada, 2011.
43. Cooper, D.; Cronje, W. Autonomous water heater control for load regulation on smart grids. In Proceedings of the IEEE International Energy Conference (ENERGYCON), Leuven, Belgium, 4–8 April 2016; pp. 1–6.
44. Xu, Z.; Diao, R.; Lu, S.; Lian, J.; Zhang, Y. Modeling of electric water heaters for demand response: A baseline PDE model. *IEEE Trans. Smart Grid* **2014**, *5*, 2203–2210. [CrossRef]
45. Epiard, X.; Prestat, B. *EDF Experience with Storage in EU and US Markets*; World Future Energy Summit: Abu Dhabi, UAE, 2017; Available online: www.issuu.com (accessed on 3 May 2024).
46. Marnell, K.; Eustis, C.; Bass, R.B. Resource study of large-scale electric water heater aggregation. *IEEE Open Access J. Power Energy* **2020**, *7*, 82–90. [CrossRef]
47. Government of Oregon. Actively Adapting to the Changing Electricitysector. 2018. Available online: https://www.oregon.gov/puc/utilities/Documents/SB978LegislativeReport-ExecSummary.pdf (accessed on 8 April 2021).
48. Mabina, P.; Mukoma, P.; Booysen, M.J. Sustainability matchmaking: Linking renewable sources to electric water heating through machine learning. *Energy Build.* **2021**, *246*, 111085. [CrossRef]

49. Laurent, J.C.; Desaulniers, G.; Malhamé, R.P.; Soumis, F. A column generation method for optimal load management via control of electric water heaters. *IEEE Trans. Power Syst.* **1995**, *10*, 1389–1400. [CrossRef]
50. Sepulveda, A.; Paull, L.; Morsi, W.G.; Li, H.; Diduch, C.P.; Chang, L. A novel demand side management program using water heaters and particle swarm optimization. In Proceedings of the 2010 IEEE Electrical Power & Energy Conference, Halifax, NS, Canada, 25–27 August 2010; pp. 1–5.
51. Kepplinger, P.; Huber, G.; Petrasch, J. *Demand Side Management via Autonomous Control-Optimization and Unidirectional Communication with Application to Resistive Hot Water Heaters*; ENOVA: Eisenstadt, Austria, 2014; Volume 8.
52. Kara, E.C.; Berges, M.; Krogh, B.; Kar, S. Using smart devices for system-level management and control in the smart grid: A reinforcement learning framework. In Proceedings of the 2012 IEEE Third International Conference on Smart Grid Communications (SmartGridComm), Tainan, Taiwan, 5–8 November 2012; pp. 85–90.
53. Diduch, C.; Shaad, M.; Errouissi, R.; Kaye, M.E.; Meng, J.; Chang, L. Aggregated domestic electric water heater control-building on smart grid infrastructure. In Proceedings of the 7th International Power Electronics and Motion Control Conference, Harbin, China, 2–5 June 2012; pp. 128–135.
54. Al-Jabery, K.; Wunsch, D.C.; Xiong, J.; Shi, Y. A novel grid load management technique using electric water heaters and Q-learning. In Proceedings of the 2014 IEEE International Conference on Smart Grid Communications (SmartGridComm), Venice, Italy, 3–6 November 2014; pp. 776–781.
55. Guo, Y.; Mahdavi, N. Machine learning method for day classification to understand thermostatically controlled load demand. In Proceedings of the 2017 IEEE Innovative Smart Grid Technologies-Asia (ISGT-Asia), Auckland, New Zealand, 4–7 December 2017; pp. 1–5.
56. Lu, N.; Katipamula, S. Control strategies of thermostatically controlled appliances in a competitive electricity market. In Proceedings of the IEEE Power Engineering Society General Meeting, San Francisco, CA, USA, 12–16 June 2005; pp. 202–207.
57. Liu, M.; Shi, Y. Model predictive control of aggregated heterogeneous second-order thermostatically controlled loads for ancillary services. *IEEE Trans. Power Syst.* **2015**, *31*, 1963–1971. [CrossRef]
58. Zong, Y.; Böning, G.M.; Santos, R.M.; You, S.; Hu, J.; Han, X. Challenges of implementing economic model predictive control strategy for buildings interacting with smart energy systems. *Appl. Therm. Eng.* **2017**, *114*, 1476–1486. [CrossRef]
59. Qu, Z.; Xu, C.; Ma, K.; Jiao, Z. Fuzzy neural network control of thermostatically controlled loads for demand-side frequency regulation. *Energies* **2019**, *12*, 2463. [CrossRef]
60. Diao, R.; Lu, S.; Elizondo, M.; Mayhorn, E.; Zhang, Y.; Samaan, N. Electric water heater modeling and control strategies for demand response. In Proceedings of the 2012 IEEE Power and Energy Society General Meeting, San Diego, CA, USA, 22–26 July 2012; pp. 1–8.
61. Gelažanskas, L.; Gamage, K. Distributed Energy Storage Using Residential Hot Water Heaters. *Energies* **2016**, *9*, 127. [CrossRef]
62. Nehrir, M.H.; Jia, R.; Pierre, D.A.; Hammerstrom, D.J. Power management of aggregate electric water heater loads by voltage control. In Proceedings of the 2007 IEEE Power Engineering Society General Meeting, Tampa, FL, USA, 24–28 June 2007; pp. 1–6.
63. Kondoh, J.; Lu, N.; Hammerstrom, D.J. An evaluation of the water heater load potential for providing regulation service. In Proceedings of the 2011 IEEE Power and Energy Society General Meeting, Detroit, MI, USA, 24–29 July 2011; pp. 1–8.
64. European Commission. Realising Value from Electricity Markets with Local Smart Electric Thermal Storage Technology, EU Commission. 2018. Available online: https://cordis.europa.eu/article/id/241249-smart-storage-heaters-benefit-households-and-business (accessed on 12 October 2022).
65. Bloess, A.; Schill, W.-P.; Zerrahn, A. Power-to-heat for renewable energy integration: A review of technologies, modeling approaches, and flexibility potentials. *Appl. Energy* **2018**, *212*, 1611–1626. [CrossRef]
66. Andrade, B.B.; Rathie, P.N. Fitting asymmetric bimodal data with selected distributions. *J. Stat. Comput. Simul.* **2016**, *86*, 3205–3224. [CrossRef]
67. EirGrid 2016–2018, Power Off and Save, EirGrid. 2021. Available online: http://www.eirgridgroup.com/how-the-grid-works/power-off-save/ (accessed on 6 February 2021).
68. Ahern, C.; Griffiths, P.; O'Flaherty, M. State of the Irish Housing stock—Modelling the heat losses of Ireland's existing detached rural housing stock & estimating the benefit of thermal retrofit measures on this stock. *Energy Policy* **2013**, *55*, 139–151.
69. Barrett, M.; Farrell, N.; Roantree, B. *Energy Poverty and Deprivation in Ireland*; Economic & Social Research Institute: Dublin, Ireland, 2022.
70. Energy Savings Trust. Measurement of Domestic Hot Water Consumption in Dwellings. 2008. Available online: https://assets.publishing.service.gov.uk/government/uploads/system/uploads/attachment_data/file/48188/3147-measure-domestic-hot-water-consump.pdf (accessed on 10 January 2023).
71. López-Rodríguez, M.A.; Santiago, I.; Trillo-Montero, D.; Torriti, J.; Moreno-Munoz, A. Analysis and modeling of active occupancy of the residential sector in Spain: An indicator of residential electricity consumption. *Energy Policy* **2013**, *62*, 742–751. [CrossRef]
72. Gerin, O.; Bleys, B.; de Cuyper, K. Seasonal variation of hot and cold water consumption in apartment buildings. 2014. Available online: https://www.irbnet.de/daten/iconda/CIB_DC27564.pdf (accessed on 5 January 2023).
73. Zhou, S.L.; McMahon, T.A.; Walton, A.; Lewis, J. Forecasting operational demand for an urban water supply zone. *J. Hydrol.* **2002**, *259*, 189–202. [CrossRef]
74. Gato, S.; Jayasuriya, N.; Roberts, P. Forecasting Residential Water Demand: Case Study. *J. Water Resour. Plan. Manag.* **2007**, *133*, 309–319. [CrossRef]

75. de Santiago, J.; Rodriguez-Villalón, O.; Sicre, B. The generation of domestic hot water load profiles in Swiss residential buildings through statistical predictions. *Energy Build.* **2017**, *141*, 341–348. [CrossRef]
76. George, D. Mech 6340 Project High Time-Step Resolution Measured Domestic Hot Water Consumption Profiles of Single Detached Canadian Homes, Dalhousie University, 13 February 2015. Available online: https://dalspace.library.dal.ca/bitstream/handle/10222/72580/George-Dane-MASc-Thesis-MECH-December-2016.pdf?sequence=5 (accessed on 3 May 2024).
77. Blokker, E.J.M.; Vreeburg, J.H.G.; van Dijk, J.C. Simulating Residential Water Demand with a Stochastic End-Use Model. *J. Water Resour. Plan. Manag.* **2010**, *136*, 19–26. [CrossRef]
78. Sborz, J.; Kalbusch, A.; Henning, E. A Review on Domestic Hot Water Consumption in Social Housing. *Water* **2022**, *14*, 2699. [CrossRef]
79. Jack, M.W.; Suomalainen, K.; Dew, J.J.W.; Eyers, D. A minimal simulation of the electricity demand of a domestic hot water cylinder for smart control. *Appl. Energy* **2018**, *211*, 104–112. [CrossRef]
80. Roux, M.; Booysen, M.J. Use of smart grid technology to compare regions and days of the week in household water heating. In Proceedings of the 2017 International Conference on the Domestic Use of Energy (DUE), Cape Town, South Africa, 4–5 April 2017; pp. 276–283.
81. Fairey, P.; Parker, D. *A Review of Hot Water Draw Profiles Used in Performance Analysis of Residential Domestic Hot Water Systems*; Florida Solar Energy Center: Cocoa, FL, USA; University of Central Florida: Orange County, FL, USA, 2004.
82. Ritchie, M.J.; Engelbrecht, J.A.A.; Booysen, M.J. A probabilistic hot water usage model and simulator for use in residential energy management. *Energy Build.* **2021**, *235*, 110727. [CrossRef]
83. Neu, O.; Oxizidis, S.; Flynn, D.; Finn, D. Utilising time of use surveys to predict domestic hot water consumption and heat demand profiles of residential building stocks. *Br. J. Environ. Clim. Change* **2016**, *6*, 77–89. [CrossRef]
84. Jordan, U.; Vajen, K. DHWcalc: Program to generate domestic hot water profiles with statistical means for user defined conditions. In Proceedings of the ISES Solar World Congress, Orlando, FL, USA, 6–12 August 2005.
85. Fischer, D.; Toral, T.R.; Lindberg, K.B.; Wille-Haussmann, B.; Madani, H. Investigation of Thermal Storage Operation Strategies with Heat Pumps in German Multi Family Houses. *Energy Procedia* **2014**, *58*, 137–144. [CrossRef]
86. Ferrantelli, A.; Ahmed, K.; Pylsy, P.; Kurnitski, J. Analytical modelling and prediction formulas for domestic hot water consumption in residential Finnish apartments. *Energy Build.* **2017**, *143*, 53–60. [CrossRef]
87. Widén, J.; Lundh, M.; Vassileva, I.; Dahlquist, E.; Ellegård, K.; Wäckelgård, E. Constructing load profiles for household electricity and hot water from time-use data—Modelling approach and validation. *Energy Build.* **2009**, *41*, 753–768. [CrossRef]
88. Popescu, D.; Serban, E. Simulation of domestic hot-water consumption using time-series models. In Proceedings of the 6th IASME/WSEAS International Conference On Heat Transfer, Thermal Engineering and Environment, Rhodes, Greece, 20–22 August 2008; pp. 20–22.
89. Kepplinger, P.; Huber, G.; Petrasch, J. Field testing of demand side management via autonomous optimal control of a domestic hot water heater. *Energy Build.* **2016**, *127*, 730–735. [CrossRef]
90. Kamal, S. Energy Follow Up Survey: Fuel Poverty. 2018. Available online: https://assets.publishing.service.gov.uk/government/uploads/system/uploads/attachment_data/file/1018737/efus-fuel-poverty.pdf (accessed on 10 January 2023).
91. SEAI. Energy in the Residential Sector—2018 Report. 2018. Available online: https://www.seai.ie/resources/publications/Energy-in-the-Residential-Sector-2018-Final.pdf (accessed on 10 January 2024).
92. Element Energy. Bottom-up Analysis of Fuel Poverty in Ireland. 2015. Available online: https://assets.gov.ie/76593/331e2543-a5a1-4057-a902-d0e2a8827d50.pdf (accessed on 12 March 2022).
93. Nestor, R. *Social Impact Assessment—SEAI Programmes Targeting Energy Poverty*; Department of Public Expenditure and Reform, Ed.; Irish Government Economic and Evaluation Service: Dublin, Ireland, 2020.
94. CSO 2019, Survey on Income and Living Conditions (SILC) 2019, CSO, viewed 29 March 2021 2020. Available online: https://www.cso.ie/en/releasesandpublications/ep/p-silc/surveyonincomeandlivingconditionssilc2019/ (accessed on 4 April 2024).
95. CSO. Census of Population—Preliminary Results, Central Statistics Office. 2022. Available online: https://www.cso.ie/en/releasesandpublications/ep/p-cpr/censusofpopulation2022-preliminaryresults/housing/#:~:text=A%20total%20of%202,124,590%20permanent,(-9%25)%20to%20166,752 (accessed on 24 September 2022).
96. O'Doherty, C. Families Live in Constant Cold as Energy Poverty Forces Heat Rationing. The Irish Independent, Independent New and Media: Dublin, Ireland, 2021.
97. Gold Standard. Carbon Pricing: Setting an Internal Price on Carbon. 2021. Available online: https://www.goldstandard.org/blog-item/carbon-pricing-setting-internal-price-carbon (accessed on 15 May 2024).
98. Government of Ireland. *An Overview of Carbon Pricing*; Parliamentarty Budget Office, Ed.; Houses of the Oireachtas: Dublin, Ireland, 2019; Available online: https://data.oireachtas.ie/ie/oireachtas/parliamentaryBudgetOffice/2019/2019-07-04_an-overview-of-carbon-pricing_en.pdf (accessed on 15 May 2024).
99. Government of Ireland. Public Spending Code A Guide to Economic Appraisal: Carrying Out a Cost Benefit Analysis. In *Public Spending Code, A Guide to Evaluating, Planning and Managing Public Investment*; Government of Ireland: Dublin, Ireland, 2019. Available online: https://www.gov.ie/en/publication/public-spending-code/ (accessed on 12 January 2022).
100. European Commission. EU Emissions Trading System (EU ETS). 2005. Available online: https://ec.europa.eu/clima/policies/ets_en (accessed on 12 January 2021).

101. Government of Ireland 2023, Climate Action Plan, Department of Environment, Climate and Communications. Available online: https://www.gov.ie/en/publication/7bd8c-climate-action-plan-2023/ (accessed on 15 January 2023).
102. Lawlor, D.; Visser, A. Energy Poverty in Ireland. 2022. Available online: https://data.oireachtas.ie/ie/oireachtas/libraryResearch/2022/2022-03-04_l-rs-note-energy-poverty-in-ireland_en.pdf (accessed on 15 May 2024).
103. Hohne, P.A.; Kusakana, K.; Numbi, B.P. A review of water heating technologies: An application to the South African context. *Energy Rep.* **2019**, *5*, 1–19. [CrossRef]

sustainability

MDPI

Article

# Energy Performance Analysis of the Renovation Process in an Italian Cultural Heritage Building

**Nikolaos Ziozas [1], Angeliki Kitsopoulou [1], Evangelos Bellos [1], Petros Iliadis [1], Dimitra Gonidaki [1], Komninos Angelakoglou [1], Nikolaos Nikolopoulos [1], Silvia Ricciuti [2] and Diego Viesi [2,*]**

[1] Centre for Research and Technology Hellas (CERTH), Chemical Process and Energy Resources Institute (CPERI), Aigialeias 52, 15125 Marousi, Greece; n.ziozas@certh.gr (N.Z.); a.kitsopoulou@certh.gr (A.K.); e.bellos@certh.gr (E.B.); iliadis@certh.gr (P.I.); d.gonidaki@certh.gr (D.G.); angelakoglou@certh.gr (K.A.); n.nikolopoulos@certh.gr (N.N.)

[2] Fondazione Bruno Kessler (FBK), Center for Sustainable Energy (SE), Via Sommarive 18, 38123 Povo, TN, Italy; sricciuti@fbk.eu

* Correspondence: viesi@fbk.eu

**Abstract:** Renovating buildings with cultural heritage significance is an important step toward achieving sustainability in our cities. The benefits are not only energy-related but also encompass social aspects that make these renovations a high apriority. The present work investigates the renovation process of a cultural heritage building in the Municipality of Trento in Italy, specifically focusing on achieving energy savings and renewable energy integration by implementing various renovation actions. These renovation actions include improvements to the building envelope, such as roof insulation and window replacements. Additionally, the renovation actions for active systems involve the installation of a ground-source heat pump for heating/cooling coupled with a borehole thermal energy storage system, which is an innovative technology for the renovation of cultural heritage buildings. The electrical systems of the building are upgraded through the addition of standard rooftop photovoltaics, innovative building-integrated photovoltaics (shingles), and the installation of an LED lighting system. The baseline and the renovation scenarios are studied using the dynamic simulation tool INTEMA.building, written in the programming language Modelica. This tool simulates both the building envelope and the energy systems with a high level of detail, using advanced control systems and adjustable time steps. According to the simulation analysis, the primary energy demand is reduced by 30.49%, the final energy demand by 36.74%, and the net electricity demand by 8.72%. Results from this study can be useful to interested stakeholders (e.g., building owners, architects, construction companies, public agents, and urban planners) dealing with the renovation of cultural heritage and protected buildings. Also, the results can be exploited for estimating energy savings by applying advanced renovation strategies for cultural heritage buildings.

**Keywords:** renovation; cultural heritage; geothermal heat pump; historical building; energy retrofit

**Citation:** Ziozas, N.; Kitsopoulou, A.; Bellos, E.; Iliadis, P.; Gonidaki, D.; Angelakoglou, K.; Nikolopoulos, N.; Ricciuti, S.; Viesi, D. Energy Performance Analysis of the Renovation Process in an Italian Cultural Heritage Building. *Sustainability* **2024**, *16*, 2784. https://doi.org/10.3390/su16072784

Academic Editors: Reza Daneshazarian, Elena Lucchi, Tianyi Chen and Wen Zhang

Received: 30 January 2024
Revised: 7 March 2024
Accepted: 21 March 2024
Published: 27 March 2024



## 1. Introduction

The decarbonization of the incumbent worldwide energy regime due to the climate crisis's escalation over the last decade highlights the contribution of the building sector to the associated total energy consumption and greenhouse gas emissions as well [1]. Characteristically, the building sector is responsible for approximately 15% of the world's total $CO_2$ emissions and about one-third of the total global final energy consumption [2]. Respectively, buildings account for 40% of the total energy consumption in the European Union and 36% of energy-related greenhouse gas emissions [3]. For that reason, the European Union has set building decarbonization at the forefront of its energy policy agenda, acknowledging energy efficiency measures as a core pillar towards a net-zero buildings pathway [4]. Under the energy efficiency obligation's schemes, European Union member countries should follow an average annual energy savings rate of 1.49% from 2024 to 2030 and an annual renovation rate of 3% for their public buildings [3]. Additionally, the integration of renewable energy sources, as well as the commercialization of high-technology solar energy systems would positively contribute to the decarbonization of energy systems for the future of the European community [5].

However, the outdated stock of the European Union's buildings challenges the commitment to energy efficiency directives, as over 60% of European buildings were built before 1980 [6]. Over 70% of the building stock is aged in the southern European countries (i.e., Italy, Greece, Spain, Cyprus, etc.) [6]. This status quo indicates the need for applied energy retrofitting interventions to enhance the poor energy performance of these buildings due to their outdated energy systems, structural elements, and materials [7]. Nonetheless, over 25% of these European aged buildings constitute historical or so-called cultural heritage buildings, challenging the integration of energy-efficient interventions [8].

Therefore, it is crucial to investigate proper sustainable technologies and retrofitting interventions to improve their energy performance without compromising their cultural and social idiosyncrasies [9].

### 1.1. Literature Review

Typical energy renovation interventions for historical buildings concern the insulation of roofs and walls, the installation of renewable energy technologies (i.e., solar panels, geothermal pumps, etc.), the integration of upgraded heating, ventilation, and air conditioning (HVAC) systems and respective control systems, and the replacement of lighting and electrical equipment [10]. Nevertheless, the unique feature of historical buildings' renovation is the accomplishment of tradeoffs between energy-efficient/renewable interventions and the preservation of their cultural and historical heritage. Arumägi et al. [11] implemented a renovation strategy consisting of insulation and HVAC interventions in a historical wooden apartment, resulting in a primary energy reduction of up to 65% and economic energy savings of up to 50%. Ardente et al. [12] found that building envelope enhancement was the most energy-efficient measure in their retrofit of six historical buildings. Another interesting retrofitting intervention implemented by Fonseca et al. [13] concerned the integration of an intelligent lighting system in public schools, while Cohen et al. [14] proposed that ceiling and wall insulation are more cost-effective than the replacement of windows for an energy consumption metering system during the retrofit of a single-family building. The installation of ground-source heat pumps (GSHPs) was studied by Boait et al. [15] as an energy-efficient action in retrofitting historical buildings in the United Kingdom. Their results show the heat pumps' poor seasonal performance compared to similar European studies, emphasizing the importance of the building's thermal behavior and characteristics. Hens [16] concluded that enriching the building envelope and ventilation system performs better in terms of energy savings than the installation of solar boilers and PV panels in a retrofit of a historical building built in 1957.

Gravagnuolo et al. [17] approached the retrofit of four historic buildings in Salerno from a transdisciplinary viewpoint by developing a multicriteria decision-making framework to evaluate the examined renovation solutions. This innovative approach led to the co-production of knowledge among designers, stakeholders, and citizens, resulting in a democratic decision-making process and respecting the cultural heritage of the community as a whole. De Toro et al. [18] focus on evaluating design-supporting tools, such as geographic information systems, for the restoration of cultural heritage buildings. At the same time, the Green Building Council protocol in Italy proposes an evaluation process for such buildings, including their cultural status and life-cycle analysis during the retrofit intervention phase [17].

However, implementing the above energy-efficient/renewable measures requires extensive energy and architectural analyses of historical buildings through energy modeling and simulation tools, as the majority of these buildings were built from local resources and materials lacking detailed drawings and studies [19]. For example, Ascione et al. [20] model and simulate a historical building in Benevento, Italy using EnergyPlus software

(Version 7.0), providing viable energy retrofit solutions. Similarly, Harrestrup and Svendsen [21] evaluate the energy behavior of a multi-story brick building in Copenhagen by simulating the integration of exterior wall insulation before and after the renovation. Sahin et al. [22] ran simulations for a 19th-century educational center in Turkey, comparing passive and active energy-efficient measures. They conclude that energy savings of up to 34% can be reached without compromising the building's heritage value. On the other hand, Moran et al. [8] performed steady-state calculations using the Passive House Planning Package version 7.0 software for Georgian dwellings in Bath, the United Kingdom, questioning the accuracy of the occupants' thermal comfort with the associated tested heating patterns. Likewise, Zagorskas et al. [23] point out the mismatch of energy simulation software for historical buildings between actual and simulated thermal performances due to the lack of modeling accuracy for the building envelope. From another perspective, Serano et al. [24] implemented a life-cycle assessment of a historical building in Denmark, comparing the most suitable option between a restoration scenario and a renovation scenario. Their research concludes that the restoration scenario performs slightly better in terms of both energy savings and architectural style [24].

### 1.2. Research Contribution

The previous survey of the literature indicates a great interest in the renovation of buildings with cultural heritage importance due to the numerous energy- and society-related benefits. In this direction, the present work examines a deep renovation strategy of a building with cultural heritage status located in Trento, Italy [25]. The examined location is of great importance as the province of Trento possesses 700 public heritage buildings, of which 166 belong to the public building stock with a registered consumption higher than 45 $kWh/m^3y$ [26]. Moreover, the geographical location of this retrofit scenario raises scientific interest, as the research on Italian historical buildings is considered to be the most active of its kind worldwide [9]. This can be attributed to the fact that 31% of Italy's total structures were constructed before 1945, and approximately 12% were constructed before 1919 [27]. This work's novelty lies in its use of an innovative, dynamic simulation tool that allows a thorough investigation of suitable energy-efficient/renewable solutions in a building with a substantial floor area (about 5400 $m^2$). Specifically, an analysis is conducted using the INTEMA.building tool, a simulation tool developed in the Modelica programming language [28], using the Dymola interface [29]. This tool presents numerous advancements because it precisely models the control systems of a building, provides detailed simulations of both the building envelope and the energy systems, and effectively solves problems with high accuracy due to the adjustable time step of the solving procedure. Also, the value of this work lies in it being based on the renovation of a cultural heritage building with some special characteristics like a high floor area and geometrical/space limitations regarding the installation of new renovation technologies. Furthermore, the most critical limitation of this research is the conservation of the building's historical idiosyncrasies and architectural features throughout the renovation process. Another novel aspect lies in the incorporation of an innovative ground-source heat pump in the renovation, as the majority of the incumbent renovations in cultural heritage buildings are based mostly on the application of technologies like solar thermal collectors and photovoltaics [30]. The results of this work indicate the energy-related benefits of the renovation process and can serve as a guideline for future renovations in cultural heritage buildings, especially for buildings with special characteristics like the studied building.

## 2. Materials and Methods

This section includes a basic description of the studied building's baseline and renovation scenarios. The data for the building envelope and the energy systems are given, and the simulation strategy is clearly described.

### 2.1. The Studied Building—Baseline Scenario

The examined building, located and managed by the Municipality of Trento (Italy), is a cultural heritage building that serves as a pivotal center of cultural and recreational activities in the city. This building's use has changed several times over the centuries: it went from being a monastery (1235–1810) to being a hospital (1810–1970) to being a university (1970–2018) until its current use as one of the most important centers of cultural and recreational activities in the city. Specifically, the building is located in the Santa Chiara District of Trento, with a latitude of 46.063281° and a longitude of 11.124574°, with a 10° counter-clockwise deviation from the direction of north. It is partially adjacent to a gym and an auditorium on the east side and to a church on the west side. Figure 1 depicts both the entrance of the examined building and a top view of the building retrieved from a satellite photograph from Google Earth, showcasing the azimuth angles of its external walls. The design drawings and construction plans of the building, including a top view and three different cross sections, are demonstrated in Figure 2.

(a) (b)

**Figure 1.** Photograph of the examined building before the renovation: (**a**) entrance; (**b**) depiction from a satellite image with azimuth orientation angles.

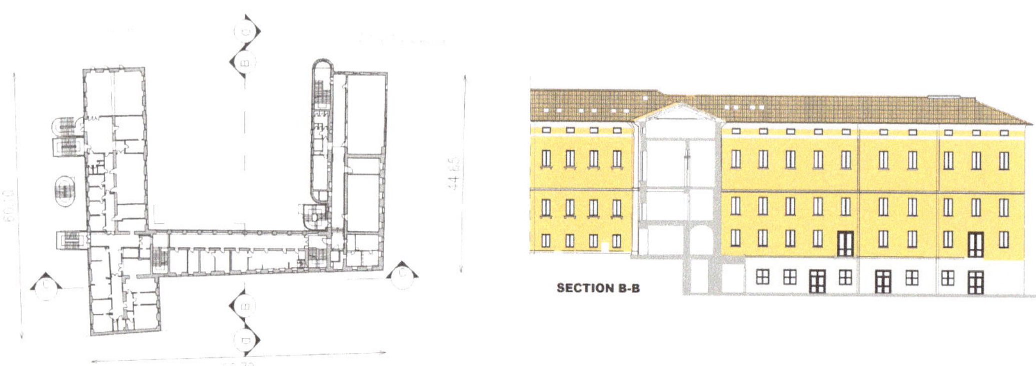


(a) (b)

**Figure 2.** *Cont.*

(c) (d)

**Figure 2.** Drawings of the examined building before the renovation: (**a**) top view with basic dimensions; (**b**) B–B cross section, (**c**) C–C cross section, (**d**) D–D cross section.

#### 2.1.1. Building Envelope Description

The building consists of five floors and a partially heated basement, with a mean total height of 20.3 m, while the mean internal height of each floor equals 3 m. Additionally, the building's conditioned floor area is calculated at 5400 m$^2$, 17% of which serves as a guesthouse. The thermal envelope of the building consists of the part of the basement that is heated, the ground floor, and four additional floors above. The basement's ground slab is in contact with the ground and has a nominal U-value of 2.2 W/m$^2$K, whereas the basement's ceiling is characterized by a nominal U-value of 1.8 W/m$^2$K. The ground floor is in contact with the basement and the two are separated by the basement ceiling. The external and basement walls are denoted by a U-value of 0.7 W/m$^2$K, while the roof construction's total thermal transmittance value is estimated at 0.5 W/m$^2$K. The building's glazing system consists of double-pane windows with a total U-value of 2.5 W/m$^2$K, including the glazing, frame, and thermal bridge effects. The absorbance of the opaque structural elements was estimated at 60% and the emittance was estimated at 80%, which are typical values according to [31]. Regarding the windows, the emittance value is estimated at 90% and the g-value is estimated to be 84%. Table 1 includes basic information regarding the detailed description of the examined building envelope.

**Table 1.** Basic data for the description of the examined building's envelope.

| Parameter | Value |
|---|---|
| **Opaque structural elements** | |
| U-value of the external walls (W/m$^2$K) | 0.70 |
| U-value of the roof (W/m$^2$K) | 0.50 |
| U-value of the basement ceiling (W/m$^2$K) | 1.80 |
| U-value of the ground slab (W/m$^2$K) | 2.20 |
| Slope of the roof (°) | 22 |
| Absorbance (%) | 60 |
| Emittance (%) | 80 |
| Slope of the roof (°) | 22 |
| **Transparent structural elements** | |
| U-value of the window (W/m$^2$K) | 2.5 |
| g-value of the window | 0.84 |
| Emittance (%) | 80 |

Table 2 gives the areas of the building's structural components for every orientation and the adjacency of the building's external walls. Specifically, the external walls are in contact with either ambient air, a neighboring building, or the ground. The sum of these three cases gives the total area of the external walls for each orientation. It is mentioned that the present building has windows in all directions, with the largest opening areas being located on the east- and south-oriented façades.

**Table 2.** Geometrical calculations of the external walls, roofs, and windows.

| Summary Table of Areas [m$^2$] | | | | | | |
|---|---|---|---|---|---|---|
| | Roofs | Windows | External Walls | External Walls in Contact with: | | |
| Orientation | Total Area | Total Area | Total Area | Ambient Air | Neighboring Building | Ground |
| North ($\gamma = -10°$) | 1157 | 134 | 1916 | 1695 | 0 | 221 |
| South ($\gamma = 170°$) | 1122 | 215 | 1704 | 1388 | 95 | 221 |
| East ($\gamma = -100°$) | 521 | 182 | 1145 | 907 | 135 | 103 |

Table 3 includes operational parameters correlated with the building's internal loads and aeration loads. The specific loads are given per conditioned area, equal to 5400 m$^2$. Additionally, the mean operating factors were estimated according to the usual occupancy of the building. Specifically, the daily operating profile of the building follows the profile illustrated in Figure 3a, which corresponds to the percentage of the maximum electricity load throughout a typical day [32]. Regarding the building's occupancy profile, its maximum occupancy is 200 occupants, while Figure 3b illustrates the daily distribution of the occupancy. The infiltration and natural ventilation rates are selected to be 0.35 and 0.7 air changes per hour (ACH), respectively.

**Table 3.** Operating parameters of the examined building for the baseline scenario.

| Parameter | Value |
|---|---|
| Mean daily occupants | 200 |
| Thermal load of the occupants (W/person) | 80 |
| Mean operating fraction of the occupants (%) | 30 |
| Specific electrical load of the lighting (W/m$^2$) | 9 |
| Mean operating fraction of the lighting (%) | 40 |
| Specific electrical load of the appliances (W/m$^2$) | 4 |
| Mean operating fraction of the appliances (%) | 50 |
| Infiltration rate (ACH) | 0.35 |
| Natural ventilation rate (ACH) | 0.7 |

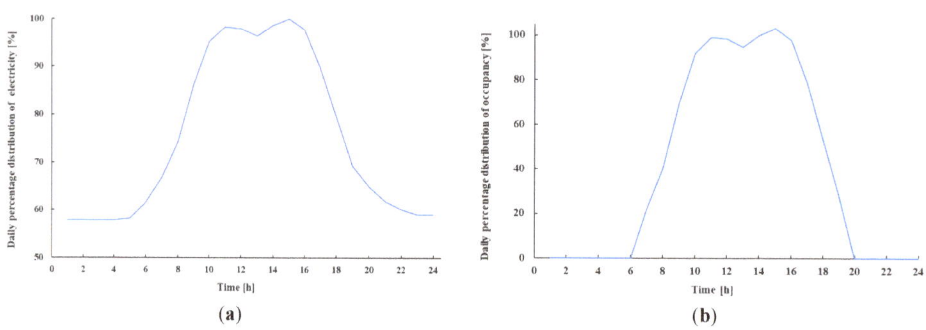


**Figure 3.** Daily distribution of (**a**) the electrical energy consumption and (**b**) the occupancy of the building [32].

### 2.1.2. Energy System Description

The space heating system of the building includes radiators, which are distributed throughout the entire building and are fed by hot water. The hot water is produced by two efficient condensing gas boilers with a nominal efficiency of 98.2% and a capacity of 534 kW and 584 kW. The distribution thermal losses are estimated to be 15% due to the relatively high distance between the gas boilers and the radiators. For the heating period, which in adherence to Italian regulations begins on the 14th of October and ends on the

16th of April, the air temperature setpoint is configured at 20 °C. The space cooling system of the building includes several air-to-air heat pumps (split units) of various capacities. The nominal efficiency of the heat pumps is estimated to be EER = 2.16. The temperature setpoint for the cooling period is set to 26 °C, whereas no specific regulations determine the time period of the cooling season. Indicatively, the need for cooling exists from the middle of May up to the end of September. For the purposes of the energy analysis and because a single heating-and-cooling system serves the entire building under the same temperature setpoints, the building is considered a single thermal zone [31].

The DHW is produced by local electrical heaters, which are located in the building's bathrooms. The DHW temperature setpoint is 48 °C, and the occupants' mean daily demand is approximately 22 L per person. A typical profile representing the daily hot water demand distribution is shown in Figure 4, which considers a variable demand distribution throughout the day that keeps up with the occupancy profile of the building. The tap water coming from the network has an annual mean temperature of around 12 °C, and a typical water temperature variation during the year was selected [33]. Electrical heaters cover the DHW demand with an efficiency of 90%, including distribution losses. Table 4 summarizes the energy system's parameters regarding the space heating and cooling and the DHW production for the baseline scenario.

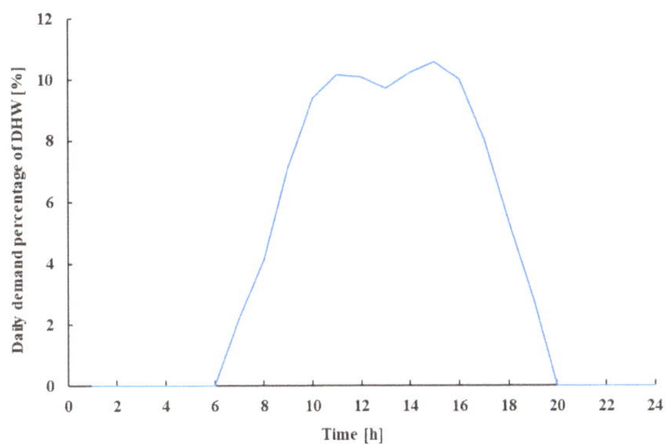


**Figure 4.** Daily dimensionless distribution of the hot water quantity demand.

**Table 4.** Energy system parameters for space heating and cooling and DHW production in the baseline scenario.

| Parameter | Value |
|---|---|
| **Heating system (condensing gas boiler)** | |
| Temperature setpoint (°C) | 20 |
| Nominal capacity of the first condensing gas boiler (kW) | 534 |
| Nominal capacity of the second condensing gas boiler (kW) | 584 |
| Boiler nominal efficiency (%) | 98.2 |
| Distribution thermal losses (%) | 15 |
| Water temperature at the boiler outlet (°C) | 80 |
| **Cooling system (air-to-air heat pumps)** | |
| Temperature setpoint (°C) | 26 |
| Nominal EER | 2.16 |
| **DHW system (electrical heaters)** | |
| Nominal efficiency, including distribution losses (%) | 90 |
| DHW specific demand (L/person) | 22 |
| DHW desired temperature (°C) | 48 |
| Mean temperature of the cold tap water (°C) | 12 |

### 2.2. The Renovation Scenario of the Examined Building

In this subsection, the renovation strategy that will be implemented regarding the renovation and transformation of the examined building into a more functional building is analytically described and segmented into specific renovation actions. The renovation strategy's targets are a decrease in the building's energy consumption, an increase in the building's renewable energy sources production, and ultimately an improvement in its energy autonomy [34]. In the renovation scenario, the building's DHW production system is not modified, whereas the building is divided into two thermal zones, served by two different energy systems for heating and cooling.

#### 2.2.1. Ground-Source Heat Pump

A reversible ground-source heat pump (GSHP) partially serves the building's heating and cooling demand using a low-global-warming potential refrigerant. The respective heat pump is coupled with a borehole heat exchanger (BHE) system. This solution has no negative visual impact on the building because of the lack of outdoor units of fossil fuel boilers or air source heat pumps, and it does not deteriorate the acoustic pollution [35] or the summer heat island effect in the urban center. The GSHP heat transfer fluid consists of a mixture of water and glycol that serves as an antifreeze agent, ensuring the system's reliability in extreme cold environmental conditions and facilitating efficient heat exchange with the ground through the BHE system.

The GSHP's technology is integrated with a fan coil system that serves both the heating and cooling needs of 40% of the treated zone in the building. For the heating period, the air temperature setpoint is configured at 20 °C. In the heating mode, the heat transfer fluid in the terminal units is characterized by inlet and outlet temperature levels of 50 °C and 40 °C, respectively, while in the BHE system, the heat transfer fluid is characterized by inlet and outlet temperature levels of −3 °C and 0 °C. Conversely, for the cooling period, there are no specific regulations concerning the period's duration. The control system of the building autonomously determines the energy system's operation, and the air temperature setpoint is configured at 26 °C. In the cooling mode, the heat transfer fluid in the terminal units is characterized by inlet and outlet temperature levels of 7 °C and 12 °C, respectively, while in the BHE system, the heat transfer fluid is characterized by inlet and outlet temperature levels of 35 °C and 30 °C. Table 5 includes technical data referring to the GSHP's operational efficiency and temperature levels, retrieved from the technological solution of Eneren [36]. The distribution losses of the energy system's configuration (on the building side) are considered to be equal to 10% for both the heating and cooling operations.

**Table 5.** Technical data of the GSHP energy system.

| Parameter | Value |
|---|---|
| **Heating mode** | |
| Nominal source inlet temperature (°C) | −3 |
| Nominal source outlet temperature (°C) | 0 |
| Nominal terminal unit inlet temperature (°C) | 50 |
| Nominal terminal unit outlet temperature (°C) | 40 |
| Nominal COP | 2.83 |
| Nominal heating capacity (kW) | 248.9 |
| Compressor isentropic efficiency (%) | 55.2 |
| **Cooling mode** | |
| Nominal source inlet temperature (°C) | 35 |
| Nominal source outlet temperature (°C) | 30 |
| Nominal terminal unit inlet temperature (°C) | 7 |
| Nominal terminal unit outlet temperature (°C) | 12 |
| Nominal EER | 3.94 |
| Nominal heating capacity (kW) | 295.1 |
| Compressor isentropic efficiency (%) | 55.2 |

2.2.2. Borehole Thermal Energy Storage System

The BHEs' field design consists of 20 m × 150 m deep parallel double-U BHEs organized in a rectangular layout with 7 m spacing between the boreholes. According to local geological and hydrogeological data, the initialization values of the ground temperature are assumed to be 14 °C, and the vertical temperature gradient is 0.03 K/m. Thermal properties and other specifications concerning the BHEs' configuration and the tubing system [36], as well as the surrounding soil, are demonstrated in Table 6.

**Table 6.** Borehole thermal energy storage system specifications.

| Borehole Parameters | Values |
|---|---|
| Ground mean temperature (°C) | 14 |
| Number of parallel boreholes (conventional grouting) | 20 |
| Borehole distance (m) | 7 |
| Borehole depth (m) | 150 |
| Radius of each borehole (m) | 0.075 |
| Thermal conductivity of filling material for grouting (W/mK) | 1.6 |
| Grouting volume [$m^3$] | 2.65 |
| **Soil Characteristics** | |
| Thermal conductivity (W/mK) | 2.30 |
| Heat capacity (MJ/$m^3$K) | 2.16 |
| Geothermal heat flow (W/$m^2$) | 0.06 |
| **Tubing System Specifications** | |
| Radius of the tubes (mm) | 16 |
| Thickness of the tubes (mm) | 3 |

For the province of Trento, several studies have focused on the geo-exchange potential of vertical closed-loop systems, using Spatial and Geographical Information System tools [37], experimental geothermal monitoring, numerical modeling, and multi-scale/multi-method geological surveys. It is worth mentioning that the most recent Provincial Energy–Environmental Plan has highlighted the strategic role of sector coupling among large-scale hydroelectric production and the electrification of thermal and transport demands (heat pumps, electric mobility). This strategic role of heat pumps in local decarbonization scenarios has been modeled not only on a provincial scale but also on a local scale [38].

2.2.3. Building-Integrated Photovoltaic System

Building-integrated photovoltaic (BIPV) shingles are slim and flexible bituminous-based lightweight photovoltaic (PV) shingles suitable for applications in complicated building roof geometries. Their cells are made of triple-junction thin-film amorphous silicon that absorbs blue, green, and red light through three separate layers, producing energy with direct and diffuse light. Bypass diodes between the cells allow each module to produce energy even when it is partially in the shade. Table 7 lists the data concerning a PV module of 144 W nominal maximum power production obtained from the manufacturer data sheets of Tegosolar [39] referring to standard test conditions. The respective technology is installed on the south roof of the neighboring gym, covering a roof area of 305 $m^2$. In total, the PV panels installed are equal to 140 panels of 20.16 kW maximum power production. The PV field is characterized by an azimuth angle of $-10°$ and a tilt angle of $10°$.

Table 7. Technical data specifications of the BIPV system.

| Parameter | Value |
|---|---|
| Module length (m) | 5.412 |
| Module width (m) | 0.373 |
| Number of cells | 22 |
| Open-circuit voltage—$V_{oc}$ (V) | 46.2 |
| Short-circuit current—$I_{sc}$ (A) | 5.3 |
| Maximum power point (W) | 144 |
| Temperature coefficient of $I_{sc}$ (%/°C) | 0.1 |
| Temperature coefficient of $V_{oc}$ (%/°C) | −0.38 |

#### 2.2.4. Insulation of the Roof

The building's roof is covered with a layer of insulation, characterized by a thickness of 18 cm and a thermal conductivity value of 0.036 W/mK. This renovation action decreases the roof's thermal transmittance value from 0.5 W/m$^2$K to 0.2 W/m$^2$K. The insulation will be installed on the interior surface of the building, so there will not be any visual change in the building's exterior.

#### 2.2.5. Replacement of the Windows

The building's windows are replaced with double-glazed windows, characterized by a total thermal transmittance value of 1.3 W/m$^2$K and a heat gain coefficient of 0.6. The new windows will be selected to avoid any visual change to the building's exterior; therefore, similar colors and materials to the existing ones will be selected.

#### 2.2.6. Installation of Standard PV Panels

A total roof surface of 300 m$^2$ is covered with standard PV panels, characterized by a panel surface of 1.76 m$^2$. These panels are installed on the neighboring auditorium's roof, with a south orientation (−10°) and an inclination of 10°. The PV field's maximum power production is equal to 57 kW. Table 8 gives the technical data of the Aleo PV panels [39] with 330 W nominal maximum power production used in the present analysis.

Table 8. Technical characteristics of the PV panel system.

| Parameters | Values |
|---|---|
| Panel surface (m$^2$) | 1.76 |
| Open-circuit voltage—$V_{oc}$ (V) | 40.8 |
| Short-circuit current—$I_{sc}$ (A) | 10.7 |
| Maximum power point (W) | 330 |
| Temperature coefficient of $I_{sc}$ (%/°C) | +0.05 |
| Temperature coefficient of $V_{oc}$ (%/°C) | −0.29 |

#### 2.2.7. Installation of an LED Lighting System

The entire building's lighting system is replaced with an LED lighting system, decreasing the specific lighting load from 9 W/m$^2$ to 5 W/m$^2$, and therefore reducing the building's electricity consumption through its lighting.

### 2.3. Basic Mathematical Modeling

The thermal transmittance (U-value) of the structural components (wall, roof, ground slab, etc.) is given as follows [40]:

$$U = \frac{1}{h_{out} + \sum\left(\frac{t_i}{k_i}\right) + h_{in}} \tag{1}$$

where (i) represents the different layers of each structural component. The heat convection coefficients were selected according to ISO 6946:2017 [41], and they are $h_{in}$ = 7.7 W/m$^2$K and $h_{out}$ = 25 W/m$^2$K.

The thermal transmittance (U-value) of windows is given as follows [40]:

$$U = \frac{U_{window} \cdot A_{window} + U_{glass} \cdot A_{glass}}{A_{window} + A_{glass}} \quad (2)$$

In the previous equation, the frame and window parameters (i.e., area and thermal transmittance) are used.

The fuel energy demand of the boiler ($Q_b$) is calculated by using the heating load ($Q_{heat}$), the boiler efficiency ($\eta_{boiler}$), and the distribution thermal loss efficiency ($\eta_{dis,boiler}$):

$$Q_b = \frac{Q_{heat}}{\eta_{boiler} \cdot \eta_{dis,boiler}} \quad (3)$$

The condensing boiler's nominal efficiency is selected to be ($\eta_{boiler}$ = 98%), and the distribution thermal loss efficiency is selected to be ($\eta_{dis,boiler}$ = 85%). More details about the boiler modeling with INTEMA.buidling can be found in Ref. [33].

Regarding the split units (air-to-air heat pumps) for cooling in the baseline scenario, the nominal EER was selected at 2.16 because they are old units, and the respective electricity consumption is calculated as follows:

$$P_{el,cool,split} = \frac{Q_{cool}}{EER_{split}} \quad (4)$$

For the renovation scenario with the geothermal heat pump, the nominal heating COP is selected to be 2.83 and the nominal cooling EER is selected to be 3.94. Also, the distribution losses from the heat pumps to the fan coils are estimated by using the distribution thermal loss efficiency ($\eta_{dis,fancoil}$), which is assumed to be 90%. This distribution efficiency is greater than the case of the boiler with radiators because the fan coils operate at lower temperatures. In these cases, the electricity consumption for heating and cooling is calculated as below:

$$P_{el,heat,geoth} = \frac{Q_{heat}}{\eta_{dis,fancoil} \cdot EER_{geoth}} \quad (5)$$

$$P_{el,cool,geoth} = \frac{Q_{cool}}{\eta_{dis,fancoil} \cdot EER_{geoth}} \quad (6)$$

A detailed modeling of the ground-source heat pump with the INTEMA.building tool can be found in Ref. [42].

The electricity production of the PV panels ($P_{el,PV}$) is calculated by using the electrical efficiency of the PV panels ($\eta_{el}$), the PV panels' area ($A_{PV}$), and the incident solar irradiation on the PV panels ($G_T$):

$$P_{el,PV} = \eta_{el} \cdot A_{PV} \cdot G_T \quad (7)$$

The modeling of the PV panels is based on the equivalent circuit of a two-diode model cell, and more details of this can be found in Refs. [43,44]. Moreover, this model has been used in our previous work with INTEMA.building [33].

*2.4. Simulation Strategy*

The simulation software used for the current numerical study is INTEMA.building, a customized dynamic simulation tool developed using the Modelica language [28] and the Dymola modeling and simulation environment, including a graphical user interface and a solver [29]. The distinctive feature of this tool is rooted in its capacity to simulate complicated and detailed energy system configurations with accuracy through the utilization and combination of simpler and modular subcomponents. The simulations are realized with adjustable time steps to account for strongly fluctuating variables in the analysis, such as those triggered by the implementation of control systems. This method guarantees a highly realistic framework for the acquired results. For the present analysis, the basic timestep is set to be 1 h, wherein the solver user is the Dassl solver [45] with $10^{-4}$ tolerance. The INTEMA.building tool has been used in various studies investigating the thermal performance of buildings and the operation of their energy systems [33,42,46] by the authors of this paper. Additionally, the respective tool has also been compared with the TRNSYS 18 simulation software [33] and validated according to the European Standard EN15265 [47].

The baseline scenario includes one energy system (ESC 1), which is depicted in Figure 5. On the other hand, the renovation scenarios use two energy systems (ESC 1, ESC 2): the old system in Figure 5 and the new one in Figure 6. Practically, the renovation scenario includes a partial renovation of the energy system because there is not enough capacity in the ground-source heat pump to serve the entire building.

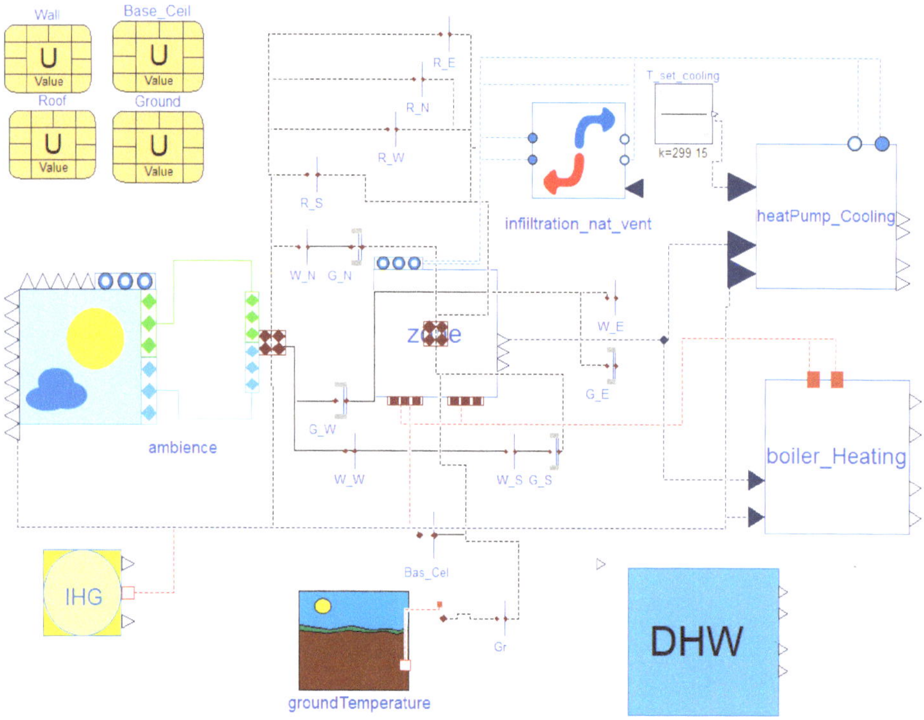


**Figure 5.** Configuration of ESC 1 coupled with the examined building in the Modelica environment.

The layout of the building with a space heating system, a cooling system, and the DHW system for the baseline scenario is depicted in Figure 5. More specifically, the essential components that comprise the total system are depicted, as well as their basic connections. In the baseline scenario, the entire treated floor area of the building is served by a single heating system composed of condensing gas boilers coupled with the radiators distributed throughout the entire building, a single cooling system composed of air-to-air heat pump split units, and DHW production systems composed of local electrical heaters. The respective heating, cooling, and DHW system combination is referred to as energy system configuration 1 (ESC 1).

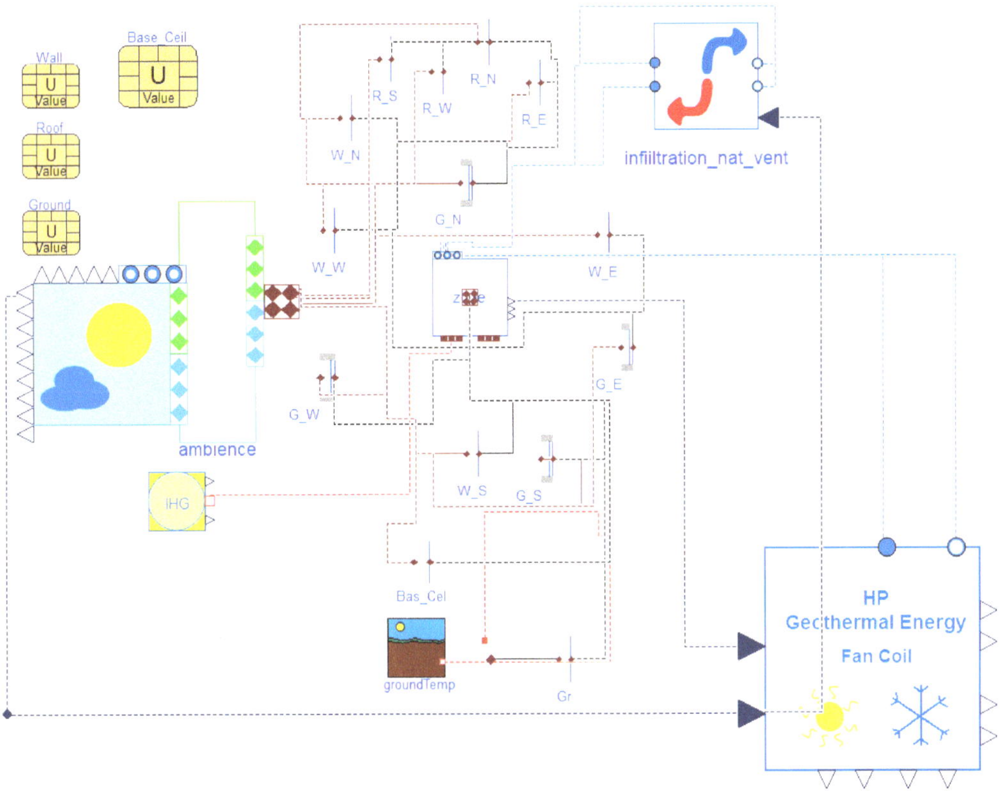


**Figure 6.** Configuration of ESC 2 coupled with the examined building in the Modelica environment.

Practically, Figure 5 includes the building envelope and the energy systems. The building envelope is modeled by using the central component of the thermal zone where the energy balances are conducted in every time step. The external structural components (walls, roof, windows, etc.) are coupled with the thermal zone in order to simulate the energy exchange between the thermal zone and the ambient air. There is also a specific component for infiltration/ventilation, a specific component for the internal loads, and a component for the heat exchange with the ground. Regarding the energy systems, there are components for the DHW, a boiler for covering the heating loads, and a conventional heat pump for covering the cooling loads.

Conversely, in the renovation scenario, the building is served by a combination of heating and cooling systems, whereas no alterations are realized regarding the DHW production system. In the respective scenario, 60% of the building-covered floor area is served by the heating and cooling system of the baseline scenario, namely the gas boiler system coupled with radiators for heating and the heat pump split units for cooling. The remaining 40% of the building's treated floor area is served by the integration of GSHPs and BHEs coupled with the fan coil system, serving both the building's heating and cooling needs, referred to as energy system configuration 2 (ESC 2). Figure 6 depicts the configuration of 40% of the building served by ESC2 developed in the Modelica environment. Specifically, Figure 6 shows the new thermal zone, which is served by one energy system, which is the innovative GSHP system. The rest of the components regarding the building envelope are similar to the baseline case, but they have been properly adjusted to the examined renovated thermal zone.

Additionally, Figure 7a illustrates the integration of the developed components of GSHP system and the fan coil system, as well as their basic connections, while Figure 7b represents the component of the water-to-water reversible heat pump connected with the component of the geothermal field of the BHEs. Specifically, Figure 7a illustrates the connection of the heat pump with the fan coil and the total configuration including circulating pumps, a control unit, thermometers, and piping components. Figure 7b shows the connection of the heat pump with the ground boreholes, and it includes a thermometer and a flowmeter in the given depiction.

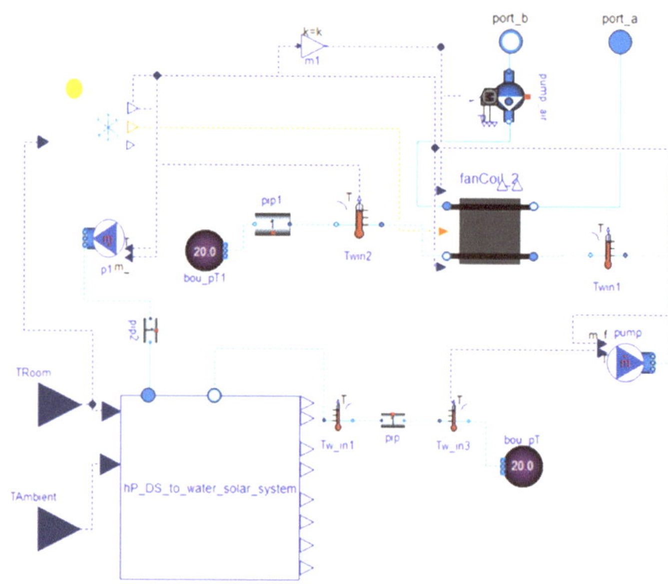


(**a**)

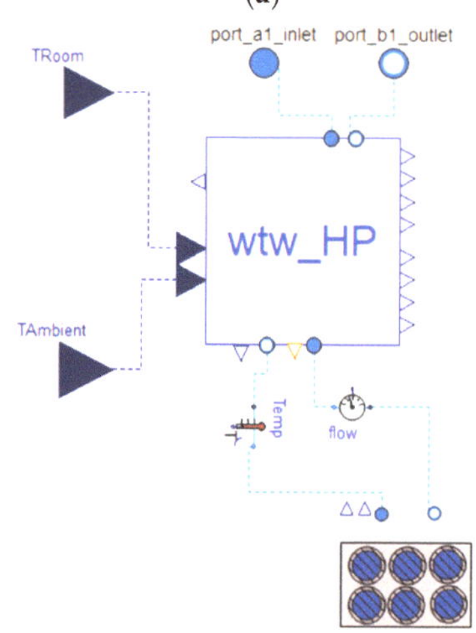


(**b**)

**Figure 7.** Detailed configuration of ESC 2: (**a**) GSHP coupled with the fan coil system, and (**b**) water-to-water reversible heat pump coupled with the BHE system.

## 3. Results and Discussion

This section includes the simulation results for the baseline scenario and the investigation of the renovation scenarios. Also, a comparison of both scenarios in terms of energy is given in the last part of this section.

### 3.1. Energy Analysis of the Baseline Scenario

Firstly, the results of the dynamic simulations regarding the baseline scenario are presented. Specifically, Figure 8a depicts the heating energy load and cumulative energy demand of the examined building, while Figure 8b shows the boiler's power and cumulative energy demand during the heating season. According to the simulation results, the yearly demand for fuel consumed by the natural gas boiler is calculated to be 914,352 kWh, covering a heating load of 762,866 kWh. Moreover, Figure 9a depicts the cooling energy load and cumulative energy demand, while Figure 9b illustrates the electrical load and cumulative electrical energy demand of the building. The calculated yearly cooling demand is 55,762 kWh, while the respective consumed electricity is 25,816 kWh. It is important to highlight that the cooling needs are significantly lower than the heating needs because the examined location presents relatively low temperature levels, especially during the winter. Additionally, Figure 10 depicts the indoor and ambient air temperatures during the year. The temperature setpoint of 20 °C and the desired thermal comfort conditions are not satisfied during short periods in May and September due to the definition of the six-month heating season (14 October–16 April).

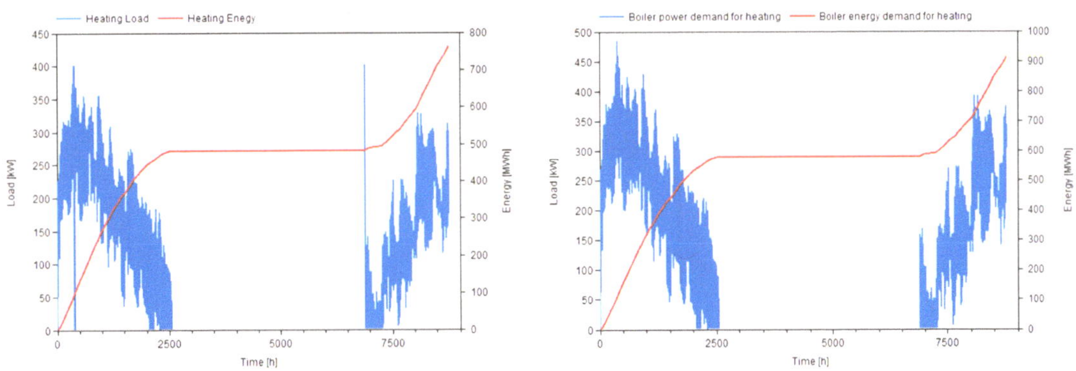


(a) (b)

**Figure 8.** (**a**) Heating energy load and cumulative heating energy demand, (**b**) Boiler power and cumulative boiler energy demand for heating.

As far as the DHW production system is concerned, Figure 11a depicts the DHW energy load and cumulative energy demand, whereas Figure 11b illustrates the electrical power and cumulative electricity consumption of the local electrical DHW heaters. The calculated yearly energy demand for DHW is 67,241 kWh, while the respective electricity demand for its preparation is 74,653 kWh. Moreover, the yearly demand for DHW is around 1606 m$^3$ of water. These results refer to the baseline as well as the renovation scenario, since no alterations have been realized regarding the building's DHW production system. Finally, Figure 12 depicts the electricity power and cumulative energy demand for covering the yearly needs of the appliances and the lighting. This quantity is computed to be 267,502 kWh.

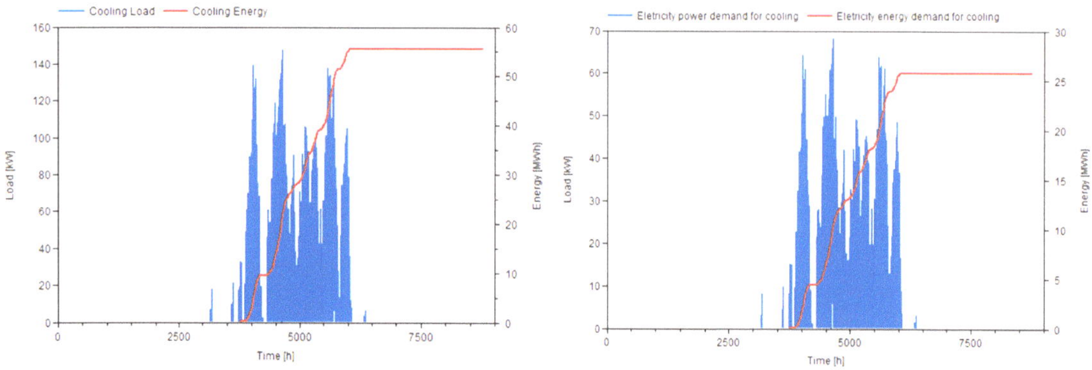


(a) (b)

**Figure 9.** (**a**) Cooling energy load and cumulative heating energy demand. (**b**) Heat pump's electrical power and cumulative electrical energy demand for cooling.

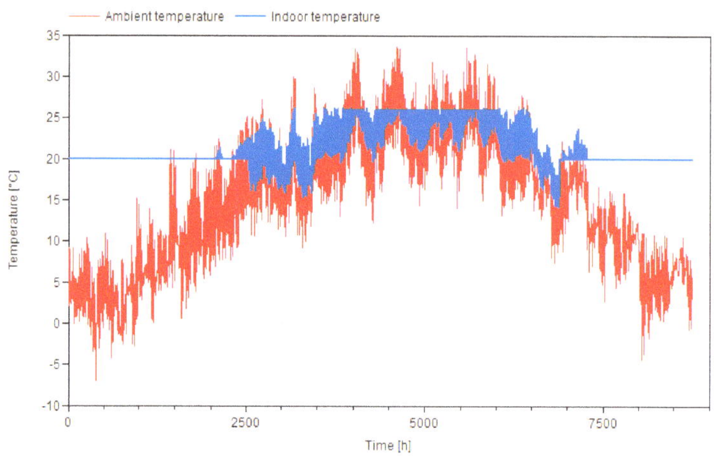


**Figure 10.** Indoor air temperature and ambient air temperature levels.

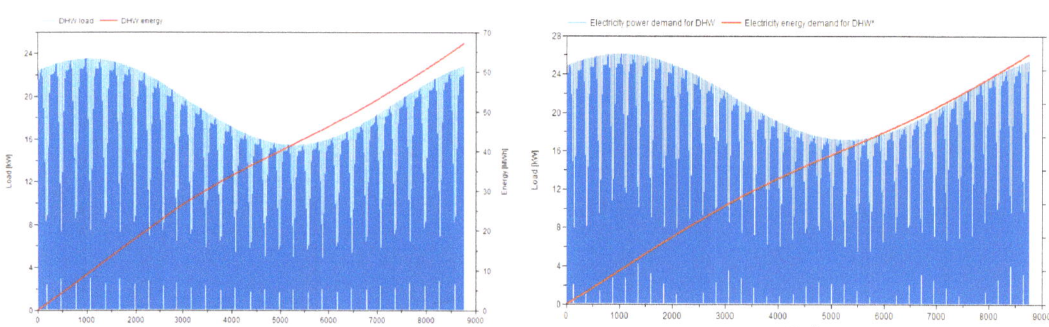


(a) (b)

**Figure 11.** (**a**) DHW energy load and cumulative DHW energy demand. (**b**) Electrical power and cumulative electrical energy demand for DHW production.

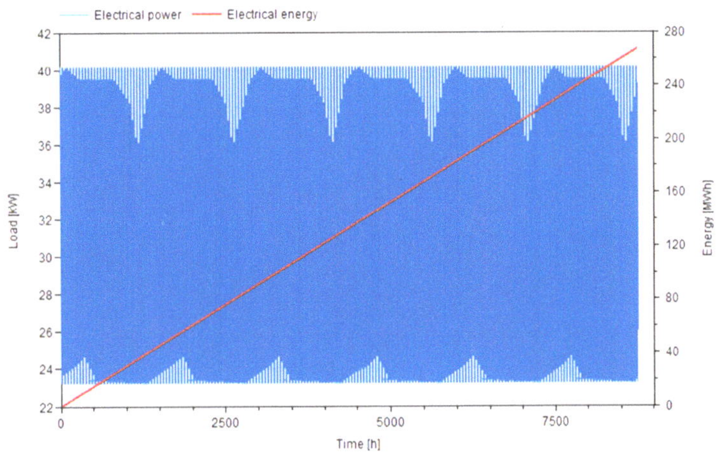


**Figure 12.** Electricity power and cumulative energy demand for covering the needs of the appliances and the lighting in the baseline scenario.

### 3.2. Energy Analysis of the Renovation Scenario

In the present subsection, the results of the dynamic simulations regarding the renovation scenario are discussed. Firstly, regarding the heating season, Figure 13a depicts the entire building's heating load and cumulative heating energy load, a value that is calculated at 658,079 kWh. Specifically, for the building zone served by the heating systems of ESC 1, corresponding to 60% of the building's total treated floor area, the cumulative heating energy demand is calculated to be 396,841 kWh, whereas for the building zone served by the systems of ESC 2, this respective value is equal to 261,238 kWh. Additionally, Figure 13b illustrates the natural gas boiler's power and yearly energy demand, a value which is calculated to be 475,325 kWh, while Figure 13c depicts the electrical power and yearly electricity demand of the GSHPs for heating, which is found to be 102,603 kWh.

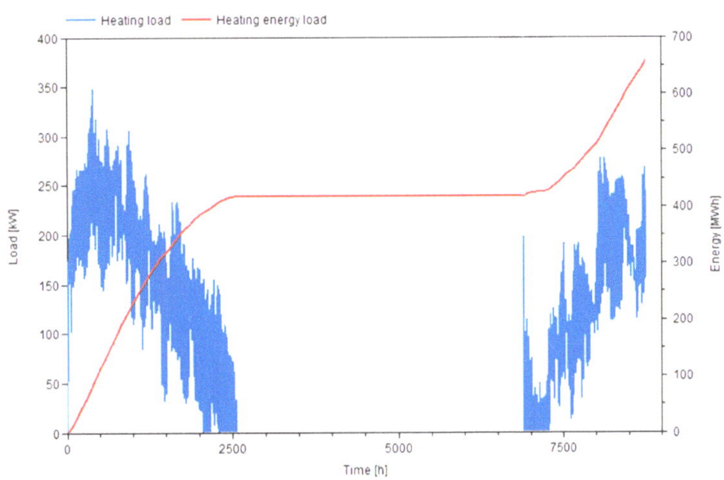


**(a)**

**Figure 13.** *Cont.*

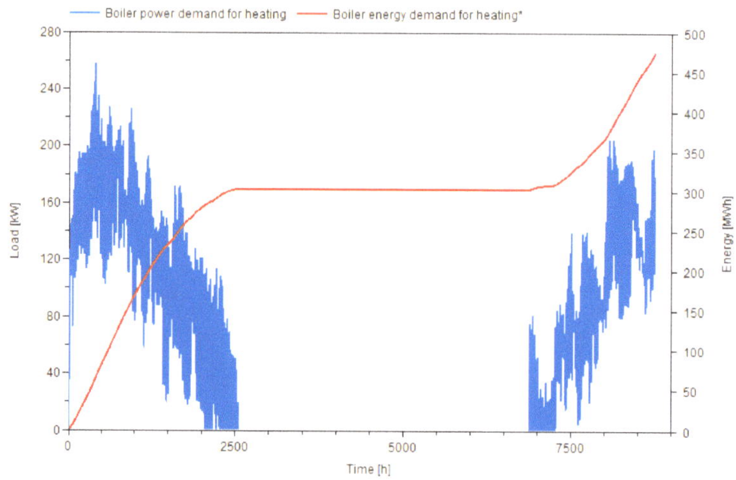


(b)

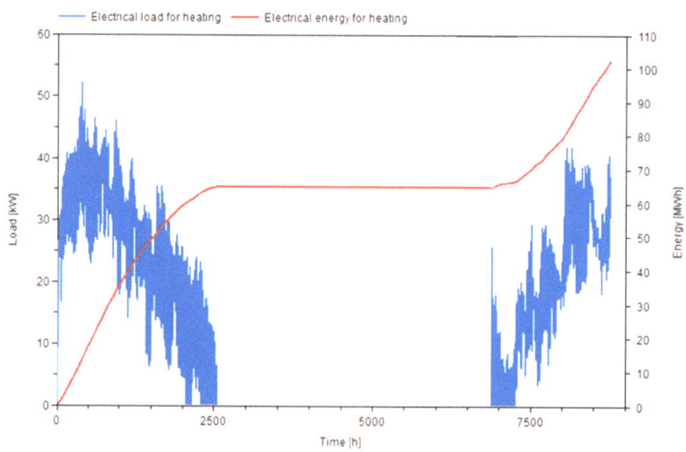


(c)

**Figure 13.** (**a**) Heating load and cumulative heating energy load of the entire building. (**b**) Boiler power and cumulative boiler energy demand for heating in ESC 1. (**c**) Electrical power and electrical cumulative energy demand for heating in ESC 2.

Moreover, as far as the building's cooling demand is concerned, Figure 14a depicts the entire building's cooling load and cumulative cooling energy load, which is calculated to be 52,126 kWh. Regarding the ESC 1 cooling scenario, the cumulative cooling energy load is calculated to be 33,859 kWh, while according to Figure 14b, the corresponding electricity consumption of the air-to-air heat pump split units is calculated at 15,676 kWh. On the other hand, for the ESC 2 cooling scenario, the cumulative cooling energy load is calculated at 18,267 kWh, while according to Figure 14b, the corresponding electricity consumption of the air-to-air heat pump split units is calculated at 5166 kWh. Additionally, Figure 15 depicts the ambient air and building zones' indoor air temperature levels.

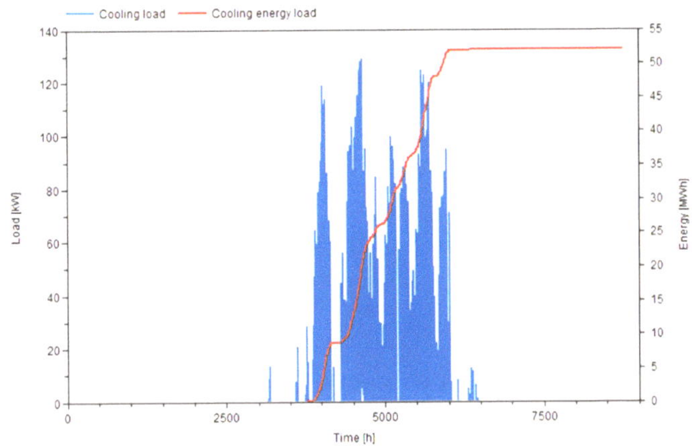


(**a**)

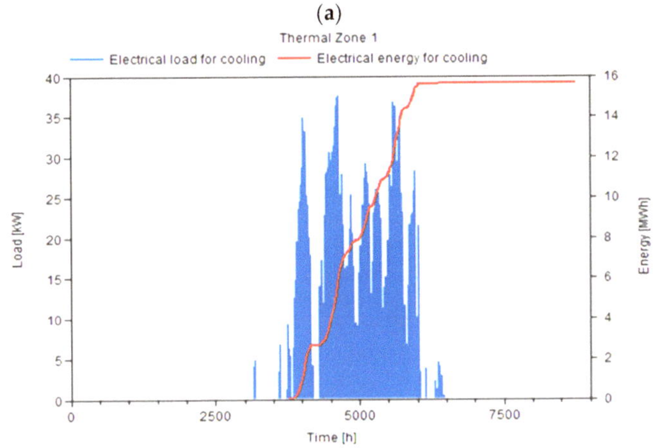


(**b**)

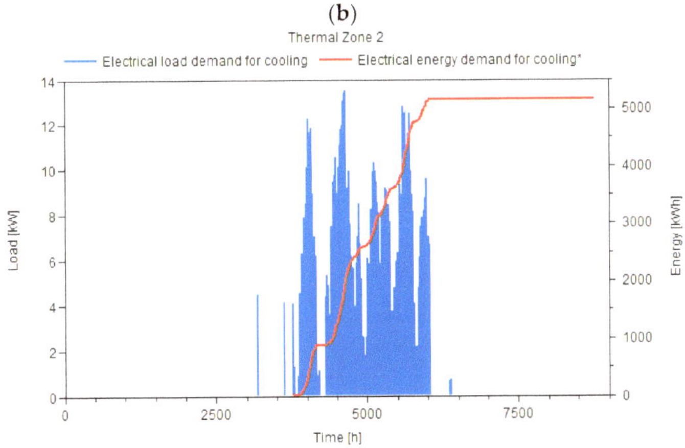


(**c**)

**Figure 14.** (**a**) Cooling load and cumulative cooling energy load of the entire building. (**b**) Electrical power and cumulative electrical energy demand for cooling in ESC 1. (**c**) Electrical power and cumulative electrical energy demand for cooling in ESC 2.

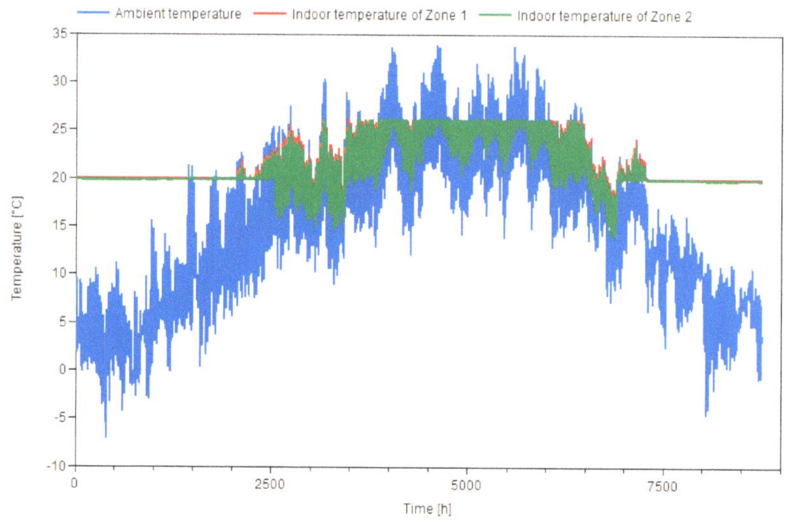


**Figure 15.** Ambient temperature and indoor temperature of building zones served by ESC 1 and ESC 2.

The installation of the LED lighting system affects the building's electricity demand for appliances and lighting, according to Figure 16. Specifically, the cumulative electricity demand for these purposes is calculated at 236,915 kWh. As far as the electricity production is concerned, according to the calculations and Figure 17a, the total electricity produced by the BIPV field installed on the neighboring gym is equal to 24,293 kWh. Additionally, Figure 17b illustrates the electricity production from the standard PV field installed in the neighboring auditorium, calculated at 74,832 kWh.

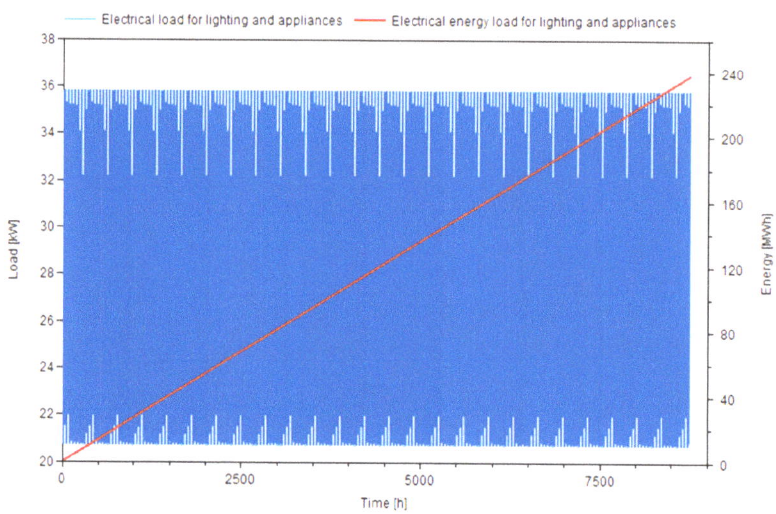


**Figure 16.** Electricity power and cumulative energy demand for covering the needs of the appliances and the lighting in the renovation scenario.

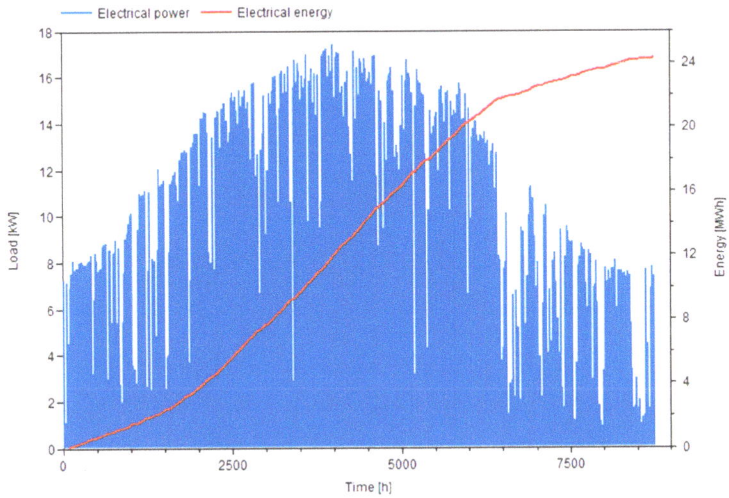


(**a**)

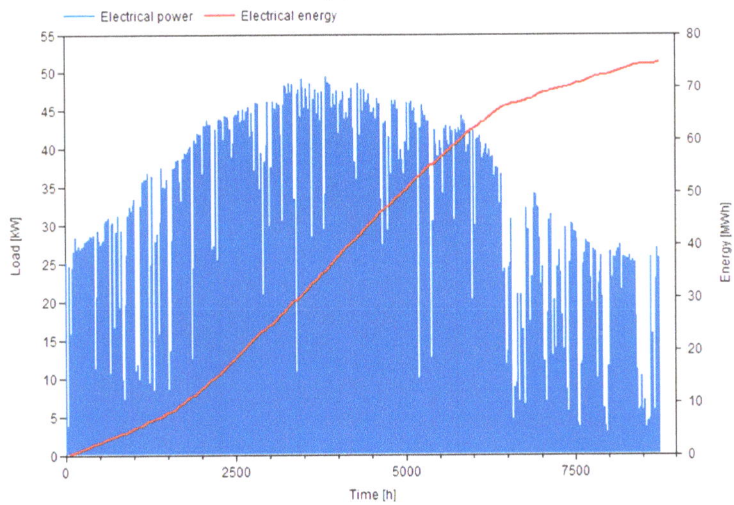


(**b**)

**Figure 17.** Electrical power and electrical energy produced by (**a**) the BIPV system and (**b**) the standard PV field.

### *3.3. Discussion*

This section compares and briefly discusses the simulation results regarding the baseline and renovation scenarios. The purpose of the renovation strategy for the examined building is to decrease its thermal loads and improve the thermal comfort conditions of its occupants. To achieve that, various standards, as well as technologically advanced solutions, are implemented. The energy simulation analysis conducted in this report allows for the comparison between the two states of the building and the prediction of the energy savings induced by the renovation actions. In Table 9, the basic yearly energy demands of each examined scenario, as well as the energy efficiency indexes of the studied heating and cooling systems, are given.

**Table 9.** Comparison of basic energy demands between the baseline and renovation scenarios on a yearly basis.

| Parameters | Baseline [kWh] | Renovation [kWh] | Difference [%] |
|---|---|---|---|
| Heating energy demand | 762,866 | 658,079 | −13.74% |
| Cooling energy demand | 55,762 | 52,126 | −6.52% |
| Energy demand for DHW | 67,241 | 67,241 | 0% |
| Electrical energy for appliances and lighting | 267,502 | 236,915 | −11.43% |
| Natural gas boiler energy demand for heating | 914,352 | 475,325 | −48.02% |
| Electricity consumption for heating | 0 | 102,603 | - |
| Electricity consumption for cooling | 25,816 | 20,842 | −19.27% |
| Electricity consumption for DHW | 74,653 | 74,653 | 0% |
| Electricity production of BIPVs | 0 | 24,293 | - |
| Electricity production of standard PV panels | 0 | 74,832 | - |
| Total electricity production | 0 | 99,125 | - |
| Total gross electricity demand | 367,971 | 435,013 | 18.22% |
| Total net electricity demand | 367,971 | 335,888 | −8.72% |
| Total primary energy demand | 1,650,294 | 1,147,101 | −30.49% |
| Total final energy demand | 1,282,323 | 811,213 | −36.74% |

In the baseline scenario, the heating energy demand is calculated at 762,866 kWh, whereas in the renovated state, the heating demand of the building is calculated at 658,079 kWh. This decrease in the heating energy load of the building is combined with the improvement of the living conditions and is equal to 13.74%. In addition, the natural gas boilers' demand is greatly decreased. More specifically, in the baseline scenario, the natural gas boilers' energy demand for heating is calculated to be 914,352 kWh, while for the renovation scenario, this value is equal to 475,325 kWh, which corresponds to a 48.02% reduction. This is explained by the fact that, in parallel with the renovation actions, in the baseline scenario, the heating system in ESC 1 serves the entire building, whereas, in the renovation scenario, the respective heating system serves only 60% of the studied building. Simultaneously, the building's cooling energy loads are also reduced. Precisely, the cumulative cooling energy demand for the baseline scenario is calculated at 55,762 kWh, whereas for the renovation scenario, this value is reduced to 52,126 kWh, indicating a decrease of 6.52%. Regarding the total electricity consumption for cooling, this value is calculated at 25,816 kWh for the baseline scenario. For the renovation scenario, though, applying a combination of the cooling systems in ESC 1 and ESC 2, the total electricity consumption required for cooling is reduced to 20,842 kWh.

Furthermore, in the renovation scenario, the building is equipped with two PV systems, which assist in the reduction of the building's net electricity consumption—specifically, the total electricity produced by the BIPV system and standard PV field aggregates to 99,125 kWh. The electricity production that stems from the installation of the RES systems and the electricity savings achieved with the LED lighting system outweigh the electricity demand of the building's energy systems for heating and cooling in the renovation scenario, resulting in a decrease in the net electricity consumption of the building. Specifically, the building's net electricity consumption is computed to be 335,888 kWh for the renovation scenario, a value indicating an 8.72% decrease compared to the baseline scenario. Finally, as far as the primary energy demand is concerned, for Italy, the primary energy factors are selected at 2.0 for electricity [48] and at 1.0 for natural gas. The renovation strategy applied in the examined building is predicted to result in a 30.49% reduction in the building's primary energy demand.

Moreover, the energy-related renovation of cultural heritage buildings leads to sideline gains as well. Specifically, the increase in the thermal comfort conditions inside the building and reduction in the thermal bridges can decrease the internal humidification in the walls/roof. Also, the high peak temperatures in the summer can lead to thermal stresses and consequently damage the external appearance of the building.

Considering the architectural, legislative, and physical limitations of the present deep renovation, this research proposes a significant reduction of 30.49% in the building's primary energy demand compared to the problematic EU energy efficiency renovation yearly rate of 0.2% [49]. Moreover, the examined renovation strategy respects the historical status of the building and the respective surrounding area, reducing the total primary energy to 50 kWh/m$^3$ per year: a fact that places the examined building among the 40% of public buildings in Trento with low energy consumption [26].

The significant reductions in the building's total primary energy demand (−30.49%) and total final energy demand (−36.74%) align with Italy's national energy objectives, aiming to reduce the total primary energy consumption by 24–43% in the period of 2020–2030 compared to the PRIMES 2007 scenario [50,51]. Moreover, according to the results of the present study, the electricity produced by PV technologies integrated into the examined building covers 22.8% of the building's electricity needs. Taking into consideration the aerothermal and geothermal contributions to the heating and cooling of the building (40% of the building's treated floor area is served by the integration of GSHPs and BHEs), the overall share of renewable energy sources in the building's energy consumption satisfies the national objectives (for the 2020–2030 period, these objectives expect a 17–30% share of the gross final consumption of energy to come from renewable energy sources).

## 4. Conclusions

The renovation of buildings with cultural heritage significance is a critical aspect in the design of decarbonized sustainable cities. These renovations have both energy-related and social benefits and are characterized as urgent due to the intense climate crisis and the aforementioned unsustainable incumbent building stock.

The present analysis focuses explicitly on an energy-related investigation into the renovation process of a cultural heritage building in the Municipality of Trento in Italy. The selection of this location and this building status constitutes a significant research insight, as Italy, specifically Trento, showcases a critical issue regarding its stock of energy-inefficient historical buildings [52]. Moreover, the historical heritage profile of the examined building challenges the integration of a deep renovation strategy due to architectural and legislative limitations and the need to preserve the building's historical timeline.

This investigation was conducted with the novel dynamic simulation tool INTEMA.building, which is based on the Modelica programming language. The renovation strategy involves a combination of passive and active actions. Below, the most important conclusions of this energy analysis are briefly presented:

- The results indicate that even in a quite challenging building typology (an old protected building, with different uses and common areas), there are solutions that can have a very positive impact in terms of energy performance.
- The building envelope renovation actions discussed in this study reduce the heating and cooling loads of the building by about 13.74% and 6.52%, respectively.
- In the renovation scenario, a reduction of approximately 48% in the natural gas consumption is achieved through the use of a ground-source heat pump.
- The total electricity produced by the PV and the BIPV systems is calculated at 99.1 MWh, covering 22.8% of the building's electricity needs in the renovation scenario.
- The implementation of the renovation strategy is calculated to result in a 30.49% reduction in the primary energy demand, a 36.74% reduction in the final energy demand, and an 8.72% decrease in the net electricity demand.

In the future, an evaluation of the social benefits of the present renovation strategy can be conducted. Additionally, the economic and environmental benefits can be calculated to obtain an overall view of the renovation process of the present case study.

**Author Contributions:** N.Z.: methodology, software, investigation, writing—original draft preparation; A.K.: methodology, software, investigation, writing—original draft preparation; E.B.: conceptualization, methodology, software, investigation, writing—original draft preparation; P.I.: conceptualization, methodology, writing—original draft preparation; D.G.: software, investigation, writing—original draft preparation; K.A.: conceptualization, supervision, writing—original draft preparation; N.N.: supervision, writing—original draft preparation; S.R.: resources, supervision, writing—original draft preparation; D.V.: resources, supervision, writing—original draft preparation. All authors have read and agreed to the published version of the manuscript.

**Funding:** This work has been carried out as part of the European Union's Horizon Europe program framework under grant agreement No. 101069610 (InCUBE—An INClUsive toolBox for accElerating and smartening deep renovation).

**Institutional Review Board Statement:** Not applicable.

**Informed Consent Statement:** Not applicable.

**Data Availability Statement:** Data are available upon request.

**Conflicts of Interest:** The authors declare that there are no conflicts of interest.

## Nomenclature

| | |
|---|---|
| A | Area, $m^2$ |
| COP | Heating mode coefficient of performance |
| E | Energy, kWh |
| EER | Cooling mode coefficient of performance/Energy Efficiency Ratio |
| g | Solar heat gain factor for the windows |
| $G_T$ | Incident solar irradiation, $W/m^2$ |
| $h_{in}$ | Inside convection coefficient, $W/m^2K$ |
| $h_{out}$ | Outside convection coefficient, $W/m^2K$ |
| I | Current, A |
| k | Thermal conductivity, W/mK |
| $P_{el}$ | Electrical energy, kWh |
| PV | Photovoltaic |
| Q | Thermal load, W |
| t | Layer thickness, m |
| U | Thermal transmittance, $W/m^2K$ |
| V | Voltage, V |
| **Subscripts and Superscripts** | |
| b | Boiler |
| cool | Cooling |
| dis | Distribution thermal losses |
| el | Electricity |
| frame | Window's opening frame |
| geoth | Geothermal |
| glass | Window's glass |
| heat | Heating |
| max | Maximum |
| oc | Open-circuit |
| sc | Short-circuit |
| split | Heat pump split units |
| **Greek symbols** | |
| $\gamma$ | Azimuth angle, ° |
| $\eta$ | Efficiency |
| **Abbreviations** | |
| ACH | Air changes per hour |
| BHE | Borehole heat exchanger |
| BIPV | Building-integrated photovoltaic |
| DHW | Domestic hot water |
| ESC | Energy system configuration for heating, cooling, and DHW purposes |
| GSHP | Ground-source heat pump |
| HVAC | Heating, cooling, and air-conditioning |

## References


1. Lidelöw, S.; Örn, T.; Luciani, A.; Rizzo, A. Energy-Efficiency Measures for Heritage Buildings: A Literature Review. *Sustain. Cities Soc.* **2019**, *45*, 231–242. [CrossRef]
2. Zhang, Y.; Teoh, B.K.; Wu, M.; Chen, J.; Zhang, L. Data-Driven Estimation of Building Energy Consumption and GHG Emissions Using Explainable Artificial Intelligence. *Energy* **2023**, *262*, 125468. [CrossRef]
3. European Parliament and the Council of the European Union. Directive (EU) 2023/1791 of the European Parliament and of the Council of 13 September 2023 on energy efficiency and amending Regulation (EU) 2023/955. Available online: https://eur-lex.europa.eu/eli/dir/2023/1791/oj (accessed on 11 January 2024).
4. Zhu, X.; Zhang, X.; Gong, P.; Li, Y. A Review of Distributed Energy System Optimization for Building Decarbonization. *J. Build. Eng.* **2023**, *73*, 106735. [CrossRef]
5. Kataray, T.; Nitesh, B.; Yarram, B.; Sinha, S.; Cuce, E.; Shaik, S.; Vigneshwaran, P.; Roy, A. Integration of Smart Grid with Renewable Energy Sources: Opportunities and Challenges—A Comprehensive Review. *Sustain. Energy Technol. Assess.* **2023**, *58*, 103363. [CrossRef]
6. Semprini, G.; Gulli, R.; Ferrante, A. Deep Regeneration vs Shallow Renovation to Achieve Nearly Zero Energy in Existing Buildings. *Energy Build.* **2017**, *156*, 327–342. [CrossRef]
7. Assimakopoulos, M.-N.; De Masi, R.F.; Fotopoulou, A.; Papadaki, D.; Ruggiero, S.; Semprini, G.; Vanoli, G.P. Holistic Approach for Energy Retrofit with Volumetric Add-Ons toward nZEB Target: Case Study of a Dormitory in Athens. *Energy Build.* **2020**, *207*, 109630. [CrossRef]
8. Moran, F.; Blight, T.; Natarajan, S.; Shea, A. The Use of Passive House Planning Package to Reduce Energy Use and $CO_2$ Emissions in Historic Dwellings. *Energy Build.* **2014**, *75*, 216–227. [CrossRef]
9. Cho, H.M.; Yun, B.Y.; Yang, S.; Wi, S.; Chang, S.J.; Kim, S. Optimal Energy Retrofit Plan for Conservation and Sustainable Use of Historic Campus Building: Case of Cultural Property Building. *Appl. Energy* **2020**, *275*, 115313. [CrossRef]
10. Ma, Z.; Cooper, P.; Daly, D.; Ledo, L. Existing Building Retrofits: Methodology and State-of-the-Art. *Energy Build.* **2012**, *55*, 889–902. [CrossRef]
11. Arumägi, E.; Kalamees, T. Analysis of Energy Economic Renovation for Historic Wooden Apartment Buildings in Cold Climates. *Appl. Energy* **2014**, *115*, 540–548. [CrossRef]
12. Ardente, F.; Beccali, M.; Cellura, M.; Mistretta, M. Energy and Environmental Benefits in Public Buildings as a Result of Retrofit Actions. *Renew. Sustain. Energy Rev.* **2011**, *15*, 460–470. [CrossRef]
13. Fonseca, D.J.; Bisen, K.B.; Clark Midkiff, K.; Moynihan, G.P. An Expert System for Lighting Energy Management in Public School Facilities. *Expert Syst.* **2006**, *23*, 194–211. [CrossRef]
14. Cohen, S.; Goldman, C.; Harris, J. Energy Savings and Economics of Retrofitting Single-Family Buildings. *Energy Build.* **1991**, *17*, 297–311. [CrossRef]
15. Boait, P.J.; Fan, D.; Stafford, A. Performance and Control of Domestic Ground-Source Heat Pumps in Retrofit Installations. *Energy Build.* **2011**, *43*, 1968–1976. [CrossRef]
16. Hens, H. Energy Efficient Retrofit of an End of the Row House: Confronting Predictions with Long-Term Measurements. *Energy Build.* **2010**, *42*, 1939–1947. [CrossRef]
17. Gravagnuolo, A.; Angrisano, M.; Bosone, M.; Buglione, F.; De Toro, P.; Fusco Girard, L. Participatory Evaluation of Cultural Heritage Adaptive Reuse Interventions in the Circular Economy Perspective: A Case Study of Historic Buildings in Salerno (Italy). *J. Urban Manag.* **2024**, *13*, 107–139. [CrossRef]
18. De Toro, P.; Nocca, F.; Renna, A.; Sepe, L. Real Estate Market Dynamics in the City of Naples: An Integration of a Multi-Criteria Decision Analysis and Geographical Information System. *Sustainability* **2020**, *12*, 1211. [CrossRef]
19. Akkurt, G.G.; Aste, N.; Borderon, J.; Buda, A.; Calzolari, M.; Chung, D.; Costanzo, V.; Del Pero, C.; Evola, G.; Huerto-Cardenas, H.E.; et al. Dynamic Thermal and Hygrometric Simulation of Historical Buildings: Critical Factors and Possible Solutions. *Renew. Sustain. Energy Rev.* **2020**, *118*, 109509. [CrossRef]
20. Ascione, F.; De Rossi, F.; Vanoli, G.P. Energy Retrofit of Historical Buildings: Theoretical and Experimental Investigations for the Modelling of Reliable Performance Scenarios. *Energy Build.* **2011**, *43*, 1925–1936. [CrossRef]
21. Harrestrup, M.; Svendsen, S. Full-Scale Test of an Old Heritage Multi-Storey Building Undergoing Energy Retrofitting with Focus on Internal Insulation and Moisture. *Build. Environ.* **2015**, *85*, 123–133. [CrossRef]
22. Şahin, C.D.; Arsan, Z.D.; Tunçoku, S.S.; Broström, T.; Akkurt, G.G. A Transdisciplinary Approach on the Energy Efficient Retrofitting of a Historic Building in the Aegean Region of Turkey. *Energy Build.* **2015**, *96*, 128–139. [CrossRef]
23. Zagorskas, J.; Paliulis, G.M.; Burinskienė, M.; Venckauskaitė, J. Energetic Refurbishment of Historic Brick Buildings: Problems and Opportunities. *Sci. J. Riga Tech. Univ. Environ. Clim. Technol.* **2013**, *12*, 20–27. [CrossRef]
24. Serrano, T.; Kampmann, T.; Ryberg, M.W. Comparative Life-Cycle Assessment of Restoration and Renovation of a Traditional Danish Farmer House. *Build. Environ.* **2022**, *219*, 109174. [CrossRef]
25. Roman, O.; Farella, E.M.; Rigon, S.; Remondino, F.; Ricciuti, S.; Viesi, D. From 3D Surveying Data to Bim to Bem: The Incube Dataset. *Int. Arch. Photogramm. Remote Sens. Spat. Inf. Sci.* **2023**, *XLVIII-1-W3-2023*, 175–182. [CrossRef]

26. Andreucci, M.B. Linking Future Energy Systems with Heritage Requalification in Smart Cities. On-Going Research and Experimentation in the City of Trento (IT). *TECHNE—J. Technol. Archit. Environ.* **2018**, *1*, 87–91. [CrossRef]
27. Fabbri, K.; Zuppiroli, M.; Ambrogio, K. Heritage Buildings and Energy Performance: Mapping with GIS Tools. *Energy Build.* **2012**, *48*, 137–145. [CrossRef]
28. Modelica. Available online: https://modelica.org/ (accessed on 1 December 2023).
29. Dymola. Available online: https://www.3ds.com/products/catia/dymola (accessed on 1 December 2023).
30. Lucchi, E. Renewable Energies and Architectural Heritage: Advanced Solutions and Future Perspectives. *Buildings* **2023**, *13*, 631. [CrossRef]
31. Greek Technical Chamber. TOTEE 20701-1 Technical Guidelines on Buildings' Energy Performance 2017. Greek Technical Chamber: Athens, Greece, 2017.
32. Thomas, D.; Bagheri, A.; Feldheim, V.; Deblecker, O.; Ioakimidis, C.S. Energy and Thermal Comfort Management in a Smart Building Facilitating a Microgrid Optimization. In Proceedings of the IECON 2017—43rd Annual Conference of the IEEE Industrial Electronics Society, Beijing, China, 29 October–1 November 2017; pp. 3621–3626.
33. Bellos, E.; Iliadis, P.; Papalexis, C.; Rotas, R.; Mamounakis, I.; Sougkakis, V.; Nikolopoulos, N.; Kosmatopoulos, E. Holistic Renovation of a Multi-Family Building in Greece Based on Dynamic Simulation Analysis. *J. Clean. Prod.* **2022**, *381*, 135202. [CrossRef]
34. Arandelovic, B.; Musil, R. The Renovation, Rehabilitation and Adaptation of Historical Heritage Buildings in Public Ownership. The Case of Historical Buildings of Exceptional Importance in Vienna. *Build. Environ.* **2023**, *246*, 110937. [CrossRef]
35. Masullo, M.; Pellegrino, R.; Scorpio, M.; Maffei, L. Auditory and Visual Impact of Split Systems on the Façade of Historical Buildings. *Appl. Acoust.* **2021**, *178*, 107997. [CrossRef]
36. newwave-media.it, N. srl- GSP Pompe di calore geotermiche polivalenti. Eneren. Available online: https://eneren.it/prodotti/gsp/ (accessed on 11 January 2024).
37. Viesi, D.; Galgaro, A.; Visintainer, P.; Crema, L. GIS-Supported Evaluation and Mapping of the Geo-Exchange Potential for Vertical Closed-Loop Systems in an Alpine Valley, the Case Study of Adige Valley (Italy). *Geothermics* **2018**, *71*, 70–87. [CrossRef]
38. Viesi, D.; Mahbub, M.S.; Brandi, A.; Thellufsen, J.Z.; Østergaard, P.A.; Lund, H.; Baratieri, M.; Crema, L. Multi-Objective Optimization of an Energy Community: An Integrated and Dynamic Approach for Full Decarbonisation in the European Alps. *Int. J. Sustain. Energy Plan. Manag.* **2023**, *38*, 8–29. [CrossRef]
39. Tegola Canadese. Available online: https://www.aleo-solar.de/en (accessed on 10 January 2024).
40. *Technical Guidelines of Technical Chamber of Greece*; TEE: Athens, Greece, 2020.
41. *ISO 6946:2017*; Building Components and Building Elements—Thermal Resistance and Thermal Transmittance—Calculation Methods. ISO: Geneva, Switzerland, 2017. Available online: https://www.iso.org/standard/65708.html (accessed on 9 January 2024).
42. Kitsopoulou, A.; Zacharis, A.; Ziozas, N.; Bellos, E.; Iliadis, P.; Lampropoulos, I.; Chatzigeorgiou, E.; Angelakoglou, K.; Nikolopoulos, N. Dynamic Energy Analysis of Different Heat Pump Heating Systems Exploiting Renewable Energy Sources. *Sustainability* **2023**, *15*, 11054. [CrossRef]
43. Duffie, J.; Beckman, W.; Blair, N. Design of Photovoltaic Systems. In *Solar Engineering of Thermal Processes, Photovoltaics and Wind*; John Wiley & Sons, Ltd.: Hoboken, NJ, USA, 2020; pp. 760–788. ISBN 978-1-119-54032-8.
44. Maoucha, A.; Djeffal, F.; Arar, D.; Lakhdar, N.; Bendib, T.; Abdi, M.A. An Accurate Organic Solar Cell Parameters Extraction Approach Based on the Illuminated I-V Characteristics for Double Diode Modeling. In Proceedings of the 2012 First International Conference on Renewable Energies and Vehicular Technology, Nabeul, Tunisia, 26–28 March 2012; pp. 74–77.
45. Hairer, E.; Norsett, S.P.; Wanner, G. *Solving Ordinary Differential Equations I*; Springer Series in Computational Mathematics; Springer: Berlin/Heidelberg, Germany, 1993; Volume 8, ISBN 978-3-540-56670-0.
46. Bellos, E.; Iliadis, P.; Papalexis, C.; Rotas, R.; Nikolopoulos, N.; Kosmatopoulos, E.; Halmdienst, C. Dynamic Investigation of Centralized and Decentralized Storage Systems for a District Heating Network. *J. Energy Storag.* **2022**, *56*, 106072. [CrossRef]
47. *EN 15265:2007*; Energy Performance of Buildings—Calculation of Energy Needs for Space Heating and Cooling Using Dynamic Methods—General Criteria and Validation Procedures. European Standards: Brussels, Belgium, 2007. Available online: https://standards.iteh.ai/catalog/standards/cen/3b7d56e1-21c8-4f7f-8fe0-eb9a39c80893/en-15265-2007 (accessed on 13 December 2023).
48. Balaras, C.A.; Dascalaki, E.G.; Psarra, I.; Cholewa, T. Primary Energy Factors for Electricity Production in Europe. *Energies* **2023**, *16*, 93. [CrossRef]
49. Senior, C.; Salaj, A.T.; Vukmirovic, M.; Jowkar, M.; Kristl, Ž. The Spirit of Time—The Art of Self-Renovation to Improve Indoor Environment in Cultural Heritage Buildings. *Energies* **2021**, *14*, 4056. [CrossRef]
50. Integrated National Energy and Climate Plan. | UNEP Law and Environment Assistance Platform. Available online: https://leap.unep.org/en/countries/it/national-legislation/integrated-national-energy-and-climate-plan (accessed on 5 March 2024).

51. Ministry of the Environment and Energy Security, Italian National Integrated Plan for Energy and Climate—June 2023. Available online: https://commission.europa.eu/system/files/2023-07/ITALY%20-%20DRAFT%20UPDATED%20NECP%202021%202030%20(1).pdf (accessed on 5 March 2024).
52. Anelli, D.; Tajani, F. Valorization of Cultural Heritage and Land Take Reduction: An Urban Compensation Model for the Replacement of Unsuitable Buildings in an Italian UNESCO Site. *J. Cult. Herit.* **2022**, *57*, 165–172. [CrossRef]

sustainability

*Article*

# Passive Energy Conservation Strategies for Mitigating Energy Consumption and Reducing $CO_2$ Emissions in Traditional Dwellings of Peking Area, China

**Liang Xie [1], Lai Fan [2,*], Dayu Zhang [2] and Jixin Liu [1]**

1 School of Architecture and Art, Central South University, Changsha 410075, China; liangxie@csu.edu.cn (L.X.); 221311001@csu.edu.cn (J.L.)
2 School of Architecture and Urban Planning, Beijing University of Civil Engineering and Architecture, Beijing 100044, China; zhangdy@bucea.edu.cn
* Correspondence: 1108130320001@stu.bucea.edu.cn; Tel.: +86-15701658035

**Abstract:** Within China, brick dwellings stand as archetypal relics of traditional habitation, embodying a "living fossil" status. The sustainability of these dwellings is contingent upon the integration of energy-conservation strategies. This study scrutinized and empirically assessed a representative dwelling in the Peking area. Using numerical simulations, the impact on energy consumption of factors such as insulation and glazing type, external wall thickness, insulation thickness, and solar energy utilization was evaluated. The outcomes reveal that introducing external thermal insulation—specifically, expanded polystyrene panels with a thickness of 60 mm and 40 mm for the roof and exterior walls, respectively—along with a sunspace of depth 1.5 m yielded superior energy efficiency. Additionally, substituting conventional roofing with solar tiles exhibited a potential annual electricity generation coupled with an annual solar radiation conversion efficiency of 17%. Collectively, these strategies induced a substantial reduction in annual energy consumption. This study presents tailored energy-conservation measures and provides design decision support for architects' practical recommendations on thermal environment control of passive traditional dwellings in the Peking area.

**Keywords:** passive energy conservation; energy consumption; $CO_2$ emissions; traditional dwelling

**Citation:** Xie, L.; Fan, L.; Zhang, D.; Liu, J. Passive Energy Conservation Strategies for Mitigating Energy Consumption and Reducing $CO_2$ Emissions in Traditional Dwellings of Peking Area, China. *Sustainability* **2023**, *15*, 16459. https://doi.org/10.3390/su152316459

Academic Editors: Elena Lucchi, Tianyi Chen and Wen Zhang

Received: 16 October 2023
Revised: 21 November 2023
Accepted: 22 November 2023
Published: 30 November 2023



## 1. Introduction

### 1.1. Previous Studies

Numerous research and analytical studies have consistently indicated that energy consumption associated with buildings comprises more than 30% of global energy consumption, a figure that escalates to approximately 35–40% within developed nations [1]. To specify, a significant proportion of energy, ranging from 30% to 60%, is dedicated to enhancing the indoor thermal conditions within buildings [2]. Over the last 15 years, China has witnessed a 2.24-fold increase in its overall carbon emissions from the construction sector, demonstrating an average annual growth rate of 5.92%, surging from 2.234 billion tons of $CO_2$ to 4997 million tons of $CO_2$ [3]. Given the swift expansion of China's economy, it becomes imperative for the Chinese government to exercise control over internal emissions and curtail building energy consumption to a maximum threshold of 1.1 billion tons of standard coal (representing 23%). This effort is not only vital for ecological conservation but also crucial for maintaining a healthy and comfortable indoor environment [4–6].

In confronting the multifaceted challenges of global climate warming and energy crisis, it becomes imperative to magnify our focus on energy conservation embedded within architectural paradigms. Buildings, notably dwellings, command a significant portion of global energy consumption, illuminating a tangible trajectory for enhanced energy streamlining and diminished environmental impact. While much of the discourse gravitates towards the sleek silhouettes of modern skyscrapers and the prowess of technologically advanced edifices, the traditional folk homes of northern China, with their profound architectural lineage, beckon for commensurate contemplation. Northern Chinese dwellings, with their robust brick masonry, courtyards, and intricate woodwork, reflect a deep architectural legacy. Designed for a region marked by cold, dry conditions, these homes emphasize heat conservation and wind protection. The severe winters highlight the need for energy efficiency. Given this architectural heritage, it is essential to harness their energy potential, aligning with global sustainability goals and maintaining their relevance in a modernizing world.

The global push towards sustainability and energy conservation has been rapidly gaining traction in the field of architecture and urban planning. Numerous studies have delved into the energy-efficient retrofitting of traditional dwellings, with a primary focus on enhancing the performance of existing architectural materials. Among these, research on passive retrofitting for Tibetan traditional residences has pinpointed potential passive modifications, underscoring the paramount importance of cultural and architectural preservation [5]. The energy performance of new Yaodong dwellings in China's Zaoyuan village showcases the efficacy of traditional heavy envelopes combined with improved insulation techniques, resulting in significantly reduced energy consumption [7]. The concern for preserving historical features while improving energy performance is evident in Korea's modern historic buildings [8]. Using specialized retrofit technologies, such as vacuum insulation panels (VIP), the study highlights an effective method to enhance energy performance without compromising historical integrity. Concurrently, EU countries are recognizing the potential of retrofitting attic spaces to maximize urban density without the necessity of new construction [9]. Here, reflective insulation and low emissivity paint treatments demonstrate the potential to alter the thermal behavior of roofs. A study has been conducted to greenify traditional dwellings in the cold region (Zone B) of central Hebei. This study involved research on typical traditional dwellings, and quantified energy-saving measures were derived through optimization. The research findings recommend an EPS insulation layer thickness of 100 mm for the roof and 60 mm for the external walls. However, it is important to note that this study exclusively analyzed modern residential buildings in the area and did not take historical structures into account [10].

On the other hand, solar energy utilization stands as a pivotal topic in residential energy conservation. Notably, the use of passive solar energy offers novel strategies to enhance thermal comfort and energy efficiency in addressing the energy concerns of traditional homes [11,12]. One such strategy involves the addition of sunrooms on the southern side of houses to mitigate heat loss during winter and optimize solar utilization, catering to both energy conservation and thermal comfort needs [13–16]. Research has been initiated on the photovoltaic potential and feasibility in areas sensitive to natural and heritage concerns. When assessing rooftops' photovoltaic potential with heritage compatibility in mind, there is a discernible reduction in conversion efficiency. However, there is a notable decrease in overall carbon emissions, and the investment return rate is promising. Further, studies on rooftop photovoltaic technologies for new residences in developing countries indicate that smaller solar panels offer greater adaptability [17]. Overall, comprehensive research on northern Chinese brick dwellings remains underexplored, especially considering their unique heritage value and climatic characteristics. It suggests a notable research opportunity for exploring their potential in energy conservation and photovoltaic applications.

### 1.2. Motivation

In northern China, including Beijing, Hebei, Shanxi, and Tianjin, traditional brick-built dwellings are quite common. The materials and forms of these traditional brick-built dwellings are similar, and research on energy-efficient renovations of traditional brick-built dwellings in the Beijing region can be effectively applied and replicated elsewhere. These brick residences of northern China epitomize the essence of Chinese culture. They embody a specific type of housing, characterized by local representation and distinctiveness, largely conceptualized and constructed autonomously by the native inhabitants. However, with the ongoing surge of modernization and rapid urban development, many of these brick homes fall short of contemporary living comfort standards, teetering on the verge of abandonment or substitution. Due to the obsolescence of materials and techniques, their innate energy conservation benefits may be compromised in today's setting. It is, therefore, pivotal to rejuvenate these buildings, ensuring their historical and cultural significance remains intact while meeting contemporary living and energy-efficiency criteria. The following questions then arise:

- What strategies can enhance the thermal efficiency of these traditional brick houses while preserving their architectural integrity?
- Given the abundance of solar radiation in the Peking region, how can we effectively harness this resource to reduce indoor energy consumption?
- How much energy efficiency can these strategies offer, and how economically feasible and implementable are they?

This study seeks to address these knowledge gaps by offering a thorough examination of the thermal and energy consumption dynamics of traditional brick dwellings in Peking. It also introduces energy-efficient renovation strategies for traditional brick houses in Beijing, considering both economic and environmental factors, marking a significant step in this direction. Further material is divided into several parts. Thus, in Section 3.4.1, we delve into the impacts of three strategies—envelope structure optimization, solariums, and rooftop solar energy generation—on indoor comfort and energy efficiency. Section 3.4.2 presents the examination of the carbon emission rates of these strategies, as well as the energy conversion rates of solar tiles made of different materials, comparing photovoltaic panel materials with electricity consumption. Finally, in Section 3.4.3, the study explores the investment payback period from an implementation standpoint. This research aims to enhance the understanding of architectural energy conservation for Peking's traditional brick houses, providing valuable data support for local governments and researchers interested in photovoltaic systems for northern brick residences.

## 2. Materials and Methods

### 2.1. Geographical Location and Climatic Characteristics

The subject of this study is a brick-and-tile house located in Cuandixia Village, Mentougou District, Peking, China. Mentougou District is situated to the west-southwest of central Peking, between longitudes 115°25′00″ and 116°10′07″ as well as latitudes 39°48′34″ and 40°10′37″. The region is a transitional area from the North China Plain to the Mongolian Plateau, with elevations higher in the northwest and lower in the southeast. Figure 1 displays the location, layout, and surroundings of the residence. The district experiences a mid-latitude continental monsoon climate: dry and windy springs, hot and rainy summers, cool and humid autumns, and cold and dry winters. The annual average high temperature is around 20 °C, with an average low of 9 °C. The frost-free period lasts about 200 days annually, with ample sunshine averaging 2470 h per year. Precipitation peaked at 970.1 mm in 1977 and hit a low of 377.4 mm in 1997, averaging about 600 mm annually.

As shown in Figure 1, Cuandixia Village, where the brick-and-tile house is located, is a traditional Chinese village at an altitude of 650 m, covering an area of 5.33 square kilometers. Built along the mountain contours, the village sprawls out fan-shaped on both sides. It boasts 74 sets of Qing Dynasty residences, totaling 689 rooms. These historic homes are invaluable cultural relics, designated as the sixth batch of national key cultural heritage sites in Peking. They play a significant role in the study of China's architectural history.

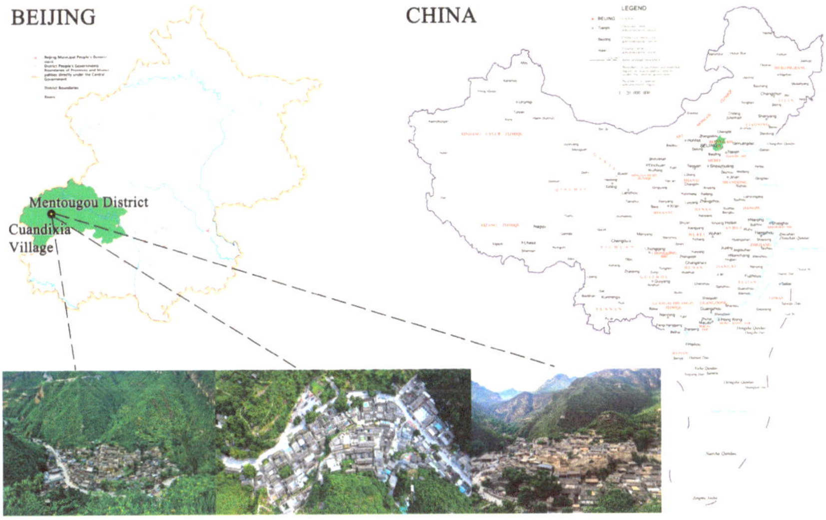


**Figure 1.** Location of brick dwellings in China and the surrounding environment.

*2.2. In-Field Experimental Campaign*

The measurements for this experiment were conducted throughout the year. We collected data on both indoor and outdoor air temperatures, humidity, wind speed, and solar radiation. The setup of the measurement instruments is depicted in Figure 2. We considered the room's practical applications and its frequency of occupation. Table 1 provides detailed specifications of all the measurement equipment utilized, ensuring compliance with the GBT 50785-2012 standard. Temperatures and humidity levels, both indoor and outdoor, were measured at a height of 0.6 m, centered within the room and the courtyard respectively. For spaces less than 16 square meters, special consideration was given due to their size. Given the room's layout, outdoor temperature measurements were solely taken from the southern direction. For a comprehensive layout of the testing points, refer to Figure 2.

*2.3. The Characteristics of Brick Dwellings*

Brick dwellings in the region have prominently evolved by harnessing local materials from the neighboring mountains. Residents meticulously shaped these large stone slabs, incorporating them into the foundation, walls, roofs, and other structural parts of the dwellings based on specific requirements. Additionally, the village's roads, stairs, and various public amenities have been constructed from these stone slates. Compared to other rural housing structures, brick dwellings provide notable thermal benefits—ensuring warmth during winter and a cooling effect in the summer. Table 1 details the distinctive features of these brick dwellings.

**Figure 2.** On-site testing instruments.

**Table 1.** Dwelling character.

| **Walls** | **Doors and Windows** | **Roof** |
|---|---|---|
|  |  |  |
| Considering the construction materials employed, the walls are primarily composed of brick, with a thickness of 370 mm. These walls serve a dual purpose: they offer insulation while also promoting ventilation. A plaster coat completes the exterior finish of the wall. | One side of the courtyard is designated as the front. Besides a brief segment of the windowsill wall, the majority of this facade features wooden doors and windows. All rooms within the compound have their doors and windows oriented towards the inner courtyard. The exterior and lateral walls are constructed robustly, often devoid of windows or furnished with only diminutive ones. This architectural approach compromises the building's air-tightness, considerably impacting the indoor climate during winter. | Each roof features a double-sloped design reminiscent of a sharp mountain peak, ensuring optimal drainage efficiency. A buffer layer between the roof and ceiling effectively mitigates the notable temperature fluctuations between winter and summer, thereby elevating indoor comfort. The roof is constructed with petite green tiles, known for their durability and reduced long-term maintenance costs. Furthermore, these tiles exhibit exceptional insulation and waterproofing properties. |
| Item | construction | U-value (w/m$^2$) |
| Roof | 20 mm tile | 1.02 |
| Ground | 100 mm plain concrete | 2.92 |
| Windows | Wooden frame single glazing (3 mm) | 6.257 |
| Brick dwelling | 200 mm brick | 0.72 |
| Interior wall | 30 mm Wooden Board | 5.621 |

## 2.4. Numerical Analysis

### 2.4.1. Simulation Model Setting and Validation

Cuandixia Village, nestled in the central region of Jaitang's northwest narrow valley, is a traditional village situated to the west of Peking. With an altitude of 650 m, the village spans an area of 5.33 square kilometers. As mandated by relevant regulations, historical structures must preserve their original height, mass, external facade, and color. Any retrofitting initiatives should be undertaken with expert guidance and in collaboration with local authorities. The ensuing strategies have garnered the endorsement of both residents and experts [18].

Given the architectural functionalities, the indoor thermal conditions of the slate dwelling can be augmented via pertinent improvement measures. This not only fosters energy conservation but also aligns with the thermal comfort preferences of the local inhabitants [19].

The Design Builder-7.0.2.006 software was deployed for strategy validation and numerical simulation, leveraging the Energy Plus engine—a tool frequently recognized in academic circles for its precision [20]. This study pinpointed optimization techniques through quantitative simulations and then authenticated the energy consumption enhancements annually under both heating and cooling conditions. Additionally, improvements in the indoor thermal environment during periods without heating or cooling were verified for the ultimate optimized solution [21]. The model's accuracy has been substantiated both by simulations and preceding studies conducted by our team [2,18–22].

While the simulation program's reliability has been previously affirmed, its precision was further scrutinized by juxtaposing it with prototype measurements, especially given the intricacies of the solar system [23]. To ensure output accuracy, identical prototype characteristics were meticulously incorporated into the computerized model [24–26]. Concurrently, Root Mean Square Error (RMSE) was employed to discern the alignment between values simulated using standard climate files obtained from the Energy Plus official repository and empirical values. The governing equations are delineated in Equation (1).

$$RMSE = \frac{1}{\overline{E}_{mea}} \sqrt{\frac{\sum_{i=1}^{n} (M_{simulation,i} - S_{mea,i})^2}{n}} \quad (1)$$

where $M_{simulation,i}$ represents the data simulated at the experimental point, $E_{mea,i}$ represents the data measured at the experimental point, $\overline{E}_{mea}$ represents the mean measured data of all selected points, and n represents the number of detections.

#### 2.4.2. The Parameter Settings of the Improvement Strategy

Simulation results, depicted in Figure 3, reveal that the building's exterior walls, roofs, and windows serve as primary conduits for heat exchange. Thus, enhancing the insulative properties of these structural components offers a promising avenue for energy conservation. Additionally, we aim to incorporate solar housing designs, introduce solar tile roofing, and employ other techniques to augment the available energy for the residential area.

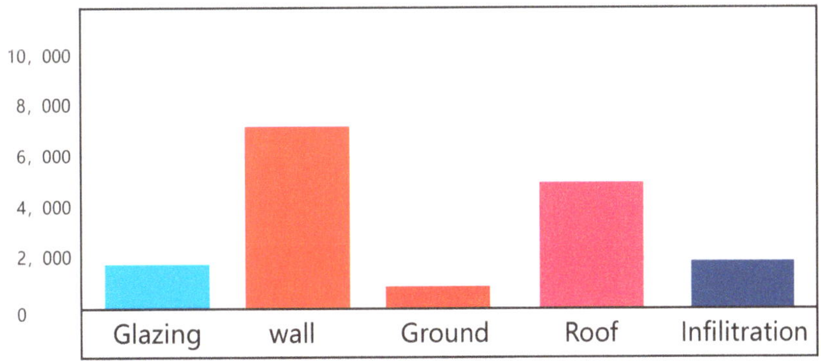


**Figure 3.** Heat Balance (kWh).

1. Enhancement of Openings

The redesigned openings aim to mitigate the effects of prevailing winter winds on the bedroom, thereby reducing indoor heat loss. Additionally, this design improves natural lighting and ventilation. Incorporating glass with superior insulation properties, as outlined in Table 2, further elevates indoor thermal comfort.

2. Optimization of the Enclosure Structure

The building's envelope structure is vital in optimizing its indoor thermal energy consumption. The brick dwelling's external walls and roof currently lack sufficient thermal insulation, compromising the indoor thermal environment. The internal surface temperature within the dwelling primarily hinges on the heat transfer properties and thickness of the insulating material utilized in the structural layer. Ideally, materials with minimal heat transfer coefficients and increased thickness would enhance the indoor thermal environment.

**Table 2.** Glazing type.

| Glazing NO. | Window Glazing Type | Total Solar Transmission | U-Value (W/M$^2$K) | Light Transmission |
|---|---|---|---|---|
| $G_1$ | Single clear (6 mm) (SC) | 0.81 | 6.1 | 0.744 |
| $G_2$ | Sgl LoE (e2 = 0.2) Clr 6 mm | 0.72 | 3.779 | 0.811 |
| $G_3$ | Dbl LoE (e2 = 0.1) Tint 6 mm/13 mm Air | 0.38 | 1.761 | 0.44 |
| $G_4$ | Trp LoE (e2 = e5 = 0.1) Clr 3 mm/13 mm Air | 0.474 | 0.982 | 0.661 |
| $G_5$ | Single reflective (6 mm) (SR) | 0.28 | 5.1 | 0.72 |
| $G_6$ | Double reflective-D (6 m/13 mm Air) (DR) | 0.7 | 2.7 | 0.69 |

As illustrated in Figure 4, the proposed renovation strategy for the external wall and roof involves augmenting an external insulation layer of a specified thickness. Through careful architectural design, the airtightness of the external wall can be enhanced. Detailed parameters of this insulation layer can be found in Table 3.

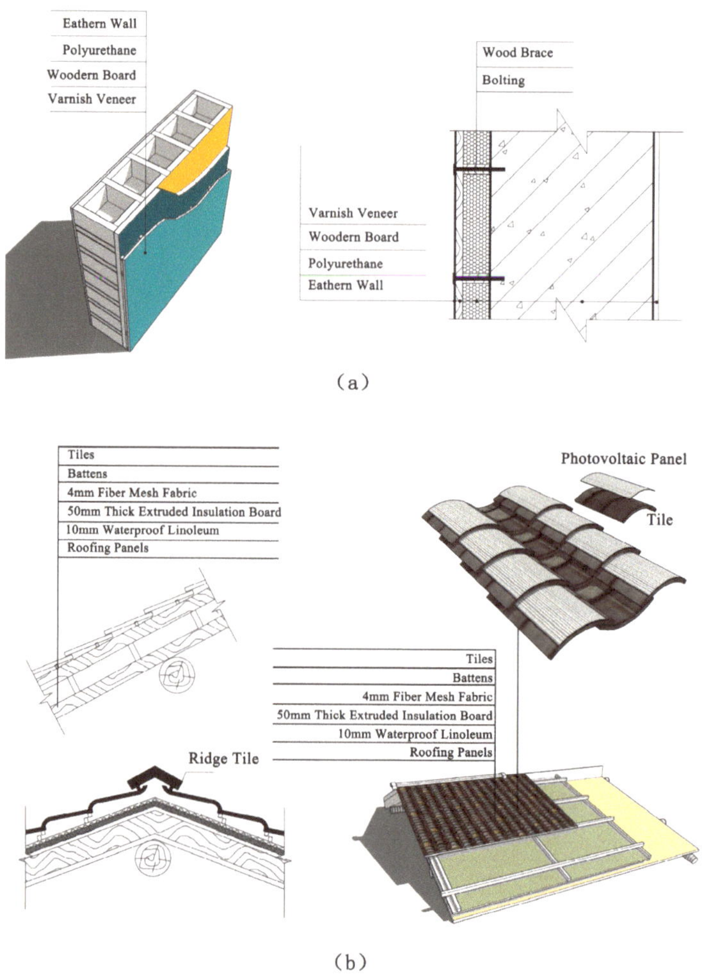


**Figure 4.** (**a**) Exterior wall renewal method; (**b**) Roof renewal method.

**Table 3.** Insulation type of roof and external wall.

| NO. | Insulation Type | Heat Conductivity W (m·k) | Specific Heat Capacity KJ/(kg·K) | Density (kg/m$^3$) |
|---|---|---|---|---|
| $I_1$ | Woods—fir, pine | 0.12 | 1380 | 510 |
| $I_2$ | Extruded polystyrene (XPS) | 0.03 | 1400 | 35 |
| $I_3$ | Aluminum Foil | 0.22 | 2700 | 130 |
| $I_4$ | Insulation Paint | 0.03 | 896 | 1240 |

- Winter Heating Scenario

In this scenario, the building is constructed to be airtight, allowing air exchange solely through infiltration. The interior air temperature is managed by the heating system. The integration of various insulation materials, as illustrated in Figure 4, fosters a more thermally accommodating indoor environment.

- Summer Mixed Cooling Scenario

During this period, the buildings adopt a hybrid approach that combines active cooling with natural ventilation. Specifically, if the external air temperature drops below the indoor temperature but remains above the natural ventilation threshold (12 °C in this study), natural ventilation is employed to cool the interior. If this cooling is insufficient, an active cooling system intervenes to maintain the indoor temperature setpoint at 26 °C. Natural ventilation rates are influenced by pressure differences at openings due to wind or stack effects, window opening sizes, and the schedule of window operations. The Energy Plus Airflow Network model provides hourly airflow rates and pertinent thermal environment parameters. By effectively leveraging outdoor air in this mixed-mode operation, cooling energy consumption is minimized.

- Heating and Cooling Systems

The primary fuel for heating is electricity, boasting a heating system coefficient of performance (COP) of 2.8. Auxiliary energy, used by pumps, which consumes 3.26 kWh per unit building area. Cooling is facilitated by a constant volume direct expansion system powered by electricity, with a COP of 3.8. Setpoint temperatures for heating, cooling, and ventilation are 16 °C, 26 °C, and 12 °C, respectively. All indoor areas are subjected to heating or cooling, except the sunspace of the newly constructed brick dwelling.

3. Integrated Sunspace

As illustrated in Figure 5, the sunspace, connected to the main building, warms the interior in two primary ways:

(1) By facilitating the transfer of warm air into the internal spaces via natural convection through shared windows and doors.
(2) By augmenting the heat transfer resistance between indoor spaces and the external environment.

However, this study does not model the natural convective heat transfer, mainly because of the bidirectional airflow in the natural ventilation mode. Natural convection arises whenever a temperature disparity exists between the sunspace and internal areas. Properly regulating window and door openings to ensure they are operational only when the sunspace is warmer than internal spaces is challenging. If the sunspace is cooler, natural convection may inadvertently reduce the warmth of interior spaces. In many instances, particularly during winter heating, the internal space's temperature, managed by the heating system, often surpasses that of the sunspace. As a result, heat may be lost through natural convective exchanges. In winter, the sunspace is considered an attached room with no convective heat transfer to the interior. During the mixed cooling scenario in summer, the sunspace is naturally ventilated, exempt from air conditioning.

**Figure 5.** The dwelling model (**a**) plan and section of brick dwelling, (**b**) brick dwelling model.

## 4. Roof solar energy utilization

Building Integrated Photovoltaic (BIPV) systems and Building Integrated Solar Thermal (BIST) systems integrate photovoltaic modules directly into building envelope components, such as roofs or façades [27–29]. This study favors solar tiles for solar energy harvesting primarily because they preserve the original architectural aesthetics of the structure.

Buildings harnessing renewable energy sources alleviate the burden on conventional energy generators and often contribute to reduced greenhouse gas emissions. To fully harness available resources, multiple systems are typically combined. This synergy has culminated in a novel technology that merges photovoltaic with solar thermal capabilities: BIPV/T. Before implementing these systems, careful consideration of the building's location and orientation is essential; shading impacts on photovoltaic or solar thermal arrays can significantly affect energy yields.

In this research, BIPV tiles either replace all or a portion of traditional roofing tiles. These are typically assembled in modular configurations, presenting a viable retrofitting solution for roofs. The design and form of these tiles can vary—some mimic the appearance of curved ceramic tiles, which, due to their curved structure, might not provide optimal surface efficiency, as depicted in Figures 4b and 6. We modeled four solar energy harnessing methods using solar tiles in place of the original small green roof tiles over an area of 66.68 $m^2$ at a 30° inclination. The specifications of these solar tiles are detailed in Table 4.

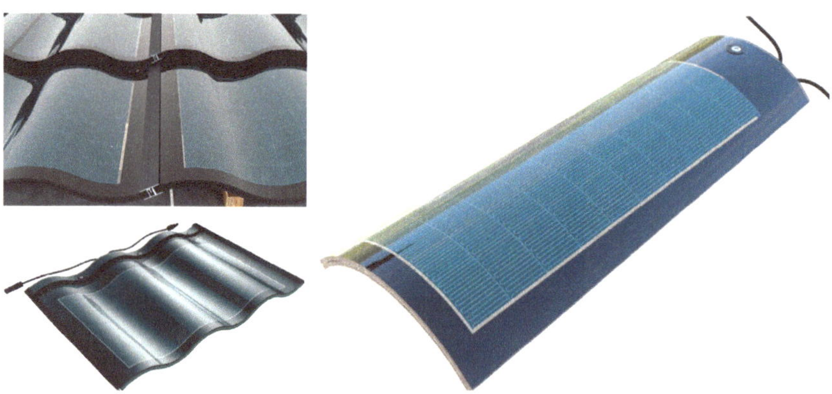

**Figure 6.** Photovoltaic tiles.

**Table 4.** Solar photovoltaic panel type.

| Photovoltaic NO. | Solar Photovoltaic Panel Type | Module Current at Max Power (A) | Module Voltage at Max Power (V) | Short Circuit Current (A) | Open Circuit Voltage (A) |
|---|---|---|---|---|---|
| $P_1$ | ASE 300-DFG/50 | 5.6 | 50.5 | 6.2 | 60 |
| $P_2$ | BP Solar 380 | 4.55 | 17.6 | 4.8 | 22.1 |
| $P_3$ | Shell S115 | 4.2 | 26.8 | 4.7 | 32.8 |
| $P_4$ | Uni Solar PVL-128 | 3.88 | 33 | 4.8 | 47.6 |

### 2.5. The Net Present Value (NPV)

The Net Present Value (NPV) serves as a dynamic instrument employed to evaluate the feasibility of the PV project. When NPV is greater than zero (NPV > 0), it signifies that the projects will yield a gain if they meet the stipulated criteria for benchmark yield. In simpler terms, an investment in this project is economically viable. The calculation of NPV is performed as follows: Equation (2):

$$\mathrm{NPV} = V_0 + \sum_{t=1}^{n} \frac{V_t}{(1+d)^t} \tag{2}$$

The formula for calculating the Net Present Value (NPV) involves several key components: $V_0$ represents the cash flow value at present, Vt signifies the cash flow value at the end of period "t," "n" denotes the number of periods, which, in this project, corresponds to the warranty period of the PV system (25 years)—maintenance, funded by the owner, becomes necessary after more than 25 years of use, and its actual operational lifespan may surpass 35 years, and finally, "d" represents the discount rate, set at 2.5% in this study.

In this research, we have chosen to incorporate the Net Present Value (NPV) as a financial analysis tool specifically within the context of photovoltaic (PV) system studies. This decision is informed by the well-established and cost-effective nature of maintenance structure renewal technology, making NPV a suitable means to assess the economic viability of various PV systems.

### 2.6. Sensitivity Analysis

#### 2.6.1. Approach and Method

The sensitivity analysis aimed to discern the model's response, as simulated by Energy Plus, to fluctuations in design factors and their ensuing effect on thermal comfort. Broadly, this analysis illuminates which combination of factors profoundly influences building performance. Sensitivity analyses can be categorized into local and global sensitivities [30]. Given that modifications in one parameter's output could concurrently influence other parameters, a global sensitivity approach was adopted for this study, comprehensively accounting for the mutual interactions and influences of all design variables. Regression analysis further elucidated parameter sensitivity, identifying primary influencing variables. To conserve computational resources and expedite processing, the Monte Carlo analysis frequently employed in thermal comfort studies was adopted [31].

#### 2.6.2. Sampling and Statistics

Three prominent sampling techniques are associated with Monte Carlo analysis: simple random sampling, stratified sampling, and Latin Hypercube Sampling (LHS). This study adopted the LHS method, grounded in the principle of stratification. This technique ensures an equitable division of the sample space, effectively preventing redundancy within specific regions. The LHS procedure involves the following:

(1) Evenly splitting the [0,1] interval into 200 distinct sub-intervals. (2) Conducting independent random sampling within each sub-interval, ensuring only a single random value is drawn per interval. This process yields a set of 200 random values corresponding to each parameter. (3) Constructing a random matrix comprising 200 rows and 10 columns

by amalgamating these random values. (4) Mapping each random value to its respective range to obtain the authentic parameter values, resulting in a matrix of these genuine values. (5) Formulating the final sample set of input parameters.

Given that the recommended LHS sample size typically ranges between 1.5 and 10 times the total input variables [32], this research initially formulated 100 LHS samples. However, to bolster the precision and validation of the sensitivity analysis, 200 LHS samples were eventually utilized. The integrated sampling tools of designbuilder facilitated the generation of data sets for the sensitivity analysis.

In terms of sensitivity measurements, both the Standardized Rank Regression Coefficient (SRRC) and the Partial Rank Correlation Coefficient (PRCC) were employed. The SRRC evaluates the linear influence of design parameters, allowing for the isolation of mutual effects among independent variables and a direct assessment of the impact of independent variables on dependent ones. In contrast, the PRCC can disentangle the interrelationships between independent variables. A higher SRRC value indicates an increased sensitivity of the parameter: a positive SRRC signifies that an increase in the parameter corresponds with a rise in the dependent variable's value, while a negative SRRC denotes the opposite.

#### 2.6.3. Independent Variables and Predictor Parameters (Energy Saving)

Drawing from pertinent literature [33,34] and the attributes of the chosen building, this study shortlisted nine design variables for sensitivity scrutiny. These encompassed an array of continuous, uniform, and discrete variable types, each characterized by diverse values and materials. Table 5 delineates these design elements, with subsequent details for discrete variables exhibited in Tables 2–4. Predominantly, the predictive mean vote (PMV) [35] serves as a reliable metric for forecasting indoor thermal comfort, validated for both naturally ventilated and air-conditioned edifices. However, certain studies suggest its potential underestimation of discomfort in colder settings [15,36]. The Design Builder-7.0.2.006 software, facilitating parametric studies, was employed to streamline simulation runs for each architectural design. This tool, interfacing seamlessly with Energy Plus, can dynamically tweak parameters within a building model, executing performance simulations. Post-simulation results were integrated into the Statistical Package for the Social Sciences (SPSS) for a nuanced sensitivity analysis.

**Table 5.** Input variables for sensitivity analysis.

| Category | Building Parameters | Unit | Variable Names | Probability Density Functions | Sampling Range | Variation Step |
|---|---|---|---|---|---|---|
| Thermal insulation | Insulation type of wall | - | $X_1$ | Discrete | $I_1$, $I_2$ | Table 1 |
| | Insulation type of roof | | $X_2$ | Discrete | $I_3$, $I_4$ | Table 1 |
| Thermal mass | Insulation thickness of the roof | mm | $X_3$ | Continuous uniform | [0,45] | 5 |
| | Insulation thickness of the wall | mm | $X_4$ | Continuous uniform | [0,200] | 20 |
| Thermal mass parameters | Thickness of roof | mm | $X_5$ | Continuous uniform | [200] | - |
| | Thickness of wall | mm | $X_6$ | Continuous uniform | [200,2000] | 200 |
| | Glazing type | - | $X_7$ | Discrete | [$G_1$, $G_4$] | Table 3 |
| Type of PV system | Long axis azimuth | - | $X_8$ | Discrete | [$P_1$, $P_4$] | Table 4 |
| Sunspace structural | Width | m | $X_9$ | Continuous uniform | [1.2, 1.8] | 0.1 |

## 3. Results and discussion

### 3.1. Field Experiment Result

Figure 7a illustrates notable outdoor temperature oscillations observed during the summer season, ranging from 15.3 °C to 33.7 °C. In stark contrast, the indoor temperature remains remarkably stable, exhibiting variations within a narrower range of 21.7 °C to 25.2 °C, with an average of 23.4 °C. It is noteworthy that the mean relative humidity during the summer period is depicted as 80%, as depicted in Figure 7a. Meanwhile, Figure 7b brings attention to the substantial fluctuations in outdoor relative humidity, which vary between 75% and 90% throughout the summer season. These fluctuations are concurrently accompanied by varying wind speeds ranging from 0 m/s to 4 m/s. The empirical data suggest that, despite the relatively favorable outdoor temperature conditions during the summer, maintaining indoor temperatures within the boundaries of comfort necessitates the incorporation of supplementary cooling equipment.

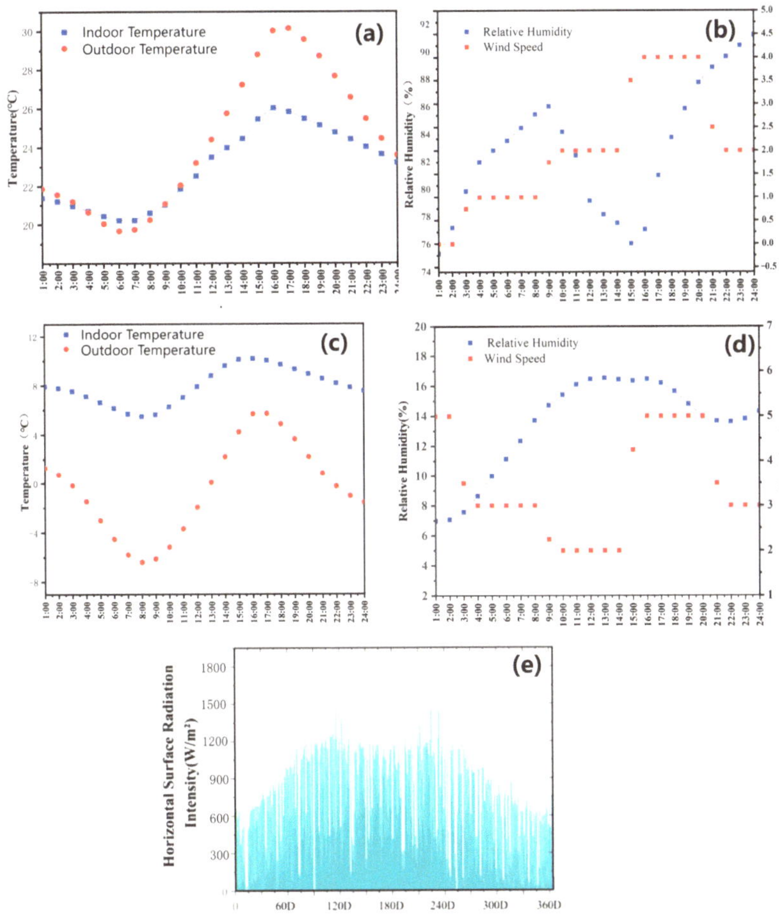


**Figure 7.** (**a**) Indoor and outdoor temperature in summer; (**b**) Indoor and outdoor relative humidity in summer; (**c**) Indoor and outdoor temperature in winter; (**d**) Indoor and outdoor relative humidity in winter; (**e**) The annual total solar radiation of the Peking area.

In the context of winter conditions, Figure 7c presents an analysis of temperature variances within a brick house. Notably, the outdoor temperature experiences pronounced fluctuations, exhibiting a peak difference of 5.6 °C. The average outdoor temperature is recorded at −4.1°C, with its lowest point at −5.1 °C observed around 7:00 and its highest point reaching 5.5 °C at 14:05. In contrast, the average indoor temperature during the winter season attains a maximum of 7.9 °C, peaking at 10.1 °C approximately at 14:50, and reaching its lowest point of 5.7 °C between 7:00 and 8:00, with a lag of approximately one hour compared to the outdoor minimum temperature.

Finally, Figure 7d portrays a gradual fluctuation in outdoor relative humidity during the winter period, oscillating within the range of 17% to 6%. This fluctuation can be attributed to the porous nature of rammed earth construction and its inherent capability to regulate humidity levels.

During the winter season, an examination of the temperature differential between indoor air and the inner surface of the structure reveals an average of −0.23 °C, with a range spanning from −1.8 °C to 0.4 °C. Although this variance is generally negligible for the majority of the time, a conspicuous upsurge in indoor air temperature is observed during the hours between 13:00 and 18:00. This temperature shift, reaching a peak difference of 1.8 °C from the inner surface temperature, raises concerns about the potential introduction of cold radiation, which has the potential to compromise indoor thermal comfort. It is worth noting that the inner surface temperature exhibits more subtle fluctuations when compared to the fluctuations observed in the outer surface temperature, emphasizing the superior thermal retention attributes associated with rammed earth construction. Lastly, as delineated in Figure 7e, the total annual solar radiation in the vicinity of Peking approximates 1500 kWh/m$^2$, with an annual average of 140 W/m$^2$. This ample solar energy resource presents a promising opportunity for reducing indoor energy consumption through the integration of solar technologies.

### 3.2. Simulation Result

To attain a comprehensive comprehension of indoor energy consumption, our study encompassed a year-long simulation, spanning a total of 8760 h. Figure 8a vividly illustrates the indoor temperature dynamics, fluctuating between 1.1 °C and 29.5 °C, with an annual mean of 15.9 °C. In sharp contrast, the outdoor temperature exhibits pronounced oscillations, ranging from −14.1 °C to 36.9 °C. Notably, the average outdoor temperature stands at 12.6 °C lower than its indoor counterpart.

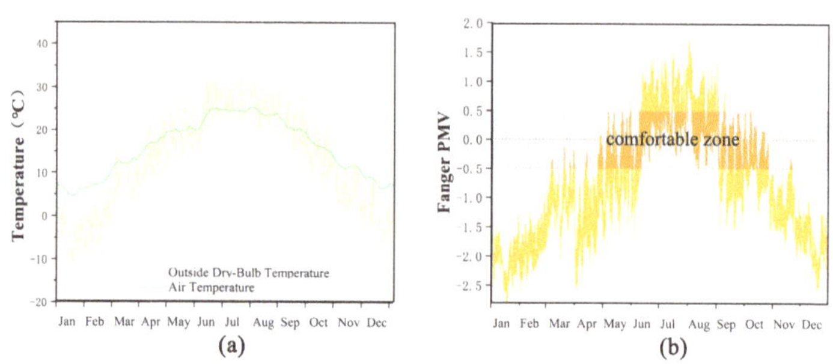


**Figure 8.** (**a**) Annual hourly temperature of the bedroom; (**b**) Annual hourly PMV value of the room.

The evaluation of thermal comfort is based on the Predicted Mean Vote (PMV), as prescribed by the Design Code for Heating, Ventilation, and Air Conditioning of Civil Buildings [37–40]. Ideally, the PMV should fall within the range of −0.5 and 0.5 for comfortable conditions. However, as depicted in Figure 8b, the room's PMV varies between −2.7 and 1.79, indicating that for a substantial portion of the year, indoor thermal conditions deviate from the comfort range. Particularly during winter, indoor thermal comfort is found to be suboptimal, leading to thermal discomfort experienced for approximately 86% of the year. Given these compelling findings, it is imperative to explore key design elements to enhance the energy efficiency of brick houses.

### 3.3. Model Validation

As shown in Figure 9, we analyzed the error between the simulated and measured values, and the error analysis index in statistics was applied, namely, relative mean bias error (RMSErel), which is illustrated in Section 2.4.1 Equation (1), where Mi is considered the measured value (Y axis), Si is the simulation value from the model (X axis), and n is the sample size; this research chose simulated data generated on the same day as the actual measurements for comparison. According to the ASHRAE guidelines, a building model is calibrated if hourly RMSE values fall below 2%. By calculation, in this model, RMSE $\leq$ 1%. That means experimental results verify the validity of the model.

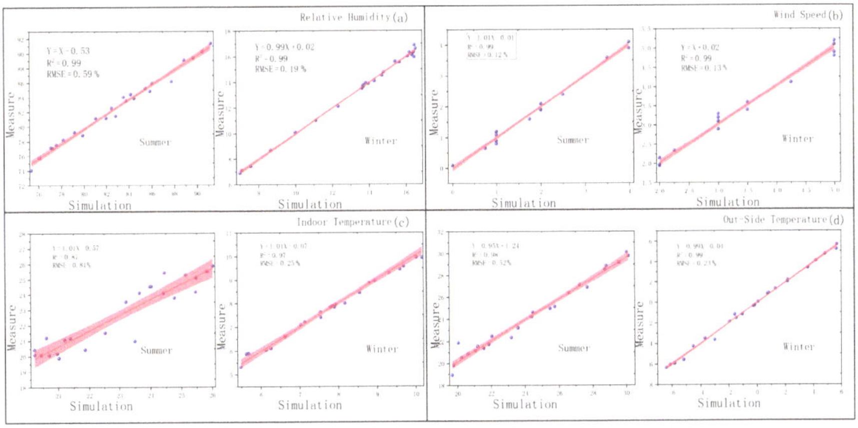

**Figure 9.** Scatter plots of S_(simulation, i) and M_(mean, i) in summer and winter (**a**) Relative Humidity; (**b**) Wind Speed; (**c**) Indoor Temperature; (**d**) Outside Temperature.

### *3.4. Sensitivity Analysis Result*

#### 3.4.1. Global Sensitivity Analysis Result

A sensitivity analysis was conducted using 200 samples, utilizing selected parameters within the SPSS-26.0 software. Two distinct metrics were employed: the standardized rank regression coefficient (SRRC) and the partial rank correlation coefficient (PRCC). Both metrics were calculated through regression methods and served as sensitivity indices. While SRRC assesses linear relationships among parameters, PRCC provides a sensitivity evaluation without incorporating the correlation effects among parameters [41].

Figure 8 illustrates the impact of SRRC and PRCC on the Predicted Mean Vote (PMV) for each parameter. The analysis highlighted that, among the nine chosen design factors, the thickness of roof insulation emerged as the most critical factor for achieving thermal comfort. Additionally, factors such as roof thickness, external wall insulation thickness, wall thickness, and the type of insulation used for roofs were identified as highly sensitive factors positively correlated with thermal comfort. Conversely, factors like glazing type and orientation exerted a negative effect, while the depth of the sunspace and the type of insulation used for external walls had minimal impact. It is noteworthy that PRCC values exceeded those of SRRC, indicating the presence of nonlinear effects.

The results from the sensitivity analysis emphasize the primacy of building insulation parameters over transparent parameters associated with the attached sunspace in ensuring thermal comfort. In pursuit of optimal comfort, both the thermal insulation of the building and its thermal storage capabilities should be primary considerations during the design phase. The PMV demonstrates improvement with a decrease in heat transfer coefficient, suggesting that the utilization of high-quality LOW-E glazing and low-emissivity windows can enhance thermal comfort, particularly during colder seasons.

Figures 10 and 11 provide in-depth insights into the outcomes of the sensitivity analysis concerning energy consumption. Notably, roof thickness emerges as the most influential determinant of energy usage, with roof insulation thickness and external wall insulation thickness following closely as significant parameters affecting energy consumption. While sunspace depth has a negative impact on energy consumption, factors such as glazing window type, wall thickness, wall insulation thickness, and roof insulation thickness continue to dominate as key contributors. A noticeable nonlinear relationship is observed between the depth of the sunspace and energy consumption, indicating that as the sunspace extends beyond 1.5 m, total energy consumption increases. This phenomenon can be attributed to the expanded glazing area within the sunspace, which, while allowing for greater solar absorption, simultaneously leads to heat loss during colder conditions. The use of high-performing glazing and insulation materials can help mitigate this heat loss but may require a higher initial investment. Therefore, finding an equilibrium depth for the sunspace, and balancing heat gain and loss becomes crucial. Additionally, when designing buildings adjacent to sunspaces, the roof's design plays an integral role in energy conservation.

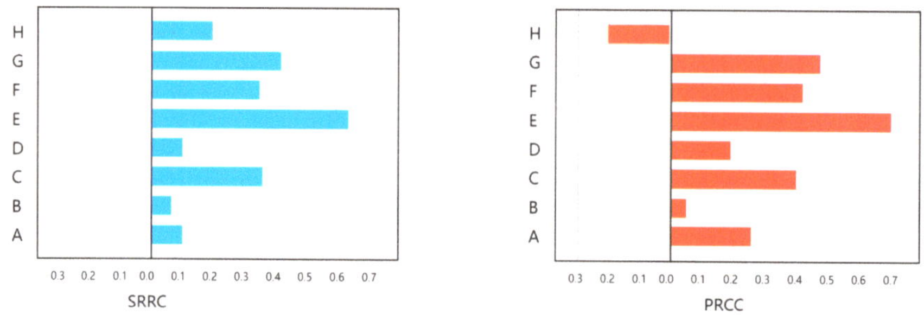


**Figure 10.** PMV global sensitivity analysis.

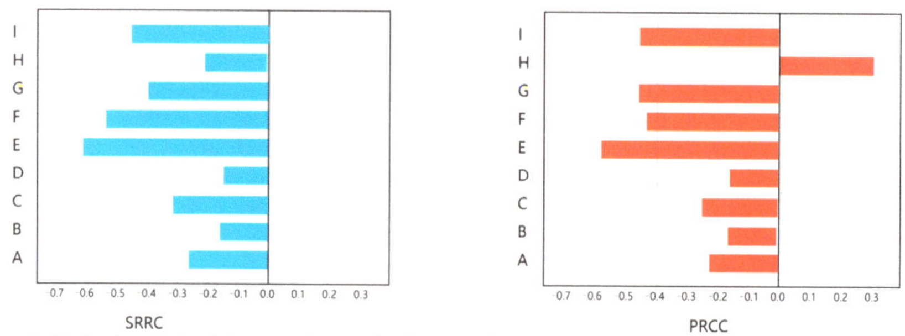


**Figure 11.** Energy Consumption Global Sensitivity Analysis.

Prior research has underscored the significance of the interplay between heat collection, storage, and insulation in passive sunspace design. This study complements and extends these insights. Components that store heat absorb energy during daylight hours, subsequently delaying temperature drops and mitigating temperature fluctuations. Effective insulation techniques can counteract heat loss from glazing windows, particularly those with high heat transfer coefficients, preserving the energy stored within materials and maximizing solar utility. It is crucial to recognize that the primary factor influencing the thermal performance of a building envelope is its insulation elements. In the absence of an internally located heat-storing roof, the wall assumes a pivotal role as both a primary insulator and thermal storage component in this context. Simultaneously, improving the quality of glass layers is indispensable for achieving optimal thermal comfort.

### 3.4.2. Single-Factor Sensitivity Analysis

Following the global sensitivity analysis discussed in Section 3.4.1, we conducted a local sensitivity analysis on parameters that demonstrated high sensitivity. The objective of this analysis was to determine the direction of influence of these parameters on both thermal comfort (indicated by PMV) and energy consumption. We also examined factors such as glazing type, wall thickness, insulation thickness, and the depth of the attached sunspace. Figure 12 provides visual representations of the results of the local sensitivity analysis, with the x-axis depicting varying levels of factors and the y-axis indicating the PMV value and annual energy consumption of the building.

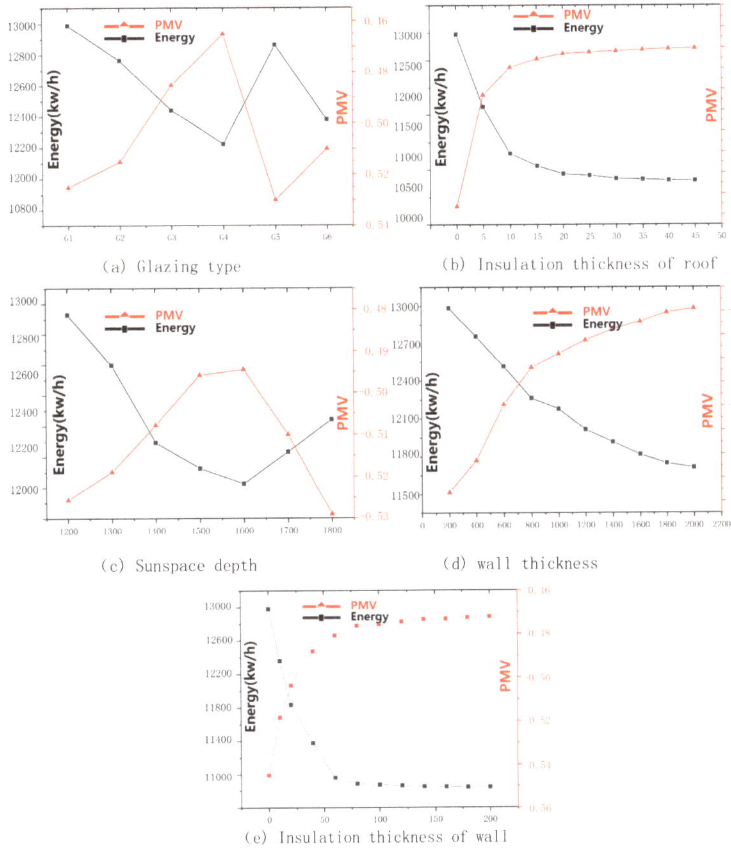


**Figure 12.** Effects of selected design parameters on thermal comfort (PMV) and annual energy consumption. (**a**) glazing type; (**b**) insulation thickness of roof; (**c**) sunspace depth; (**d**) wall thickness; (**e**) insulation thickness of wall.

As depicted in Figure 12a, the choice of roof glazing type has a profound impact on energy consumption. Specifically, selecting G4, as detailed in Table 3, leads to optimal energy savings, resulting in a yearly reduction of 12,215.15 kWh, which represents a decrease of 784.85 kW/h compared to other glazing types. Therefore, for cost-efficiency and maintaining consistent indoor PMV comfort throughout the year, G4 emerges as the preferred glazing type.

Figure 12b provides insights into the influence of roof insulation thickness on both energy consumption and comfort. It is evident that increasing insulation thickness consistently improves both aspects. Particularly, within the range of 0 mm to 15 mm, the impact of insulation thickness variation on energy conservation becomes more pronounced as the curve steepens. Beyond this range, while energy consumption continues to decrease, the reduction becomes gradual. Excessive increases in roof insulation thickness may result in higher costs without substantial energy-saving benefits. Therefore, the optimal insulation thickness range for the roof is found to be between 5 mm and 15 mm. Notably, as thickness escalates from 15 mm to 45 mm, comfort begins to deteriorate. Consequently, the 5 mm to 15 mm range is identified as the ideal design scope, yielding maximum annual energy savings of 10,205.42 kWh, which represents a reduction of 2514.58 kWh, and an increase in PMV from −0.62 to −0.43.

Figure 12c depicts the sensitivity analysis of sunspace depth. The relationship between sunspace depth and energy consumption appears to be nonlinear, with energy consumption spiking as sunspace depth exceeds 1.6 m. A deeper sunspace enhances the glazing area, allowing for more solar energy capture but also increasing the risk of heat loss through transparent sections during colder periods. While enhanced glazing and insulation materials can mitigate this heat loss, they also entail higher initial costs. An optimal sunspace depth of 1.6 m captures sufficient solar radiation in winter while avoiding the dominant wind direction. However, in the unique climatic and geographical context, harnessing solar radiation is of paramount importance, making southern orientation vital. An efficiency balance between comfort and energy consumption is achieved within depths of 1.4 m to 1.6 m, optimizing annual energy savings to 12,115.44 kWh, which is a reduction of 884.6 kWh.

In Figure 12d, we examine the influence of wall thickness on both energy consumption and comfort. It is evident that as wall thickness increases, both energy consumption and comfort improve simultaneously. Notably, within the range of 200 mm to 800 mm, the curve steepens, indicating that changes in wall thickness have a significant impact on energy savings. While continuous increases in thickness consistently lead to reduced energy consumption, the rate of reduction becomes less pronounced over time. Excessive augmentation of wall thickness may create a waste of resources and increase construction difficulties without proportionate energy savings. Consequently, the optimal design range for wall thickness is identified to be between 200 mm and 800 mm. It is important to note that while thermal comfort generally follows the trends observed in energy consumption, the comfort growth rate starts to decline as wall thickness extends from 800 mm to 1600 mm. Therefore, for achieving maximum energy efficiency, the best thickness is 800 mm, and annual savings can reach their peak at 12,255.32 kWh, representing a net reduction of 744.68 kWh.

Figure 12e provides insights into the effects of wall insulation thickness on both energy consumption and comfort. Similar to wall thickness, increasing insulation thickness enhances both energy consumption and comfort. Within the range of 0 mm to 50 mm, the curve exhibits a steep incline, indicating that variations in insulation thickness have a profound impact on energy conservation. Although energy consumption decreases with greater insulation thickness, this reduction becomes progressively less pronounced. Excessive increases in wall insulation thickness can result in waste without significant energy savings. Therefore, the ideal insulation thickness range for walls is identified to be between 20 mm and 50 mm. While thermal comfort generally aligns with the trends observed in energy consumption, it is important to note that comfort levels begin to diminish as insulation thickness expands from 60 mm to 200 mm. Consequently, optimizing insulation thickness within the range of 20 mm to 50 mm achieves the most efficient energy savings, approximately 10,915 kWh annually, representing a decrease of 1997 kWh.

The following indices were employed for the comprehensive assessment of the energy utilization of the building under investigation in this study. $E_{heating}$ denotes the annual total energy consumption per unit building area, while $E_{CO2}$ quantifies the carbon dioxide ($CO_2$) emissions per unit building area, which is expressed in Equation (3) [3].

$$E_{CO2} = E_{fuel} \cdot C_{con} \quad (3)$$

Herein, '$E_{CO2}$' signifies the carbon dioxide emissions per unit building area, measured in kilograms (kg). '$E_{fuel}$' corresponds to the total fuel consumption per unit building area, which is derived from the heating or cooling system and calculated using simulation software, and is expressed in kW·h. Additionally, '$C_{con}$' designates the carbon dioxide conversion factor, representing the quantity of carbon dioxide emitted per unit of energy consumption.

The Chinese Ministry of Housing and Urban-Rural Development has divided China into seven different regions, with Peking falling under the Northern Regional category, characterized by an emission factor of 0.55 [42]. Conducting a heat balance analysis of a building proves beneficial in elucidating the dynamics of heating energy consumption [22]. Hence, a statistical analysis of heat transfer through the building envelopes was performed to assess the heat flux emanating from the buildings. These aforementioned indices, in conjunction with $CO_2$ emissions, are pertinent to the evaluation of the energy consumption efficiency of individual buildings.

As depicted in Figure 13, upon the selection of the most suitable update strategy in Section 3.4.1, the daily carbon dioxide ($CO_2$) emissions originating from the roof exhibited a range of 5 to 25 kg on 49% of the annual calendar days. In contrast, following the retrofit, the roof's daily $CO_2$ emissions fell within the range of 0 to 5 kg for 51% of the year. The wall, pre-retrofit, contributed to daily $CO_2$ emissions ranging from 5 to 45 kg for 59% of the annual duration, while post-retrofit observations showed that daily $CO_2$ emissions in this category decreased to 5–25 kg for 50% of the year. Before the retrofit, the glazing component was responsible for daily $CO_2$ emissions within the 0 to 5 kg range for approximately 73% of the annual timeframe. However, post-retrofit, this specific range of $CO_2$ emissions only accounted for 11% of the yearly duration. This noteworthy shift in emissions can be attributed to the enhanced thermal insulation of the building envelope, resulting in significantly reduced energy consumption throughout the remaining aspects of the interior construction and air infiltration. Total annual $CO_2$ emissions have been reduced by 3799.79 kg.

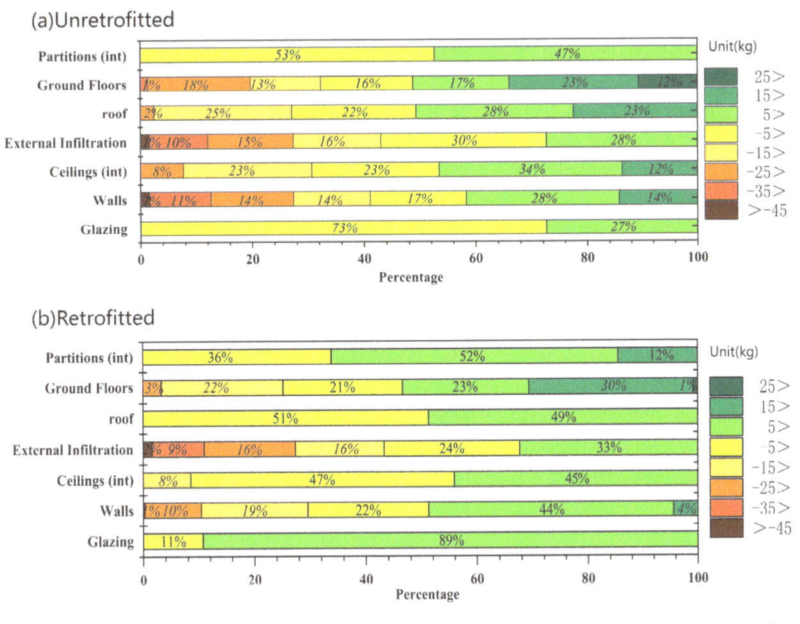


**Figure 13.** Frequency of $CO_2$ emission. (**a**) $CO_2$ emission frequency before the envelope retrofitting; (**b**) $CO_2$ emission frequency after the envelope retrofitting.

### 3.4.3. Annual Solar Tile Roof Power Generation

Figure 14a illustrates the performance of material P1, as outlined in Table 4 when utilized as a medium for renewing solar tile roofing. This material yields an annual electricity generation of 7659 kWh and demonstrates a solar radiation conversion efficiency of 10%.

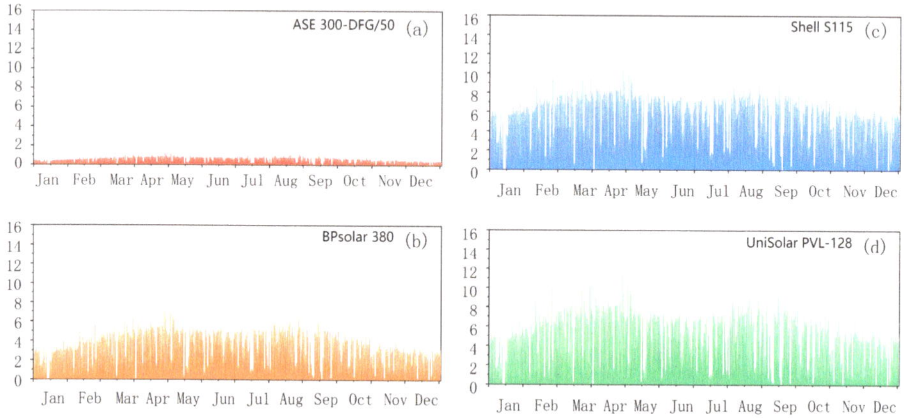


**Figure 14.** Annual Solar tile roof power generation. (**a**) ASE 300-DFG/50; (**b**) BP solar 380; (**c**) Shell S115; (**d**) Uni Solar PVL-128.

In Figure 14b, we observe the results of employing material P2, also referenced in Table 4, to renew solar tile roofing. Material P2 achieves an annual electricity output of 12,732 kWh with a remarkable conversion efficiency of 17%.

Moving on to Figure 14c, we investigate the utilization of material P3, listed in Table 4, for the renewal of solar tile roofing. This material leads to electricity production of 10,075 kWh annually, boasting a solar radiation conversion efficiency of 13%.

Lastly, in Figure 14d, we explore the performance of material P4, as detailed in Table 4, when employed for the renewal of solar tile roofing. Material P4 results in an annual electricity generation of 12,057 kWh, accompanied by a solar radiation conversion efficiency of 16%.

#### 3.4.4. PV System Economic Impact Financial

In situations involving household decision-making, financial considerations invariably take precedence, particularly when there is a need to determine the pay-back period (PBP) for an investment. This entails identifying the point at which electricity generation becomes cost-effective. For residents, it is of paramount interest to ascertain whether the installation of a photovoltaic (PV) system on tile roofing represents a sound financial investment, particularly given the potentially high initial capital outlay associated with PV panels. At present, local electricity prices stand at approximately CNY 0.588 per kilowatt-hour (kWh) at 2023 rates, with a warranty period of 25 years; the equipment incurs a modest maintenance fee over 25 years, and yet, its operational lifespan can surpass 35 years.

One critical aspect to consider is the ability to achieve equilibrium in return on investment (ROI). This equilibrium hinges on various factors, including the time value of money and the selection of nominal and real discount rates when conducting a whole-life cost appraisal (WLC).

To evaluate the economic viability of such projects, two key financial metrics, namely the pay-back period (PBP) and the net present value (NPV), were employed. These analyses serve as instrumental tools for assessing the economic impact and feasibility of implementing PV systems on tile roofing in a residential setting.

Table 6 provides information on four distinct PV panels, including details on the approximate initial system cost, which encompasses the total investment price, cost per unit of square footage, economic benefits generated during the year of electricity production, the inflation rate, and the annual return on investment amount. For the simulated residential house with a roof area of 66 square meters, the total investment of the ASE 300 system is CNY 69,010, and since the price of electricity in Peking is CNY 0.588, the annual value of solar power generation is CNY 4504, the total investment of the BP 300 system is CNY 111,220, the annual value of solar power generation is CNY 7486, the total investment of the Shell s115 300 system is CNY 85,425, the annual value of solar power generation is CNY 5924, the total investment of the Uni 128 system is CNY 101,840, and the annual value of solar power generation is CNY 7090.

**Table 6.** NPV analysis calculations for installing a PV system on the brick dwelling.

| **ASE 300-DFG/50** | | | **BP Solar 380** | | **Shell s115** | | **Uni-Solar PVL-128** | |
|---|---|---|---|---|---|---|---|---|
| Total Investment = CNY 69,010 | | | Total Investment = CNY 111,220 | | Total Investment = CNY 85,425 | | Total Investment = CNY 101,840 | |
| Unit cost = 1030 CNY/m$^2$ | | | Unit cost = 1560 CNY/m$^2$ | | Unit cost = 1275 CNY/m$^2$ | | Unit cost = 1420 CNY/m$^2$ | |
| annual profit = CNY 4504 | | | annual profit = CNY 7486 | | annual profit = CNY 5924 | | annual profit = CNY 7090 | |
| Inflation rate, d = 2.5% | | | Inflation rate, d = 2.5% | | Inflation rate, d = 2.5% | | Inflation rate, d = 2.5% | |
| Years | Cash flow (CNY) | NPV (CNY) | Cash flow (CNY) | NPV (CNY) | Cash flow (CNY) | NPV (CNY) | Cash flow (CNY) | NPV (CNY) |
| 1 | 4394.1 | −64,615.9 | 7303 | −103,917 | 5779.5 | −79,645.5 | 6917.1 | −94,922.9 |
| 2 | 4285.5 | −60,330.4 | 7123.2 | −96,793.8 | 5636.5 | −74,009 | 6745.9 | −88,177 |
| 3 | 4185.9 | −56,144.5 | 6957.2 | −89,836.6 | 5505.6 | −68,503.4 | 6589.2 | −81,587.8 |
| 4 | 4079.7 | −52,064.8 | 6780.8 | −83,055.8 | 5366 | −63,137.4 | 6422.2 | −75,165.6 |
| 5 | 3982.3 | −48,082.5 | 6618.9 | −76,436.9 | 5237.8 | −57,899.6 | 6268.7 | −68,896.9 |
| 6 | 3882.7 | −44,199.8 | 6453.5 | −69,983.4 | 5106.9 | −52,792.7 | 6112.1 | −62,784.8 |
| 7 | 3788.1 | −40,411.7 | 6296 | −63,687.4 | 4982.3 | −47,810.4 | 5963 | −56,821.8 |
| 8 | 3697.9 | −36,713.8 | 6146.2 | −57,541.2 | 4863.8 | −42,946.6 | 5821 | −51,000.8 |
| 9 | 3608.9 | −33,104.9 | 5998.4 | −51,542.8 | 4746.8 | −38,199.8 | 5681.1 | −45,319.7 |
| 10 | 3518.8 | −29,586.1 | 5848.4 | −45,694.4 | 4628.1 | −33,571.7 | 5539.1 | −39,780.6 |
| 11 | 3432.9 | −26,153.2 | 5705.8 | −39,988.6 | 4515.2 | −29,056.5 | 5403.9 | −34,376.7 |
| 12 | 3351.2 | −22,802 | 5569.9 | −34,418.7 | 4407.8 | −24,648.7 | 5275.3 | −29,101.4 |
| 13 | 3266.1 | −19,535.9 | 5428.6 | −28,990.1 | 4295.8 | −20,352.9 | 5141.4 | −23,960 |
| 14 | 3189.8 | −16,346.1 | 5301.7 | −23,688.4 | 4195.5 | −16,157.4 | 5021.3 | −18,938.7 |
| 15 | 3110.5 | −13,235.6 | 5169.9 | −18,518.5 | 4091.2 | −12,066.2 | 4896.4 | −14,042.3 |
| 16 | 3034.7 | −10,200.9 | 5043.8 | −13,474.7 | 3991.3 | −8074.87 | 4776.98 | −9265.32 |
| 17 | 2960.5 | −7240.31 | 4920.76 | −8553.94 | 3894.02 | −4180.85 | 4660.47 | −4604.85 |
| 18 | 2888.41 | −4351.9 | 4800.75 | −3753.19 | 3799.05 | −381.799 | 4546.8 | −58.04 |
| 19 | 2817.96 | −1533.94 | 4683.66 | 930.47 | 3706.39 | 3324.591 | 4435.9 | 4377.859 |
| 20 | 2749.2 | 1215.282 | 4569.43 | 5499.9 | 3615.99 | 6940.58 | 4327.71 | 8705.571 |
| 21 | 2682.2 | 3897.454 | 4457.98 | 9957.887 | 3527.79 | 10,468.37 | 4222.16 | 12,927.73 |
| 22 | 2616.8 | 6514.208 | 4349.25 | 14,307.14 | 3441.76 | 13,910.13 | 4119.18 | 17,046.91 |
| 23 | 2552.9 | 9067.138 | 4243.16 | 18,550.3 | 3357.8 | 17,267.93 | 4018.71 | 21,065.62 |
| 24 | 2490.7 | 11,557.8 | 4139.68 | 22,689.98 | 3275.91 | 20,543.84 | 3920.69 | 24,986.31 |
| 25 | 2429.9 | 13,987.72 | 4038.71 | 26,728.69 | 3196.01 | 23,739.85 | 3825.07 | 28,811.38 |

According to Figure 15, the installation of the Uni-Solar PVL-128 system is poised to be a highly favorable investment. It is expected to generate revenue close to CNY 101,840 (equivalent to USD 14,144). The system's annual electricity production is estimated at around 12,057 kWh, and it is anticipated to reach the break-even point in the 18th year, considering the residential electricity tariff of CNY 0.588. Furthermore, over its 25-year service life, the system is projected to yield a profit of CNY 28,811.38.

A summary of the monthly produced energy records is presented in Figure 16. The simulation shows that monthly power generation exceeds 600 kWh. Based on Electricity and Water Authority (EWA) data, each house consumes approximately 450 kWh per month, therefore, installing a PV system on the roof can meet all the needs of residential electricity consumption, and the excess electricity can be integrated into the grid to supply industrial electricity. According to the $CO_2$ conversion rate formula discussed in Section 3.4.3, the approximate annual reduction in $CO_2$ emissions after installation of a roof PV system is approximately 6631.35 kg.

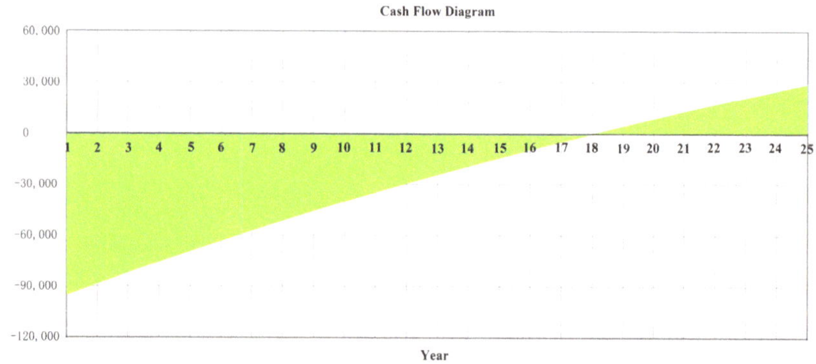


**Figure 15.** Cumulative savings over 25 years in investing in PV systems on rooftop.

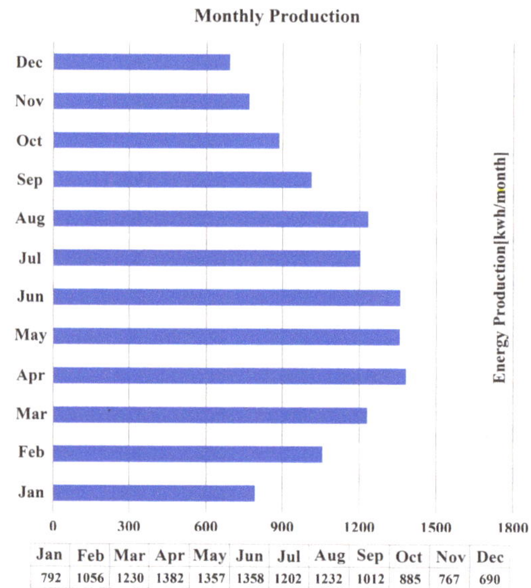


**Figure 16.** Monthly produced energy for the chosen dwelling.

## 4. Conclusions

This study employed a combination of experimental and numerical approaches to investigate mitigating energy consumption and reducing $CO_2$ emissions of a brick dwelling located in the Peking area. The following conclusions can be drawn from the findings of this research:

(1) Field experiments have revealed that the average annual solar radiation in the area amounts to 140 W/m$^2$. The outdoor temperature experiences significant fluctuations during both the winter and summer seasons. Despite some inherent thermal insulation properties of local brick dwellings, they have not been able to provide a consistently comfortable indoor living environment. Consequently, measures such as cooling and heating are necessary to enhance indoor thermal comfort.

(2) The simulation results have shown that the annual reduction in energy consumption for rooms with attached sunspaces maximum value is 884.6 kWh. Notably, there exists

a nonlinear correlation between the depth of the sunspace and energy consumption, indicating that total energy consumption increases as the depth of the sunspace exceeds 1.6 m.

(3) At a wall thickness of 800 mm, there is a reduction of 744.68 kWh in indoor energy consumption. Similarly, with the preferred thickness of roof insulation material set at 15 mm, indoor energy consumption decreases by 2514.58 kWh. Additionally, a recommended outer wall insulation material thickness of 50 mm results in a decrease of 1997 kWh in indoor energy consumption. The implementation of a well-suited brick dwelling renewal design has led to a reduction in the annual $CO_2$ emissions of a single residence by approximately 3799.79 kg.

(4) When employing Photovoltaic tiles of type BPsolar 380, the annual electricity production is estimated to reach 12,057 kWh, and the approximate annual reduction in $CO_2$ emissions after installation of a roof PV system is approximately 6631.35 kg.

This study serves as a valuable resource for architects seeking guidance on the control of energy consumption and $CO_2$ emissions within brick dwellings situated in the Peking area. Expanding our understanding of indoor thermal environments across various residential building types. However, it is important to emphasize that within this paper, the examination of parameters' impact on thermal comfort and energy consumption has been conducted separately. It is important to acknowledge that certain factors may have contrasting effects on thermal comfort and energy consumption. Subsequent endeavors will encompass extensive optimization initiatives directed toward the minimization of performance objectives, with the ultimate aim of attaining a globally optimal design that strikes a harmonious balance between economic efficiency and energy efficiency.

**Author Contributions:** Conceptualization, formal analysis, data curation, formal analysis, writing—original draft L.X.; Conceptualization, formal analysis, visualization, writing—review & editing L.F.; Funding acquisition, D.Z.; data curation J.L. All authors have read and agreed to the published version of the manuscript.

**Funding:** This work was supported by the State Key Project of the National Natural Science Foundation of China (Grant number. 51938002).

**Institutional Review Board Statement:** Not applicable.

**Informed Consent Statement:** Informed consent was obtained from all subjects involved in the study.

**Data Availability Statement:** All data generated by the author's field collection and software simulation.

**Acknowledgments:** The authors would like to thank J.L. for excellent technical support, data curation, and investigation.

**Conflicts of Interest:** The authors declare no conflict of interest.

## References

1. Kim, J.; Tzempelikos, A.; Braun, J.E. Energy Savings Potential of Passive Chilled Beams vs. Air Systems in Various US Climatic Zones with Different System Configurations. *Energy Build.* **2019**, *186*, 244–260. [CrossRef]
2. Alhaj Hasan, O.; Defer, D.; Shahrour, I. A Simplified Building Thermal Model for the Optimization of Energy Consumption: Use of a Random Number Generator. *Energy Build.* **2014**, *82*, 322–329. [CrossRef]
3. Cao, W.; Yang, L.; Zhang, Q.; Chen, L.; Wu, W. Evaluation of Rural Dwellings' Energy-Saving Retrofit with Adaptive Thermal Comfort Theory. *Sustainability* **2021**, *13*, 5350. [CrossRef]
4. Juan, X.; Ziliang, L.; Weijun, G.; Mengsheng, Y.; Menglong, S. The Comparative Study on the Climate Adaptability Based on Indoor Physical Environment of Traditional Dwelling in Qinba Mountainous Areas, China. *Energy Build.* **2019**, *197*, 140–155. [CrossRef]
5. Sun, H.; Leng, M. Analysis on Building Energy Performance of Tibetan Traditional Dwelling in Cold Rural Area of Gannan. *Energy Build.* **2015**, *96*, 251–260. [CrossRef]
6. Li, B.; Du, C.; Yao, R.; Yu, W.; Costanzo, V. Indoor Thermal Environments in Chinese Residential Buildings Responding to the Diversity of Climates. *Appl. Therm. Eng.* **2018**, *129*, 693–708. [CrossRef]

7. Zhu, X.; Liu, J.; Yang, L.; Hu, R. Energy Performance of a New Yaodong Dwelling, in the Loess Plateau of China. *Energy Build.* **2014**, *70*, 159–166. [CrossRef]
8. Yuk, H.; Choi, J.Y.; Kim, Y.U.; Chang, S.J.; Kim, S. Historic Building Energy Conservation with Wooden Attic Using Vacuum Insulation Panel Retrofit Technology. *Build. Environ.* **2023**, *230*, 110004. [CrossRef]
9. Fantucci, S.; Serra, V. Investigating the Performance of Reflective Insulation and Low Emissivity Paints for the Energy Retrofit of Roof Attics. *Energy Build.* **2019**, *182*, 300–310. [CrossRef]
10. Li, H. *Green Transformation of Thermalenvironment of Block Dwellingsin Jizhong Area Based on Dynamic Simulation*; Hebei University of Architecture: Zhangjiakou, China, 2011. [CrossRef]
11. Zhang, L.; Dong, Z.; Liu, F.; Li, H.; Zhang, X.; Wang, K.; Chen, C.; Tian, C. Passive Solar Sunspace in a Tibetan Buddhist House in Gannan Cold Areas: Sensitivity Analysis. *J. Build. Eng.* **2023**, *67*, 105960. [CrossRef]
12. Xu, J.; Yang, W.; Lu, Z.; Wu, Y.; Hou, C.; Liu, D. Quality Analysis on Indoor Thermal Comfort and Energy-Saving Improvement Strategy of Slate Dwellings, China. *Buildings* **2022**, *12*, 468. [CrossRef]
13. León, E.Z.; Barraza, C.C. Adaptability of Photovoltaic Mono-Polycrystalline Solar Panels and Photovoltaic Roof Tiles on Dwelling Roofs of Real Estate Developments. *Rev. Constr.* **2019**, *18*, 42–53. [CrossRef]
14. Odeh, S. Thermal Performance of Dwellings with Rooftop PV Panels and PV/Thermal Collectors. *Energies* **2018**, *11*, 1879. [CrossRef]
15. Földváry Ličina, V.; Cheung, T.; Zhang, H.; de Dear, R.; Parkinson, T.; Arens, E.; Chun, C.; Schiavon, S.; Luo, M.; Brager, G.; et al. Development of the ASHRAE Global Thermal Comfort Database II. *Build. Environ.* **2018**, *142*, 502–512. [CrossRef]
16. Lamsal, P.; Bajracharya, S.B.; Rijal, H.B. A Review on Adaptive Thermal Comfort of Office Building for Energy-Saving Building Design. *Energies* **2023**, *16*, 1524. [CrossRef]
17. Yao, R.; Li, B.; Liu, J. A Theoretical Adaptive Model of Thermal Comfort—Adaptive Predicted Mean Vote (APMV). *Build. Environ.* **2009**, *44*, 2089–2096. [CrossRef]
18. Timur, B.A.; Başaran, T.; İpekoğlu, B. Thermal Retrofitting for Sustainable Use of Traditional Dwellings in Mediterranean Climate of Southwestern Anatolia. *Energy Build.* **2022**, *256*, 111712. [CrossRef]
19. Hou, L.Q.; Yang, L.; Liu, D.L.; Xu, X.Y.; Liu, J.P. Research on optimization of envelope structure of traditional residential buildings in Kangding. *Build. Energy Effic.* **2016**, *3*, 43–46+50. [CrossRef]
20. Gelesz, A.; Catto Lucchino, E.; Goia, F.; Serra, V.; Reith, A. Characteristics That Matter in a Climate Façade: A Sensitivity Analysis with Building Energy Simulation Tools. *Energy Build.* **2020**, *229*, 110467. [CrossRef]
21. Wang, H.F.; Chiou, S.C. Spatial Form Analysis and Sustainable Development Research of Traditional Residential Buildings. *Sustainability* **2020**, *12*, 637. [CrossRef]
22. Saif, J.; Wright, A.; Khattak, S.; Elfadli, K. Keeping Cool in the Desert: Using Wind Catchers for Improved Thermal Comfort and Indoor Air Quality at Half the Energy. *Buildings* **2021**, *11*, 100. [CrossRef]
23. Xu, C.; Li, S.; Zhang, X. Energy Flexibility for Heating and Cooling in Traditional Chinese Dwellings Based on Adaptive Thermal Comfort: A Case Study in Nanjing. *Build. Environ.* **2020**, *179*, 106952. [CrossRef]
24. Wang, F.; Wang, S.; Cheng, B.; Wang, W. To Inhabit, Retain, or Abandon? Adaptive Utilization of Energy-Efficient Sunken Buildings by Rural Households in Shanzhou, China. *Energy Build.* **2022**, *255*, 111668. [CrossRef]
25. Becker, F.G.; Cleary, M.; Team, R.M.; Holtermann, H.; The, D.; Agenda, N.; Science, P.; Sk, S.K.; Hinnebusch, R.; Hinnebusch, A.R.; et al. Design Standard for Energy Efficiency of Residential Buildings in Hot Summer and Cold Winter Zone. *Syr. Stud.* **2015**, *7*, 37–72.
26. Huang, J.; Sun, W.; Zhang, Z.; Ling, Z.; Fang, X. Thermal Protection of Electronic Devices Based on Thermochemical Energy Storage. *Appl. Therm. Eng.* **2021**, *186*, 116507. [CrossRef]
27. Yang, T.; Wei, J.; Guo, Y.; Lv, Z.; Xu, Z.; Cheng, Z. Manipulation of Oxygen Vacancy for High Photovoltaic Output in Bismuth Ferrite Films. *ACS Appl. Mater. Interfaces* **2019**, *11*, 23372–23381. [CrossRef] [PubMed]
28. Hou, C.F.; Li, S.Q.; Li, B.; Sun, X.D. Incoherently Coupled Bright-Dark Screening-Photovoltaic Soliton Pairs in Biased Photovoltaic Photorefractive Crystals. *Acta Phys. Sin.* **2001**, *50*, 1711–1712. [CrossRef]
29. Lu, K.-Q.; Zhang, Y.-P.; Tang, T.-T.; Lu, Z.-X.; Liu, L. Steady-State Screening-Photovoltaic Spatial Solitons in the Biased Photorefractive-Photovoltaic Crystals. *Chin. Phys. Lett.* **2001**, *18*, 233–235. [CrossRef]
30. Mahar, W.A.; Verbeeck, G.; Reiter, S.; Attia, S. Sensitivity Analysis of Passive Design Strategies for Residential Buildings in Cold Semi-Arid Climates. *Sustainability* **2020**, *12*, 1091. [CrossRef]
31. Mahar, W.A.; Verbeeck, G.; Singh, M.K.; Attia, S. An Investigation of Thermal Comfort of Houses in Dry and Semi-Arid Climates of Quetta, Pakistan. *Sustainability* **2019**, *11*, 5203. [CrossRef]
32. Bayarri, M.J.; Berger, J.O.; Cafeo, J.; Garcia-Donato, G.; Liu, F.; Palomo, J.; Parthasarathy, R.J.; Paulo, R.; Sacks, J.; Walsh, D. Computer Model Validation with Functional Output. *Ann. Stat.* **2007**, *35*, 1874–1906. [CrossRef]
33. Gao, J.; Wang, Y.; Wargocki, P. Comparative Analysis of Modified PMV Models and SET Models to Predict Human Thermal Sensation in Naturally Ventilated Buildings. *Build. Environ.* **2015**, *92*, 200–208. [CrossRef]
34. Cheung, T.; Schiavon, S.; Parkinson, T.; Li, P.; Brager, G. Analysis of the Accuracy on PMV—PPD Model Using the ASHRAE Global Thermal Comfort Database II. *Build. Environ.* **2019**, *153*, 205–217. [CrossRef]
35. Enescu, D. A Review of Thermal Comfort Models and Indicators for Indoor Environments. *Renew. Sustain. Energy Rev.* **2017**, *79*, 1353–1379. [CrossRef]

36. Kim, J.; Zhou, Y.; Schiavon, S.; Raftery, P.; Brager, G. Personal Comfort Models: Predicting Individuals' Thermal Preference Using Occupant Heating and Cooling Behavior and Machine Learning. *Build. Environ.* **2018**, *129*, 96–106. [CrossRef]
37. Karunakaran, R.; Iniyan, S.; Goic, R. Energy Efficient Fuzzy Based Combined Variable Refrigerant Volume and Variable Air Volume Air Conditioning System for Buildings. *Appl. Energy* **2010**, *87*, 1158–1175. [CrossRef]
38. Wang, Y.; Zhao, F.Y.; Kuckelkorn, J.; Liu, D.; Liu, L.Q.; Pan, X.C. Cooling Energy Efficiency and Classroom Air Environment of a School Building Operated by the Heat Recovery Air Conditioning Unit. *Energy* **2014**, *64*, 991–1001. [CrossRef]
39. Zou, H.; Liu, Z.; Long, E. An Experimental Study on External Ventilation to the Heating Performance of Household Air Source Heat Pump. *Front. Energy Res.* **2021**, *9*, 785461. [CrossRef]
40. Shahsavar, A.; Salmanzadeh, M.; Ameri, M.; Talebizadeh, P. Energy Saving in Buildings by Using the Exhaust and Ventilation Air for Cooling of Photovoltaic Panels. *Energy Build.* **2011**, *43*, 2219–2226. [CrossRef]
41. Helton, J.C.; Johnson, J.D.; Sallaberry, C.J.; Storlie, C.B. Survey of Sampling-Based Methods for Uncertainty and Sensitivity Analysis. *Reliab. Eng. Syst. Saf.* **2006**, *91*, 1175–1209. [CrossRef]
42. *GB/T 51366-2019*; Standard for Building Carbon Emission Calculation. The Standardization Administration of the People's Republic of China: Beijing, China, 2019.

sustainability

MDPI

*Article*

# The Obstacles to the Growth of the Renewable Energy Industry in the European Union

**Dmytro Osiichuk**

Department of Finance, Kozminski University, 03-301 Warsaw, Poland; dosiichuk@kozminski.edu.pl

**Abstract:** The study presents a comparative analysis of the financial standing of renewable- and conventional-energy-generating companies domiciled in the European Union, and explores factors contributing to their operating performance and financing needs. The comparative analysis of electricity-generating companies demonstrates that renewable energy providers have steadily improved their financial standing and appear well-prepared to compete with conventional energy generation from fossil fuels. Relying on univariate statistical tests, we demonstrate that, at present, renewable energy companies generate a significantly higher return on assets and make larger investment outlays than their conventional peers. Univariate cross-sectional tests confirm that they also enjoy access to cheaper equity and more plentiful debt financing thanks to a shift in the priorities of financial institutions and capital markets. We posit that, besides the cumbersome permit system, the biggest challenge to the operating performance of renewable energy firms could reside in the restrictive revenue caps, price regulation, and renewable firms' exposure to spot market price fluctuations. The econometric analysis based on multivariate panel regression models demonstrates that the profitability and financing needs of renewable energy companies are more sensitive to electricity prices than those of conventional energy firms. We conclude that higher energy prices and greater reliance on market pricing mechanisms could significantly boost the performance of the renewable energy sector.

**Keywords:** renewable energy; profitability; industrial policy

**Citation:** Osiichuk, D. The Obstacles to the Growth of the Renewable Energy Industry in the European Union. *Sustainability* **2023**, *15*, 14612. https://doi.org/10.3390/su151914612

Academic Editors: Elena Lucchi, Tianyi Chen and Wen Zhang

Received: 15 September 2023
Revised: 4 October 2023
Accepted: 7 October 2023
Published: 9 October 2023



## 1. Introduction

### *1.1. The Energy Transition in the European Union*

The European Union (EU) countries have embarked on an extensive policy and regulatory update in order to accelerate the development of renewable energy as part of the broad package of measures aimed at limiting the effects of climate change. The pending and already implemented measures include revamped taxation mechanisms and carbon trading schemes, changes in trading regimes directed at carbon-intensive imports, tax reliefs, and targeted assistance to companies involved in the key industries participating in the transition towards renewables. The most ambitious project within the strategy is "Fit for 55", which aims to implement several regulatory changes precipitating a decline in the EU's carbon footprint. The capital markets and industry have followed the directions set by regulators and implemented internal mechanisms and incentive schemes designed to benefit from the ongoing transition and secure leadership positions within the overhauled industrial landscape [1].

The regulatory pressure and market shifts have already delivered some commendable results. Fueled by plentiful capital inflows, lax monetary policy, and targeted financial assistance, the industries tied to the transformation have experienced dynamic growth over the last decade [2]. Investments in renewable energy generation have become higher than in oil and gas extraction, and conventional energy generation is being gradually phased out in EU countries [3].

At the same time, some projections of greenhouse gas emissions deem the current pace of renewable energy growth insufficient to meet the targets [4] set in the policy guidelines. Therefore, an in-depth understanding of the factors that may be responsible for slowing down the renewables' growth is necessary to fine-tune the enacted policies and, if necessary, reshape them relying on market incentives rather than on regulatory pressure. In our opinion, striking the right balance between the weights of market-based mechanisms and centralized industrial policies based on the allocation of public financing may be a critical factor in shaping the trajectory of further development of the renewable energy sector.

One of the prevailing views among policymakers maintains that the renewable energy industry is primarily held back by insufficient capital inflows. The capital intensity of the industry [5] makes attaining the necessary economies of scale difficult. Since natural monopolies have historically dominated the energy industry, new entries are challenging, and transformations of the incumbents in the direction of renewable energy generation are lengthy and rigid. Across many mature markets, the energy industry is also unionized, which may further hinder the strategic pivots of the incumbents [6].

### 1.2. The Goal of the Study

The present study is structured around two principal axes. First, we compare and contrast the current financial standing of renewable energy companies and their fossil-fuel-based peers, with the aim of establishing the current context in which the sector is operating and deploying new generating capacity. We also try to quantitatively assess whether the existing regulatory framework has been successful at creating an environment favorable to renewable energy companies and their needs in terms of long-term financing. Secondly, we focus on what, in our opinion, may constitute one of the challenges to the growth of the industry: price regulation coupled with caps on the upside movement of prices in the event of shocks. The findings reported in the paper contribute to the empirical literature on the determinants of the development of the renewable energy industry and on the mechanisms of the energy market and the industrial organization of the energy sector. The findings may be of interest to regulatory bodies, which are looking for ways to shape the mechanisms underlying the energy market and accelerate the ongoing transformation of the European Union's economy in the direction of sustainable energy procurement.

Our comparative analysis of conventional and renewable energy companies in the European Union demonstrates that the problem of capital availability has mainly been solved. The findings of statistical analysis demonstrate that renewable energy companies already enjoy better access to capital and a lower cost of capital than their conventional energy generation peers. Both intermediated financing and direct capital market financing appear to greatly favor renewable companies over their conventional competitors. In the face of the growing regulatory pressure directed at accelerating the energy transformation, market discipline has forced financial institutions to adopt policies aimed at reducing exposure to carbon-intensive industries and increasing the share of renewables and other sustainable sectors in their credit portfolios.

The result of improved access to capital has been a sharp increase in the level of investment by renewable energy generation companies. Expensive greenhouse emission permits on the European market have also rendered renewable energy much more competitive by penalizing the inefficient asset base of fossil-fuel-based companies.

Overall, it appears that the critical market shift has already happened, and renewable energy is going to keep growing in the foreseeable future. At the same time, policymakers keep using direct intervention [6] instead of relying on market mechanisms, which may significantly slow down the development of the industry. Tax reliefs and subsidies remain the preferred tool for incentivizing the development of the industry and adjacent sectors. At the same time, the potential profitability and internally generated cash flows of renewable energy companies appear to be hindered by the contemplated and enacted revenue caps. Lifting the caps or abandoning the mechanism altogether, along with a systemic liberalization of the regulatory standards in the domain of permit procedures, could spur industry performance [7]. More importantly, industry development would be primarily driven by market mechanisms. This would reduce the risks of distortions in the pricing mechanisms, alter consumer incentives by rewarding the optimization of energy use [8], and possibly reduce the widespread outcry over the vast amounts of public investments, which are being directed to the development of the industry at to the expense of other sectors.

Our econometric findings derived from panel regression modeling on a sample of energy-generating companies based in European Union countries demonstrate that the financials of the renewable energy industry remain much more sensitive to the fluctuations in energy prices than those of conventional energy companies. We show that higher energy prices are not only strongly and positively associated with the level of profitability of renewable energy companies but also with their financing needs, with the newly incurred debt substituting for the weakened internally generated cash flow. Abandoning the instruments of price regulation and revenue caps and shifting the industry towards reliance on stabler energy procurement contracts could not only ameliorate the financial standing of those companies but also provide the necessary financial resources for organic growth and deleveraging. These shifts could also reduce the pressure on public finances.

This paper contributes to a broad strand of empirical literature analyzing the determinants of renewable energy generation development. By presenting a comparative analysis of renewable and conventional electricity-generating companies across the European Union, we delineate the factors that could slow down the growth of the renewable energy industry and, as a result, the transition of the European Union towards climate sustainability goals. The remainder of the paper is organized as follows: First, we present the literature review. Subsequently, we describe the research methodology and the dataset used for econometric analysis. The concluding section presents the empirical findings and policy discussion.

## 2. Literature Review

All the major decision-making and legislative bodies of the European Union appear to support a fast transition towards sustainable energy generation. This policy pivot requires institutions to navigate two major types of hurdles: sociopolitical and economic.

### 2.1. The Economic Challenges to the Development of Renewable Energy

Several specific features of the renewable energy industry shape the dynamics of the transition towards a higher share of renewables in the energy mix of EU countries. To start with, installing renewable energy generation capacity requires significant investment outlays, which can be financed either from the internally generated cash flows of energy companies or from public investment [9]. The "chicken-and-egg" problem stems from the insufficiency of cash flows of the developing renewable energy industry comprised of many newly created entities to finance the necessary investment expenditures. A bigger problem resides in the instability of renewable energy sources [10], which necessitates a redesign of the grid in order to ensure the storage or frictionless reallocation of energy depending on intermittent demand and supply fluctuations [11]. The infrastructure necessary for connecting renewable energy generating capacity into the integrated energy management system requires substantial financing. In particular, the biggest challenge appears to reside in the capacity to store redundant energy in order to cover a possible subsequent surge in electricity consumption [12].

Despite a reduction in the average expenditure involved in installing an additional unit of renewable energy generating capacity and in average production costs [13], the above factors cause renewable energy to be at a disadvantage vis-à-vis that generated by fossil-fuel-based companies. The comparison speaks in favor of renewable energy if the hidden costs associated with increased emissions of greenhouses are included in the calculations of the unit price of electricity [14]. Since burning fossil fuels is proven to cause significant negative externalities, remedial regulatory action is necessary in order to incorporate the hidden costs imposed on third parties into the market pricing mechanisms.

The European Union has done an outstanding job in developing carbon pricing mechanisms. The EU emissions trading system is the biggest functioning market for permits to emit greenhouse gases. It appears to work very well in incentivizing the market participants to change their energy mix and energy consumption patterns [15]. The spikes in the prices of greenhouse gas emission permits have been shown to carry macroeconomic implications for some of the EU countries and sometimes even threaten the competitiveness of some of the legacy industries [16]. The architecture of the market has allowed renewable energy companies to become more competitive not only by altering the pricing mechanism but also by imposing financial penalties on inefficient assets associated with increased greenhouse gas emissions. The financial markets have reacted accordingly by substantially increasing the risk premium associated with polluting assets and reducing that attached to sustainable assets [17]. As a result, renewable energy companies have enjoyed improved access to external financing through all financing vehicles.

While the reliance on the mechanism of carbon pricing to alleviate the problem of negative externalities is wholly warranted, other instruments prevalent in the energy industry may be distorting the pricing mechanisms and, as a result, the incentives of the consumers, instead of boosting the growth of the industry. The European Union spends vast amounts of money in the form of subsidies to prop up the stability of energy prices for consumers in the face of possible price shocks [18]. While subsidies undoubtedly allow sheltering the most vulnerable consumers from the effects of price hikes, they also reduce the efficiency of the pricing mechanism. As market prices are not allowed to fully and instantaneously incorporate fundamental demand- and supply-side factors, consumers may be reluctant to slash consumption, optimize consumption patterns, and invest in prosumer infrastructure; producers, in turn, may be slower in switching to cleaner generation technologies [19]. These negative incentives may be particularly detrimental to the long-term transformation of legacy industries. High energy prices represent an indirect penalty on operational activities involving excessive spending of energy, particularly that derived from non-sustainable carbon-intensive sources. Reinforcing the existing inefficiencies may result in a slowdown of the reallocation of capital from the sectors most likely to be the losers in the current transformation toward sustainable energy.

The financial situation of renewable energy companies is also complicated by the rising costs and the tightening monetary policy [20] accompanying the inflation upsurge across the EU countries. While renewable companies have enjoyed improved access to debt and equity financing through various financing vehicles, the constantly growing investment demand has caused those companies to accumulate significant amounts of interest-bearing debt. As a result, when the central banks across European countries started increasing interest rates, renewable energy companies experienced a significant increase in the cost of debt servicing. The relatively meager return on invested capital observed in the industry throughout the last decade has had to compete with the increasing cost of capital, resulting in a growing number of new investment projects in the industry being shelved for the lack of projected profitability. The rising costs embedded in the increasing levels of inflation [21] observed across the majority of European countries have also been detrimental to the bottom lines of renewable energy companies. Starting from the COVID-19 pandemic, the energy sector experienced the same problems as the rest of the real economy: the rising costs of logistics and raw materials [22]. The industry is particularly reliant on the supply of industrial metals. The resulting upsurge in the cost of installing new infrastructure and procuring new equipment, coupled with the limited ability to pass the costs along to the consumer, has created a significant challenge for industry executives and pushed them to seek external assistance.

### 2.2. The Sociopolitical Aspects of the Transition toward Renewable Energy

The observed extensive reliance on energy subsidies has important sociopolitical underpinnings [23]. Regulators, executives, and legislators react to the public outcry accompanying the fluctuation in prices in response to the ongoing energy transition. As the unfolding transformation is likely to result in serious infrastructural bottlenecks [24] and siphon substantial amounts of capital from other sectors of the economy, it is likely to contribute to price instability as well as technical problems with the grid and the stability of supply. The prevailing viewpoint among decision-makers is that cushioning any possible price fluctuations is the only way to ensure a smooth transformation without substantial public backlash. The handling of the problem of risk management is thus shifted to the energy-generating companies. The troublesome part of implementing a risk management strategy in such settings resides in the need to secure the company against the downside movements of prices. Long-term energy procurement contracts [25] could partially solve the problem by providing renewable energy companies with relatively stable cash flows and an opportunity to invest those predictable cash flows into developing operational activities. The problem is that renewable energy companies may have to rely on spot markets [26] to apportion their energy. The latter could be an advantage if companies were allowed to benefit from the upward movement of energy prices. However, the mechanisms of price regulation are likely to smooth such fluctuations. The financing shortfalls could be compensated for with financial assistance from dedicated financing vehicles. However, similar to subsidies [27], this mechanism carries a number of risks associated with the allocation of allotted financial resources. In particular, it is not always clear which criteria should be used when allocating capital or granting procurement contracts, which may result in winnowing and favoritism [28].

As the transition towards sustainable industry is pursued across the majority of developed countries, it is likely to absorb enormous amounts of capital, which would have otherwise been available to other industries. The possible resulting crowding-out effect [29] may precipitate a backlash against further growth in the use of renewables. The latter may precipitate a regulatory response aimed at curbing the development of the industry.

The relative flexibility of renewables makes them more likely to be shut off in the event of excess supply. This happened multiple times across European countries when supply significantly outstripped demand, and the resulting surplus had to be eliminated by limiting the generating capacity of renewables since fossil-fuel-generating facilities cannot be flexibly turned on and off and are used to accommodate intermittent renewable generation [30]. The integrated market has been equipped with tools to address this problem. However, it is likely to persist as long as there is no low-cost, extensively implemented solution for the temporary storage of surplus energy.

The final and possibly one of the most difficult challenges hindering the development of renewable energy generation relates to the difficulty of obtaining necessary permits, stakeholders' consent, and regulatory approvals during the deployment phase [31]. The bureaucratic procedures involved in the process of installing new generating capacities are incredibly lengthy and burdened with multiple steps, where every approval makes the project's continuation a contingency. Estimates suggest that eliminating bureaucratic barriers, particularly those related to land procurement, building permits, and integration of newly installed capacity into the energy system, could significantly accelerate the growth of the industry [32].

## 3. Database and Research Design

The present study relies on a firm-year dataset encompassing sampled listed energy companies domiciled across European Union countries. The sample consists of renewable energy companies and conventional energy companies. Notably, the former group does not distinguish between renewables from different sources, such as wind, solar, geothermal, etc., while the latter does not distinguish between companies reliant on different types of fossil fuels to generate electricity. We aggregate renewable energy companies into a single cluster and study their broad statistical properties. We did not exclude companies that conduct trading business alongside energy generation since many of the oldest electrical utilities, including the state-owned ones, have a diversified line of operating activities. The panel dataset is unbalanced and geographically heterogeneous, with companies from bigger economies prevailing in the sample composition. Overall, we studied 100 companies, all publicly traded. The country breakdown of the sample is presented in Appendix A. For the analyzed companies, we collected yearly financial data from the Refinitiv Eikon database. All nominal variables subject to empirical analysis have been scaled and winsorized at a 1% level to eliminate outliers. The period of analysis spans from 2009 to 2022, with the raw dataset comprised of 1023 firm-year observations. The definitions of variables used in the analysis are presented in Table 1. The descriptive statistics can be found in Table 2. The data on the fluctuations in electricity prices across the studied European Union countries have been collected from the dataset of the European Commission.

**Table 1.** Definitions of variables.

| Variable Name | Definition |
|---|---|
| Firm size | Natural logarithm of a firm's contemporaneous reported total assets |
| EBIT margin | The ratio of reported EBIT (earnings before interest and taxes) to total sales |
| ROA | Return on assets |
| OCF | The value of contemporaneous operating cash flows scaled by the value of total assets |
| Capex | The value of contemporaneous capital expenditures scaled by the value of total assets |
| Asset turnover | The ratio of total sales to contemporaneous total assets |
| Asset tangibility | The ratio of gross property, plant, and equipment to total assets |
| Cash | The ratio of cash and cash equivalents to total assets |
| Debt | The ratio of total debt to total assets |
| Debt issuances | The ratio of net change in total interest-bearing debt YoY to contemporaneous total assets |
| Stock issuances | The ratio of equity issuances/stock repurchases to contemporaneous total assets |
| Debt issuances binary | Binary variable equal to 1 if during a given year a company issued new debt (net change in interest-bearing debt is positive) |
| Stock issuances binary | Binary variable equal to 1 if during a given year a company issued equity |
| Net change in cash | The ratio of net change in cash to contemporaneous total assets |
| Cost of debt | The cost of debt used for the calculation of the weighted average cost of capital. It is estimated as a marginal cost of raising new debt at the time of measurement. The rate is extracted from a credit curve and is calculated as a weighted average of the short- and long-term cost of debt (Refinitiv Eikon). |
| Cost of equity | The cost of equity used for the calculation of the weighted average cost of capital. As per Refinitiv Eikon's methodology, it is calculated using the formula re = rf + beta * MRP, where re—cost of equity, rf—inflation-adjusted risk-free rate, beta—beta of the company's stock, MRP—market risk premium estimated as a difference between the expected return on a diversified market portfolio and risk-free rate (adjusted for inflation). |
| WACC | Weighted average cost of capital calculated as a weighted average of the cost of debt and equity. The methodology for calculating the cost of equity and debt is presented above. |
| Energy price | The price of electricity (EUR/kilowatt-hour), excluding taxes and levies in the band between 2500 kWh and 5000 kWh |
| Time since incorporation | The number of years elapsed since the company's date of incorporation until the end of the analyzed year |
| Renewable | Binary variable equal to 1 if a given company's primary business consists of renewable energy generation |

Source: own elaboration.

The empirical part of the paper is structured as follows: At stage one, we compare the financials of renewable and fossil-fuel-based energy companies. The comparison is performed across three principal dimensions: profitability, investment, and capital structure. The comparative analysis allows us to understand the current development context of the specific subindustries in the energy sector. We want to investigate whether renewable energy companies have sufficient internally generated and external resources to meet their investment demand and to contrast the aggregate level of investment in the industry depending on the energy generation type. To that end, we rely primarily on univariate statistical tests to compare the subsamples of companies with different asset structures. The goal of this stage is also to ascertain by the indirect means of observing the dynamics of energy companies' financials whether the existing policies aimed at boosting the development of the renewable energy sector are succeeding at channeling financial resources towards industry growth. Since the principal goal of the majority of enacted policies is to facilitate access of renewable energy companies to financial resources in order to further improve the technology and lower the unit costs, the primary key performance indicators of such policies are measures of those companies' access to external financing.

**Table 2.** Descriptive statistics.

| Variable Name | Mean | Std. Dev. | Min | Max |
|---|---|---|---|---|
| Firm size | 20.96 | 3.375 | 6.908 | 27.799 |
| EBIT margin | 0.142 | 0.206 | −0.37 | 0.546 |
| ROA | 1.571 | 6.885 | −18.62 | 15.506 |
| OCF | 0.059 | 0.072 | −0.107 | 0.27 |
| Capex | 0.068 | 0.07 | 0 | 0.355 |
| Asset turnover | 0.482 | 0.554 | 0 | 2.939 |
| Asset tangibility | 0.819 | 0.472 | 0.001 | 2.048 |
| Cash | 0.107 | 0.14 | 0 | 0.821 |
| Debt | 0.287 | 0.21 | 0 | 0.783 |
| Debt issuances | 0.022 | 0.089 | −0.17 | 0.349 |
| Stock issuances | 0.06 | 0.177 | −0.027 | 1 |
| Net change in cash | 0.017 | 0.069 | −0.08 | 0.313 |
| Cost of debt | 1.891 | 1.614 | −0.013 | 6.002 |
| Cost of equity | 5.76 | 2.923 | 1.398 | 14.068 |
| WACC | 4.329 | 2.145 | 1.13 | 10.361 |
| Energy price | 0.127 | 0.036 | 0.068 | 0.267 |
| Renewable | 0.313 | 0.464 | 0 | 1 |

Source: own elaboration.

At stage two of empirical analysis, we investigate the relationship between energy companies' operating performance and the dynamics of energy prices. The goal is to check how sensitive the companies' financials are to the market parameters, which are subject to regulatory tinkering. We also want to contrast these sensitivity indicators across the types of energy generation. We postulate that renewable energy companies may be more exposed to energy price fluctuations because of the existing procurement policies, higher exposure to spot markets, and possibly because of the caps, which may limit the possible operational benefits of upside movements in prices. In contrast, conventional energy companies may be more insulated from market movements because of legacy mechanisms of energy procurement, from which they may benefit by smoothing their contemporaneous operational performance. At this stage of analysis, we utilize multivariate panel regression models, in which the explained variables are different measures of the operating performance of the sampled companies, including EBIT margin, return on assets (ROA), and operating cash flows. The baseline regression model specification is as follows:

$$Operational\ performance_{ij} = \beta_0 + \beta_1 Energy\ price_j + \beta_2 Renewable_{ij} + \beta_3 Renewable_{ij} * Energy\ price_j + \beta_4 CONTROL_{ij} + \beta' Year_i + \beta' Country_i + \beta_5 Error_{ij}, \quad (1)$$

where $CONTROL_{ij}$ is a vector of control variables measuring firm size, asset tangibility, asset turnover, and the level of cash reserves. $\beta'$ represents a vector of regression coefficients at binary variables encoding year and country fixed effects. The primary explanatory variable is energy price, which may supposedly be associated with the firms' contemporaneous operating performance. The interaction terms comprised of a binary variable encoding renewable energy companies and a continuous variable measuring the fluctuations in energy prices should allow us to check the differences in the sensitivity of operating performance to price swings across the two subsamples of energy companies with different generation types. The models include year and country control variables.

If the energy price turns out to be correlated with firms' operating performance, then it should also be associated with their financing needs and the patterns of procurement of external financing under fluctuations of operating performance. If energy prices go down, one may reasonably expect the cash flows of a company to decline. Assuming that the company maintains a stable level of investment demand, the resulting gap in internally generated resources can be plugged with either previously accumulated financial resources or with external funding. To verify this conjecture, we model the relationship between energy price fluctuations and energy companies' demand for external financing, with the balancing figure being the change in net cash reserves. Using panel regression modeling, we quantify the associative link between energy prices and the relative sizes of stock and debt issuances year-on-year. The size of net equity and debt issuances are scaled by the value of total assets. The baseline multivariate regression model has the following specification:

$$\begin{aligned} Net\ external\ financing_{ij} = \beta_0 + \beta_1 Energy\ price_j + \beta_2 Renewable_{ij} + \beta_3 Renewable_{ij} * Energy\ price_j + \\ \beta_4 CONTROL_{ij} + \beta' Year_i + \beta' Country_i + \beta_5 Error_{ij} \end{aligned} \quad (2)$$

The definitions of variables in the equation can be found in Table 1. We run a regression of the same specification for the net change in cash over a given year scaled by total assets. We study not only the magnitude of the necessary external capital procurement but also the frequency and the relative likelihood thereof. To that end, we binary-code the instances of net debt and net equity issuances by a given company during a given year. The resulting binary variables are subject to multivariate logistic regression modeling. The explained variable is the occurrence of external capital procurement. The list of explanatory variables is the same as in Equation (2). We postulate that the sampled companies, which find themselves in need of funding to close the liquidity gap, are likely to not only issue debt and equity in larger tranches, but they may also be more likely to issue external financing with higher relative frequency compared to their peers experiencing stabler cash flows and less vulnerable to market fluctuations.

This stage of analysis should help us elucidate how energy companies compensate for intermittent fluctuations in the internally generated cash flows, which may be caused, among other things, by the fluctuations in energy prices. Companies that are more subject to market fluctuations, and as a result, may experience an elevated level of cash flow volatility, may be more reliant on external financing. Together with the results on the availability of external financing and the specificity of the capital structure of energy companies reliant on different types of energy generation, this should help us better understand the operating challenges faced by the industry in terms of cash flow planning.

At the final stage of analysis, we check how the operating performance of the sampled companies evolves over time. This is done to better understand the prospects of the studied companies in terms of their future chances of becoming sufficiently profitable to finance their investment needs and compensate for any intermittent fluctuations in operating cash flows. If the performance of renewable energy companies is found to be improving over time, this could presage a potentially faster deployment of additional energy generating capacity and, as a result, faster transition towards sustainable integrated energy systems. We are also interested to know whether the performance of conventional energy companies is deteriorating or remaining stable, as they are likely to actively participate in the ongoing transition.

## 4. Principal Findings

Table 3 presents the results of univariate tests of differences in the financials of renewable (subsample 2) and conventional (subsample 1) energy companies. The following differences stand out:

**Table 3.** Univariate tests of differences between subsamples of conventional (subsample 1) and renewable (subsample 2) energy companies.

| | Mean1 | Mean2 | dif | St Err | t Value |
|---|---|---|---|---|---|
| ROA | 2.248 | 0.023 | 2.225 | 0.398 | 5.6 *** |
| Capex | 0.057 | 0.093 | −0.037 | 0.004 | −8.8 *** |
| Debt | 0.241 | 0.387 | −0.145 | 0.011 | −12.95 *** |
| Debt issuances | 0.009 | 0.049 | −0.04 | 0.005 | −7.8 *** |
| Cost of debt | 1.708 | 2.241 | −0.533 | 0.152 | −3.5 *** |
| Cost of equity | 6.285 | 4.758 | 1.528 | 0.27 | 5.65 *** |
| WACC | 4.659 | 3.701 | 0.957 | 0.2 | 4.8 *** |

Note: The table presents the results of the Student's *t*-test for the difference of means of subsamples of conventional and renewable energy firms. The tests' statistical significance at 1% level is indicated with ***. Source: own elaboration.

Renewable energy companies exhibit significantly higher levels of investment demand than their conventional peers. The average level of capital expenditures scaled by contemporaneous total assets is 0.093 in the renewable energy subsample versus 0.0574 for conventional energy companies (sig.: 1%). In line with the expectations of the ongoing energy transition, renewable energy companies are driving the industry's investment demand.

Figure 1 demonstrates that the return on assets of renewable energy companies has been steadily improving over the last decade and thus catching up with the performance indicators of conventional energy companies. In 2018, the average return on assets generated by the sampled renewable energy companies exceeded the average among conventional energy companies. The improvement in the renewable energy companies' performance is attributable to a juxtaposition of market dynamics, whereby fossil fuels experienced a substantial increase in average prices, and concerted public assistance to the sector through various financing vehicles and mechanisms.

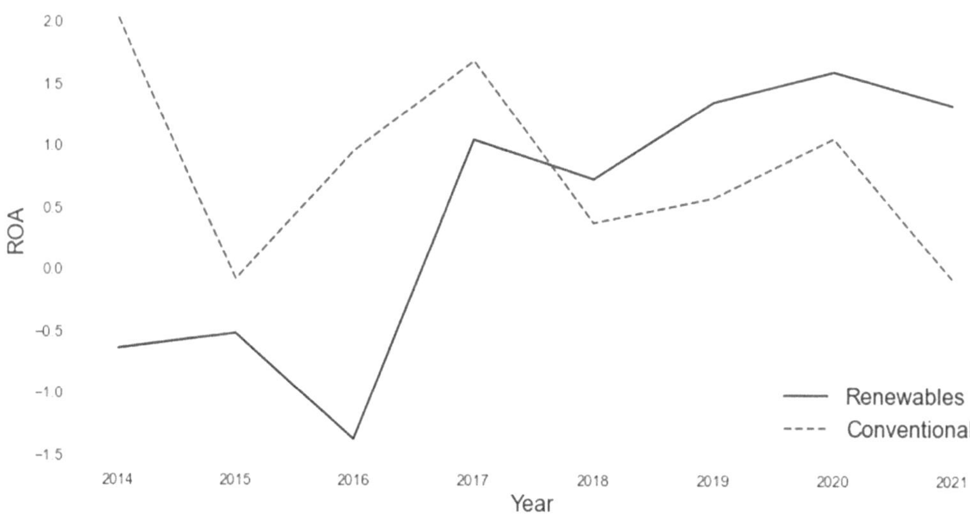


**Figure 1.** The dynamics of the average return on assets (ROA) within the subsamples of conventional and renewable energy companies. Source: own elaboration.

Importantly, renewable energy companies appear to be much more reliant on external financing than their peers. The average debt ratio of renewable energy companies is 0.387 versus 0.241 among conventional energy companies (sig.: 1%). Renewable energy companies are also evidenced to tap significantly more external debt and equity. The average size of debt issuances by renewable energy companies stands at 0.049 (in relationship to total assets). The readily available external financing appears to be one of the principal engines of the accelerated growth in renewable energy generating capacity. The cost of procuring external financing is also significantly lower among renewable energy companies. The average cost of equity is about 1.53 percentage points lower than among conventional energy companies. The weighted average cost of capital is significantly lower among renewable energy companies (Figure 2).

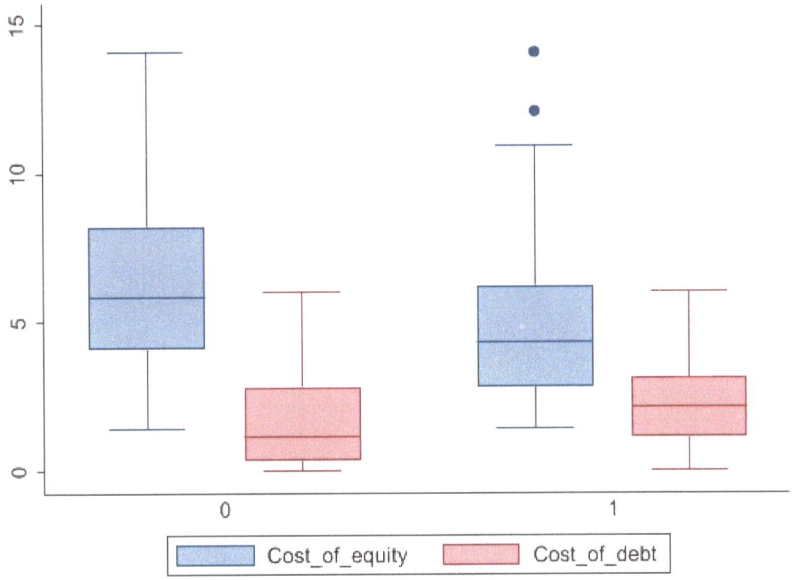


**Figure 2.** The differences in the cost of equity and cost of debt between subsamples of conventional and renewable energy companies. The $x$-axis binary-codes the subsamples of conventional (0) and renewable (1) energy companies. Source: own elaboration. The dots on the chart represent outliers defined as observations deviating from the 75th percentile of the distribution by more than 1.5 times interquartile range.

Overall, univariate analysis suggests that renewable energy companies have unimpeded access to external financing necessary for their growth and investment. However, the access to external funding favored by the transformation of financing priorities in the banking sector and on public capital markets may eventually become a burden for cash-strapped renewable energy companies, particularly under conditions of monetary tightening, whereby the management of substantial accumulated debts may become unsustainable. At the same time, it is clear that regulatory action aimed at promoting the unimpeded flow of capital to the renewable energy industry and adjacent sectors has been successful if assessed based on the inflows of cheap financing to the industry.

Despite the readily available external financing, renewable energy companies appear to remain financially constrained. The measurement of the cash flow sensitivity of investment, which is accepted as a proxy for the degree of binding financing constraints [33], suggests (Figure 3) that the investment demand of renewable energy companies is more dependent on internally generated operating cash flows than that of conventional energy companies. This may be because of the significant investment demand in the industry, which contributes to an elevated rate of reinvestment of internally generated cash flows. At the same time, it may possibly signify that renewable energy companies are still in need of more external financing to fuel further growth.

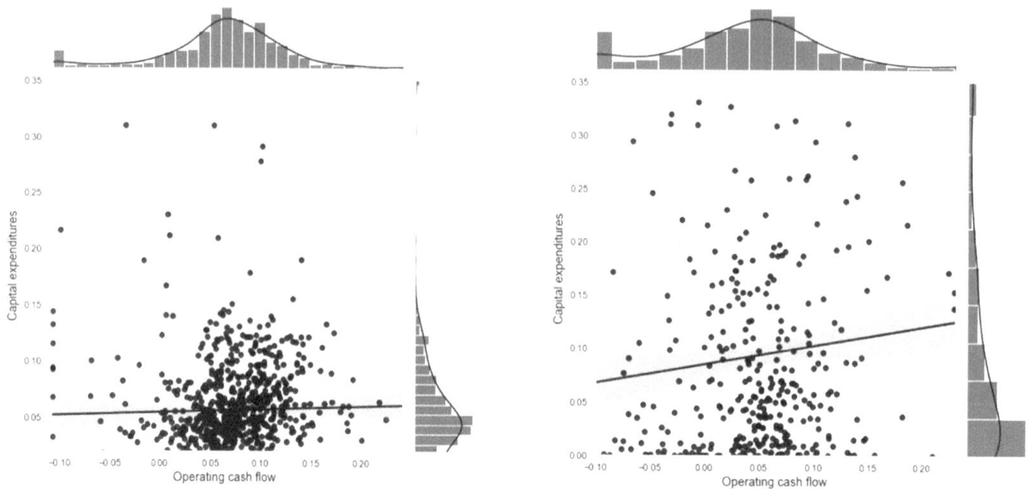


**Figure 3.** The comparison of the cash flow sensitivity of investment between subsamples of conventional (**A**) and renewable energy (**B**) companies. Source: own elaboration.

Table 4 presents the results of multivariate panel regression modeling of the nexus between the sampled firms' operating performance and the fluctuations in energy prices. First of all, the results suggest that within a multivariate setting, there are no statistically significant differences in operating performance indicators between the subsamples of renewable and conventional energy companies (the coefficient of the Renewable variable is statistically insignificant). Secondly, two measures of operating performance—EBIT margin and operating cash flows—exhibit a statistically significant (sig.: 10%) associative link with energy prices interacting with the binary variable encoding renewable energy companies. These coefficients may suggest that within a subsample of renewable energy companies, operating performance exhibits a positive link with the movements of energy prices, which may signify a relatively stronger dependence of operating performance on market fluctuations than in the case of conventional energy companies. This may be due to the fact that renewable energy companies are more exposed to the dynamics of the spot market. The higher exposure to fluctuations in energy prices could play a positive role in fostering industry growth. However, for that reason, the prices of energy should not be tampered with and instead be allowed to fluctuate in line with the changes in demand and supply. At present, the price mechanism is regulated in a way to mitigate the exposure of consumers to short-term spikes in energy prices caused by external shocks. While this may be beneficial from the standpoint of protecting industrial clients and limiting public outcry, it may deprive the industry of the benefits of upside price movements. While the trade-off between these two policy priorities is challenging to figure out, the architecture of the market could benefit from a solution based on market incentives and price signals.

Table 5 presents the results of multivariate econometric modeling of the interrelation between energy prices and the demand of energy companies for external financing. We find a statistically significant relationship between the dynamics of energy prices and debt and equity issuances by energy companies. More importantly, the difference in the size of stock issuances observed between renewable and conventional energy companies is not observable within multivariate settings. There is, however, a statistically significant difference in the relative size of debt issuances (coeff: 0.09; sig.: 1%). The energy prices are evidenced to exhibit a statistically significant associative link with the relative size of debt issuances across the studied sample of energy companies. The relationship is negative and statistically significant within the subsample of renewable energy companies as evidenced by the coefficient of the interaction term Renewable × Energy price (coeff.: −0.551; sig.: 5%). The demand for debt financing on the part of renewable energy companies appears to be increasing as energy prices decline and vice versa. This may signify that renewable energy companies are attempting to close the resulting gap in cash flows by procuring debt. This result appears to be further corroborated by the observed lack of any significant relationship between energy prices and the changes in the cash reserves of energy companies. The cash reserves, when plentiful, may serve as a buffer against the fluctuations in operating cash flows. However, within the environment of low interest rates and readily available external financing, debt may be used to cushion the fluctuations in operating performance. The models in Table 5 also demonstrate that the demand for external capital may be gradually diminishing as firms mature, as evidenced by the coefficients of the Time since incorporation variable.

**Table 4.** The relationship between energy prices and operating performance of the sampled energy firms.

| Explained Variable | EBIT Margin | ROA | OCF |
|---|---|---|---|
| Model No. | (2) | (3) | (4) |
| Firm size | 0.023 *** | 0.385 ** | 0.006 *** |
| | (3.57) | (2.07) | (3.58) |
| Asset tangibility | −0.021 | −0.808 | 0.028 *** |
| | (−1.06) | (−1.11) | (4.19) |
| Asset turnover | −0.033 * | 1.141 * | 0.024 *** |
| | (−1.76) | (1.72) | (4.03) |
| Cash | −0.177 *** | 2.071 | 0.095 *** |
| | (−2.71) | (0.78) | (3.83) |
| Energy price | −0.044 | 2.282 | −0.052 |
| | (−0.20) | (0.27) | (−0.63) |
| Renewable | −0.013 | −0.171 | −0.030 |
| | (−0.19) | (−0.07) | (−1.43) |
| Renewable × Energy price | 0.775 * | 6.452 | 0.252 * |
| | (1.80) | (0.39) | (1.67) |
| _cons | −0.325 * | −3.455 | −0.080 * |
| | (−1.68) | (−0.66) | (−1.86) |
| Country fixed effects | Yes | Yes | Yes |
| Year fixed effects | Yes | Yes | Yes |
| chi2 | 59.579 ** | 56.995 ** | 139.262 *** |

Note: the table presents the results of static panel regression modeling. The models include country and year dummies (not reported for brevity). The significance of the respective variables is denoted with asterisks: * $p < 0.1$, ** $p < 0.05$, *** $p < 0.01$ (the respective t-values are reported in parentheses beneath the coefficients). Source: own elaboration.

The results of binary logit regressions reported in Table 6 support our prior findings with regard to the relationship between energy prices and the demand for external capital on the part of renewable energy companies. To start with, we demonstrate that renewable energy companies are significantly more likely to issue debt (coeff.: 2.01; sig.: 5%) than their conventional energy peers. At the same time, this subsample exhibits a negative associative link between the selected energy prices and the likelihood of debt issuances (coeff.: −15.62; sig.: 1%), which means that tapping external debt may be more likely in the event of lower energy prices. The opposite appears to be true for equity issuances (coeff.: 18.17; sig.: 5%), which, in the case of renewable energy companies, appear to be more likely under higher energy prices. The latter may be due to the fact that higher prices may be positively associated with the stock price of the issuing companies. Therefore, issuing equity may be more financially viable under such circumstances. In the case of lower energy prices, debt appears to be the primary source of necessary external financing.

**Table 5.** The relationship between energy prices and the energy companies' demand for external financing.

| Explained Variable | Stock Issuances | Debt Issuances | Net Change in Cash |
|---|---|---|---|
| Model No. | (1) | (2) | (3) |
| Firm size | −0.010 *** | −0.001 | −0.000 |
| | (−2.67) | (−0.32) | (−0.32) |
| Asset tangibility | 0.009 | −0.030 *** | −0.004 |
| | (0.61) | (−3.05) | (−0.61) |
| Asset turnover | −0.028 ** | −0.019 ** | −0.004 |
| | (−2.39) | (−2.11) | (−0.77) |
| Cash | 0.331 *** | 0.076 * | 0.321 *** |
| | (7.44) | (1.91) | (13.40) |
| Time since incorporation | −0.001 ** | −0.001 *** | −0.000 ** |
| | (−2.42) | (−2.86) | (−2.18) |
| Energy price | 0.018 | 0.271 ** | −0.059 |
| | (0.17) | (2.15) | (−0.72) |
| Renewable | −0.047 | 0.090 *** | −0.021 |
| | (−1.07) | (2.84) | (−1.10) |
| Renewable × Energy price | 0.251 | −0.551 ** | 0.167 |
| | (0.89) | (−2.44) | (1.20) |
| _cons | 0.300 *** | 0.067 | 0.015 |
| | (2.83) | (1.07) | (0.46) |
| Country fixed effects | Yes | Yes | Yes |
| Year fixed effects | Yes | Yes | Yes |
| chi2 | 144.755 *** | 78.201 *** | 252.641 *** |

Note: The table presents the results of static panel regression modeling. The models include country and year dummies (not reported for brevity). The significance of the respective variables is denoted with asterisks: * $p < 0.1$, ** $p < 0.05$, *** $p < 0.01$ (the respective t-values are reported in parentheses beneath the coefficients). Source: own elaboration.

**Table 6.** The results of logistic regressions explaining the likelihood of debt and equity issuances by the sampled energy companies.

| Explained Variable | Debt Issuance Binary | Equity Issuance Binary |
|---|---|---|
| Model No. | (1) | (2) |
| Firm size | 0.023 | −0.292 *** |
| | (0.528) | (−4.001) |
| Asset tangibility | −1.077 *** | 1.009 ** |
| | (−4.335) | (2.354) |
| Asset turnover | −0.241 | −0.157 |
| | (−1.288) | (−0.479) |
| Cash | −1.056 | 0.523 |
| | (−1.040) | (0.282) |
| Time since incorporation | −0.005 | −0.020 *** |
| | (−1.307) | (−3.879) |
| Energy price | 8.781 ** | −2.707 |
| | (2.383) | (−0.382) |
| Renewable | 2.013 ** | −3.106 ** |
| | (2.491) | (−2.579) |
| Renewable × Energy price | −15.619 *** | 18.173 ** |
| | (−2.653) | (2.096) |
| _cons | −0.125 | 9.314 *** |
| | (−0.101) | (4.583) |
| Country fixed effects | Yes | Yes |
| Year fixed effects | Yes | Yes |
| Log-likelihood | −451.941 | −237.103 |
| chi2 | 102.097 *** | 147.562 *** |

Note: The table presents a summary of binary logit models investigating the determinants of the likelihood of a company issuing debt or equity. The significance of the respective variables is denoted with asterisks: ** $p < 0.05$, *** $p < 0.01$. Source: own elaboration.

Overall, we demonstrate that renewable energy companies appear to be more reliant on external financing, with characteristic patterns of the procurement thereof under different market conditions. The declining energy prices may be associated with inferior operating performance, forcing the companies to resort to incurring debt. At the same time, higher prices appear to favor equity issuances, which may be necessary for further financing of investments under a favorable market conjuncture. Whether such patterns have a positive or negative impact on the dynamics of the growth of the entire industry remains unclear. However, reducing the industry's reliance on external capital or at least supplementing its organic cash flows by fostering market mechanisms may inject additional resources for future growth.

The final stage of our analysis attempts to establish whether the performance of renewable energy companies is improving over time. The results of multivariate modeling of the proxies for operating performance are presented in Table 7. Only the EBITDA and EBIT margins demonstrate a statistically observable pattern over time since the companies' incorporation. There appears to be a negative statistically significant associative link (coeff.: −0.003; sig.: 1%) between the time since the firms' incorporation and the EBIT margins. The opposite appears to be true in the subsample of renewable energy companies, where we observe a positive statistically significant link between time since incorporation and the margins. This may suggest that the performance of sampled renewable energy companies is improving over time. The results should be interpreted with caution since the time frame of the analysis is too short to draw any definite conclusions.

**Table 7.** The associative link between time since the sampled firms' incorporation and their operating performance.

| **Explained Variable** | **EBIT Margin** | **ROA** | **OCF** |
|---|---|---|---|
| Model No. | (2) | (3) | (4) |
| Firm size | 0.037 *** | 0.313 * | 0.005 *** |
| | (6.99) | (1.84) | (4.02) |
| Asset tangibility | 0.008 | −0.472 | 0.029 *** |
| | (0.53) | (−0.86) | (5.52) |
| Asset turnover | −0.036 *** | 1.005 ** | 0.023 *** |
| | (−2.69) | (2.10) | (5.07) |
| Cash | −0.003 | 8.409 *** | 0.100 *** |
| | (−0.07) | (4.60) | (5.29) |
| Time since incorporation | −0.003 *** | 0.011 | 0.000 |
| | (−3.17) | (0.44) | (0.03) |
| Renewable | −0.005 | 1.476 | −0.002 |
| | (−0.10) | (1.03) | (−0.17) |
| Renewable × Time since incorporation | 0.003 *** | −0.022 | 0.000 |
| | (3.22) | (−0.92) | (0.25) |
| _cons | −0.433 *** | −3.152 | −0.077 * |
| | (−2.63) | (−0.64) | (−1.87) |
| Country fixed effects | Yes | Yes | Yes |
| Year fixed effects | Yes | Yes | Yes |
| chi2 | 126.014 *** | 135.527 *** | 223.488 *** |

Note: The table presents the results of static panel regression modeling. The models include country and year dummies (not reported for brevity). The significance of the respective variables is denoted with asterisks: * $p < 0.1$, ** $p < 0.05$, *** $p < 0.01$ (the respective t-values are reported in parentheses beneath the coefficients). Source: own elaboration.

## 5. Conclusions

The present study attempts to fulfill two tasks. The first one is to conduct a comparative analysis of the financial standing of renewable and conventional energy companies in the European Union. The second one is to analyze the associative link between the fluctuations in energy prices and the renewable energy companies' financials and external financing needs. The study demonstrates that renewable energy companies have experienced a steady improvement in their operating performance over the last decade. The current return on assets generated by renewable energy companies is significantly higher than that of conventional energy generation. At present, the sector enjoys unimpeded access to relatively cheap external capital, which in turn may be fueling the investment in new energy-generating capacity. Univariate cross-sectional statistical tests demonstrate that renewable energy companies have a lower average cost of equity (4.76% vs. 6.29% for renewable and conventional energy companies, respectively). The margins of the renewable energy businesses have also been steadily improving.

At the same time, renewable energy companies appear to remain more financially constrained than their conventional energy peers, which may be due to the steadily increasing investment demand, for which the available financing resources may be insufficient. Regression modeling demonstrates that the cash flow sensitivity of investments is significantly higher among renewable energy companies than among their conventional energy peers. Therefore, additional regulatory measures may be necessary to foster the growth of the industry and allow it to accumulate greater financial resources, including boosting the internally generated operational cash flows. We show that the performance of renewable energy companies is positively and significantly associated with the fluctuations in energy prices. The sensitivity of operational KPIs (key performance indicators) to the fluctuations in energy prices is much higher in the subsample of renewable energy companies, suggesting that they may be more exposed to the price fluctuations possibly due to their greater dependence on the spot market dynamics. The price declines appear to be associated with an increased need to procure external financing. Capital procurement appears to be following specific patterns: under lower energy prices, firms appear to be more likely to issue additional debt, while under higher prices, they exhibit a higher likelihood of issuing equity.

We posit that an additional impulse to the development of the industry may come from relaxing the regulation around the pricing mechanisms, which may allow the companies to benefit from the upside movements of prices. Fostering the market mechanisms may provide the companies with additional resources for organic development and investment. Deregulation of the permit procedures may also play an important role in spurring industry growth and speeding up the transition of the European Union towards sustainable energy generation. Further studies are necessary in order to establish the most efficient structure of energy procurement contracts and the architecture of energy markets conducive to the more dynamic growth in renewables. It remains unclear whether energy procurement through spot markets or through long-term contracts is better for spurring industry growth and providing energy companies with resources for investment. Further studies will focus on the impact of changes to the regulatory permit systems across the EU countries on the growth dynamics of the renewable energy industry.

**Funding:** This research received no external funding.

**Institutional Review Board Statement:** Not applicable.

**Informed Consent Statement:** Not applicable.

**Data Availability Statement:** The data are available from the corresponding author upon request. Restrictions on the public dissemination of data apply.

**Conflicts of Interest:** The author declares no conflict of interest.

## Appendix A

**Table A1.** The geographic breakdown of the sample.

| Country | Conventional | Renewable |
|---|---|---|
| Austria | 4.93% | 0.00% |
| Belgium | 2.07% | 0.00% |
| Bulgaria | 0.79% | 1.72% |
| Czech Republic | 3.55% | 0.00% |
| Denmark | 1.38% | 0.00% |
| Estonia | 0.00% | 0.86% |
| Finland | 2.46% | 0.00% |
| France | 9.26% | 9.46% |
| Germany | 10.54% | 12.90% |
| Greece | 2.96% | 3.87% |
| Hungary | 1.87% | 8.17% |
| Republic of Ireland | 1.28% | 1.94% |
| Italy | 9.75% | 10.97% |
| Lithuania | 5.02% | 0.00% |
| Luxembourg | 2.46% | 0.00% |
| Netherlands | 0.00% | 5.38% |
| Poland | 10.54% | 12.26% |
| Portugal | 2.46% | 1.08% |
| Romania | 4.33% | 0.00% |
| Spain | 10.44% | 17.42% |
| Sweden | 0.39% | 9.25% |
| United Kingdom | 13.50% | 4.73% |

Source: own elaboration. The table presents the geographical breakdown of the sample observations. The observations are split into subsamples of conventional and renewable energy companies.

## Appendix B

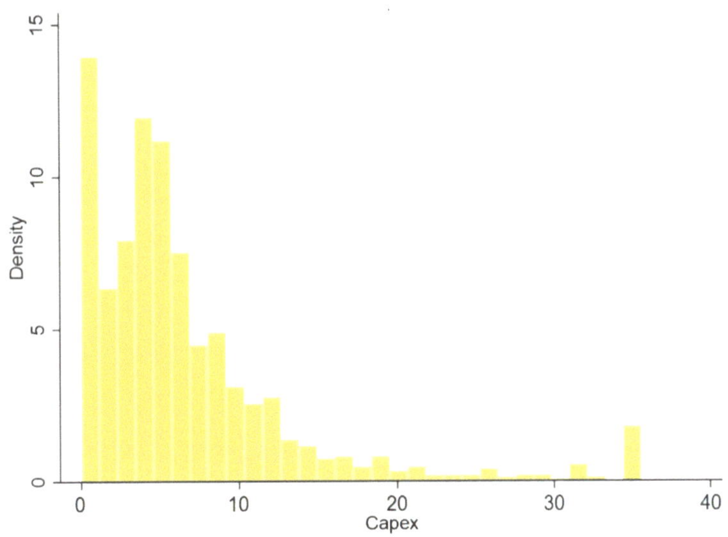


**Figure A1.** Histogram of the Capital Expenditures of the Sampled Companies (Scaled by Contemporaneous Total Assets and Winsorized), percentage points.

## References


1. Su, T.; Meng, L.; Wang, K.; Wu, J. The role of green credit in carbon neutrality: Evidence from the breakthrough technological innovation of renewable energy firms. *Environ. Impact Assess. Rev.* **2023**, *101*, 107135. [CrossRef]
2. Klinlampu, C.; Chimprang, N.; Sirisrisakulchai, J. The sufficient level of growth in renewable energy generation for coal demand reduction. *Energy Rep.* **2023**, *9*, 843–849. [CrossRef]
3. Hauenstein, C.; Braunger, I.; Krumm, A.; Oei, P. Overcoming political stalemates: The German stakeholder commission on phasing out coal. *Energy Res. Soc. Sci.* **2023**, *103*, 103203. [CrossRef]
4. Li, M.; Duan, M. Efforts-sharing to achieve the Paris goals: Ratcheting-up of NDCs and taking full advantage of international carbon market. *Appl. Energy* **2020**, *280*, 115864. [CrossRef]
5. Best, R. Switching towards coal or renewable energy? The effects of financial capital on energy transitions. *Energy Econ.* **2017**, *63*, 75–83. [CrossRef]
6. Normann, H.; Tellmann, S. Trade unions' interpretation of a just transition in a fossil fuel economy. *Environ. Innov. Soc. Transit.* **2021**, *40*, 421–434. [CrossRef]
7. Kryszk, R.; Kurowska, K.; Marks-Bielska, R.; Bielski, S.; Ezlakowski, B. Barriers and Prospects for the Development of Renewable Energy Sources in Poland during the Energy Crisis. *Energies* **2023**, *16*, 1724. [CrossRef]
8. Risch, A. Are environmental fiscal incentives effective in inducing energy-saving renovations? An econometric evaluation of the French energy tax credit. *Energy Econ.* **2020**, *90*, 104831. [CrossRef]
9. Lu, Q.; Farooq, M.; Ma, X.; Iram, R. Assessing the combining role of public-private investment as a green finance and renewable energy in carbon neutrality target. *Renew. Energy* **2022**, *196*, 1357–1365. [CrossRef]
10. Cosgrove, P.; Roulstone, T.; Magableh, S. Intermittency and periodicity in net-zero renewable energy systems with storage. *Renew. Energy* **2023**, *212*, 299–307. [CrossRef]
11. Al-Masri, H.; Dawaghreh, O.; Zachary, S. Optimal configuration of a large scale on-grid renewable energy systems with different design strategies. *J. Clean. Prod.* **2023**, *414*, 137572. [CrossRef]
12. Olabi, A. Renewable energy and energy storage systems. *Energy* **2017**, *136*, 1–6. [CrossRef]
13. Timilsina, G. Are renewable energy technologies cost competitive for electricity generation? *Renew. Energy* **2021**, *180*, 658–672. [CrossRef]
14. Varawala, L.; Hesamzadeh, M.; Dan, G.; Bunn, D.; Rosellon, J. A pricing mechanism to jointly mitigate market power and environmental externalities in electricity markets. *Energy Econ.* **2023**, *121*, 106646. [CrossRef]
15. Dechezlepretre, A.; Nachtigall, D.; Venmans, F. The joint impact of the European Union emissions trading system on carbon emissions and economic performance. *J. Environ. Econ. Manag.* **2023**, *118*, 102758. [CrossRef]
16. Rentier, G.; Lelieveldt, H.; Kramer, G. Varieties of coal-fired power phase-out across Europe. *Energy Policy* **2019**, *132*, 620–632. [CrossRef]
17. Chan, H.; Merdekawati, M.; Suryadi, B. Bank climate actions and their implications for the coal power sector. *Energy Strategy Rev.* **2022**, *39*, 100799. [CrossRef]
18. Sendstad, L.; Hagspiel, V.; Mikkelsen, W.; Ravndal, R.; Tveitstol, M. The impact of subsidy retraction on European renewable energy investments. *Energy Policy* **2022**, *160*, 112675. [CrossRef]
19. Antimiani, A.; Costantini, V.; Paglialunga, E. Fossil fuels subsidy removal and the EU carbon neutrality policy. *Energy Econ.* **2023**, *119*, 106524. [CrossRef]
20. Dottling, R.; Ratnovski, L. Monetary policy and intangible investment. *J. Monet. Econ.* **2023**, *134*, 53–72. [CrossRef]
21. Malliaris, A.; Malliaris, M. Where is the Euro Area headed? Restoration of price stability. *J. Policy Model.* **2023**, *45*, 848–863. [CrossRef]
22. Khan, K.; Khurshid, A.; Koseoglu, S. Are there bubbles in shipping freight during COVID-19? *Res. Transp. Econ.* **2023**, *100*, 101319. [CrossRef]
23. Aklin, M. Do high electricity bills undermine public support for renewables? Evidence from the European Union. *Energy Policy* **2021**, *156*, 112400. [CrossRef]
24. Denholm, P.; Hand, M. Grid flexibility and storage required to achieve very high penetration of variable renewable electricity. *Energy Policy* **2011**, *39*, 1817–1830. [CrossRef]
25. Bhagwat, P.; Marcheseli, A.; Richstein, J.; Chappin, E.; De Vries, L. An analysis of a forward capacity market with long-term contracts. *Energy Policy* **2017**, *111*, 255–267. [CrossRef]
26. Bublitz, A.; Keles, D.; Fichtner, W. An analysis of the decline of electricity spot prices in Europe: Who is to blame? *Energy Policy* **2017**, *107*, 323–336. [CrossRef]
27. Pani, M.; Perroni, C. Energy subsidies and policy commitment in political equilibrium. *Energy Econ.* **2018**, *71*, 149–160. [CrossRef]
28. Rimsaite, L. Corruption risk mitigation in energy sector: Issues and challenges. *Energy Policy* **2019**, *125*, 260–266. [CrossRef]
29. Deleidi, M.; Mazzucato, M.; Semieniuk, G. Neither crowding in nor out: Public direct investment mobilising private investment into renewable electricity projects. *Energy Policy* **2020**, *140*, 111195. [CrossRef]
30. Impram, S.; Nese, S.; Oral, B. Challenges of renewable energy penetration on power system flexibility: A survey. *Energy Strategy Rev.* **2020**, *31*, 100539. [CrossRef]
31. Fridolfsson, S.; Tangeras, T. A reexamination of renewable electricity policy in Sweden. *Energy Policy* **2013**, *58*, 57–63. [CrossRef]

32. Juszczyk, O.; Juszczyk, J.; Juszczyk, S.; Takala, J. Barriers for Renewable Energy Technologies Diffusion: Empirical Evidence from Finland and Poland. *Energies* **2022**, *15*, 527. [CrossRef]
33. Fazzari, S.; Hubbard, G.; Petersen, B.; Blinder, A.; Poterba, J. Financing Constraints and Corporate Investment. *Brook. Pap. Econ. Act.* **1988**, *1*, 141–206. [CrossRef]

sustainability

MDPI

Article

# Research on the Spatial Pattern of Carbon Emissions and Differentiated Peak Paths at the County Level in Shandong Province, China

Xinyu Han [1,2], Peng Qu [3], Jiaqi Wu [4], Beile Su [5], Ning Qiu [1,2] and Lili Zhang [5,*]

1 School of Architecture and Urban Planning, Shandong Jianzhu University, Jinan 250101, China; 24097@sdjzu.edu.cn (X.H.); qiuning22@sdjzu.edu.cn (N.Q.)
2 Yellow River Institution, Shandong Jianzhu University, Jinan 250101, China
3 Shandong Urban and Rural Planning Design Institute Co., Ltd., Jinan 250013, China; sdghqp@163.com
4 School of Architecture, Tsinghua University, Beijing 100084, China; wujq22@mails.tsinghua.edu.cn
5 School of Art, Shandong Jianzhu University, Jinan 250101, China; 202006101080@stu.sdjzu.edu.cn
* Correspondence: 24142@sdjzu.edu.cn; Tel.: +86-130-3171-5060

**Abstract:** In the pursuit of China's carbon peak and carbon neutrality objectives, county-level areas assume a pivotal role in orchestrating diverse initiatives for low-carbon development. However, empirical evidence is limited. This paper aims to fill this gap by exploring the driving factors of carbon peak and carbon peak path at the county level, using Shandong Province as a case study. Employing data related to economic development, industrial structure, land utilization, energy consumption, and emission characteristics, a principal component analysis (PCA) was utilized to extract the following five driving factors of carbon peak: green transformation, urbanization, industrial construction, energy consumption, and environmental constraints. Subsequently, K-means clustering identified five cluster areas: (1) agricultural transformation pending area, (2) low-carbon lagging area, (3) industrial transformation area, (4) low-carbon potential areas, and (5) low-carbon demonstration area. Based on these areas, this study further elucidates spatial combination models of carbon peak within the urban system, spanning central cities, coastal cities, resource-based cities, and agricultural cities. The paper enhances comprehension of the integral role county-level areas play in achieving China's carbon reduction objectives. By providing nuanced insights into diverse developmental trajectories and spatial interactions, the study contributes to effective low-carbon strategy formulation. The findings underscore the importance of considering specific county attributes in urban areas to devise precise optimization strategies and trajectories, ultimately facilitating the realization of carbon peak goals.

**Keywords:** carbon emission; driving factors; differentiated peak paths; cluster analysis; county; Shandong

**Citation:** Han, X.; Qu, P.; Wu, J.; Su, B.; Qiu, N.; Zhang, L. Research on the Spatial Pattern of Carbon Emissions and Differentiated Peak Paths at the County Level in Shandong Province, China. *Sustainability* **2023**, *15*, 13520. https://doi.org/10.3390/su151813520

Academic Editors: Elena Lucchi, Tianyi Chen and Wen Zhang

Received: 9 August 2023
Revised: 7 September 2023
Accepted: 7 September 2023
Published: 9 September 2023



## 1. Introduction

The greenhouse effect and ecological environmental issues resulting from carbon emissions have presented significant challenges to sustainable development in recent years [1,2]. China, in particular, has experienced rapid industrialization and urbanization since the 1980s, resulting in substantial socioeconomic progress. However, this growth has been accompanied by a continuous increase in the scale and intensity of carbon emissions [3,4]. China has held the title of the world's largest $CO_2$ emitter since 2007, contributing to a staggering 28% of global carbon emissions in 2019 [5]. Consequently, China's forthcoming endeavors to curtail emissions will play a pivotal role in the global pursuit of limiting global warming to 1.5 °C [6]. Acknowledging the escalating importance of carbon reduction, China has established ambitious targets, aiming to peak $CO_2$ emissions before 2030 and achieve carbon neutrality by 2060 [7]. Given regional variances, burden-sharing complexities, policy innovation, and the enhancement of local competitiveness and development, individual provinces have emerged as key actors in realizing the dual carbon strategy within this strategic framework [8,9]. Therefore, it is imperative to gain a scientific understanding of the spatial disparities and evolutionary trends in regional carbon emissions, accounting for the diverse development objectives and priorities of different stakeholders. This knowledge is vital for formulating effective strategies for carbon control and emission reduction, particularly in light of the dual challenges posed by economic growth and future "dual carbon" targets.

In response to the increasingly pressing global issue of climate change, which has garnered significant international attention, scholars have made substantial progress in addressing the challenges of carbon reduction. In the early stages of this research, a primary focus was on accurately predicting the scale of carbon emissions. This involved the development of diverse measurement methods and frameworks to assess and quantify various aspects related to carbon emissions [10–13]. Subsequent studies delved into regional carbon emission measurements, using them to comprehensively examine the multifaceted factors contributing to variations in carbon emissions. Researchers, for instance, constructed various frameworks to investigate the role of carbon emissions, analyzing driving factors such as population, per capita GDP, energy consumption intensity, and carbon intensity [13]. Notably, these studies identified an inverted 'U' curve relationship between GDP and carbon emissions, signifying diminishing marginal benefits of carbon emissions as the economy reaches a certain level of growth [14]. Analyses of energy consumption characteristics across different countries have revealed that energy intensity plays a significant role in controlling carbon emissions [15]. Additionally, researchers have identified population growth as a driver of carbon emissions scale, particularly in urban areas with significant population agglomeration [16]. Panel statistics analysis has underscored the pivotal role of the secondary industry in urban carbon emissions, emphasizing the optimization of industrial structure as a crucial pathway toward achieving low-carbon development [17]. Furthermore, various measures, such as the promotion of industrial structure rationalization, the adoption of innovation development strategies, and the enhancement of technological innovation, have proven to be effective in reducing carbon emissions [18].

As China's carbon reduction efforts have garnered global attention, an increasing volume of research has been dedicated to understanding various aspects of carbon emissions within the country. This includes studying carbon emissions from different industries, regions, and energy consumption patterns [19]. Additionally, scholars have delved into the exploration of the technological, policy, and economic factors necessary for transitioning China's energy production and consumption towards a low-carbon trajectory [20,21]. Moreover, researchers have examined the establishment and development of carbon market mechanisms in China [21,22] and evaluated the effectiveness and impact of China's carbon reduction policies [23,24]. Many of these studies have relied on regional and provincial panel data to offer empirical evidence through heterogeneity analysis across provinces and thus support provincial energy transition and industrial upgrading policies. As carbon reduction research continues to advance in terms of spatial and temporal granularity, an increasing number of studies are examining the spatial distribution and evolutionary characteristics of regional carbon emissions from a space–time perspective. These endeavors aim to unveil the effects and influences of human activities on carbon emissions by scrutinizing factors such as city size, economic development patterns, level of international cooperation [25], urban land use, industrial structure, consumption behavior [26], and the selection of action paths for reducing carbon emissions [27].

In summary, research on carbon emissions at the provincial level has provided valuable insights into regional characteristics and underlying influences. Additionally, studies examining spatiotemporal patterns and evolution across multiple regions, scales, and timeframes are gaining prominence. However, it is crucial to recognize two key limitations. Firstly, using cities as spatial analysis units might overlook variations in socioeconomic and industrial development within cities and counties. While micro-county units are considered more appropriate for studying resource utilization and administration, research focused on them remains scarce, potentially restricting our comprehension of local carbon emission dynamics. Secondly, the absence of comprehensive analysis on dominant factors driving carbon peak scenarios hampers the classification of regional carbon peak types and formulation of differentiation paths for carbon emission peak types. Addressing these limitations represents a significant enhancement to the current research within urban contexts. This effort will lead to a deeper understanding of carbon emissions at a more intricate and granular level, thereby facilitating the development of effective carbon reduction strategies. Furthermore, a comprehensive analysis of the predominant factors influencing carbon emissions will provide substantial support for categorizing, formulating, and executing emission peak trajectories across diverse regions.

In this study, Shandong Province serves as an illuminating case study, offering a comprehensive microcosm of substantial significance. Being one of China's economic powerhouses, Shandong exemplifies the intricate interplay between economic growth and its corresponding environmental consequences. The province's unique composition encompasses a wide range of industrial, agricultural, and residential activities across various counties, including urban, suburban, and rural areas. These activities work in synergy to contribute to its overall carbon emissions. The unveiling of the intricate carbon emission patterns within this distinctive geographic context is pivotal, not only for crafting localized mitigation strategies but also for yielding invaluable insights into broader trends. The primary focus of this paper is to comprehensively address the closely interrelated influencing factors of the carbon peak pathway. To achieve this, we establish an evaluation index system for carbon peak trends, which incorporates key elements such as population, economy, land usage, and carbon emissions. The framework is established using principal component analysis (PCA) and k-means cluster analysis. Specifically, two key questions will be answered: (1) What are the key factors driving carbon peak? (2) What is the spatially differentiated pattern of carbon peak features, and how does the typical spatial combination model manifest in different types of cities? (3) What are the pathways to achieve different types of carbon peaks, considering their varying levels of development and differences in industrial structure?

## 2. Methodology and Data

### 2.1. Study Area and Data Source

Shandong Province, situated along the eastern coast of China, was selected as the study area. It is important to note that Shandong has consistently reported carbon emissions per capita and intensity figures that surpass the national average. As a province with a large population and considerable economic importance, Shandong stands out due to its diverse industrial landscape, high levels of energy consumption, and significant agricultural activities, all of which contribute significantly to its substantial carbon emissions. The administrative division of Shandong Province consists of 17 prefecture-level cities and 139 county-level administrative divisions, encompassing both counties and county-level cities.

This study utilized four data sources. Firstly, vector data for administrative divisions were acquired from the map of Shandong (http://www.sdmap.gov.cn/ (accessed on 8 August 2023)). Secondly, socioeconomic data primarily from the 2018 Shandong Statistical Yearbook, along with relevant statistical yearbooks and communiqués at the municipal level, were supplemented. District gross national product (GDP) was revised using data from the fourth economic census. Thirdly, data on land use status and change were based on the latest results of the third land survey. Fourth, data on carbon emissions in 2017 were obtained by processing energy data from the China Bureau of Statistics using the apparent energy consumption estimation method. Finally, county-level energy consumption data for the year 2017 was sourced from the study conducted by Chen et al. (2022). (https://figshare.com/articles/dataset/City-_and_county-level_spatio-temporal_energy_consumption_and_efficiency_datasets_for_China_from_1997_to_2017/19196780/1 (accessed on 8 August 2023)) [28].

## 2.2. *Data and Methods*

### 2.2.1. Hotspot Analysis of Carbon Emissions: Getis–Ord Statistic

We initially performed a normality test on carbon emissions at the county level in 2017. The results of this test indicated that carbon emissions did not conform to a normal distribution. This deviation from normality made them suitable for the application of the Getis–Ord Gi* statistics. Subsequently, we employed the Getis–Ord statistic to identify hotspots of carbon emissions. This statistical test is designed to determine whether clusters of emissions are statistically significant. The resulting z-scores signify either high or low values of neighboring features. High z-scores do not necessarily indicate statistically significant hotspots. To be classified as a statistically significant hotspot, a feature should exhibit both a high z-score and be surrounded by other features with high z-scores. A positive z score indicates a hotspot, and the larger the z scores, the more intense the clustering, while a z score indicates a cold spot and the smaller the z score indicates a more intense clustering of low values (cold spot). The Getis–Ord General G and Getis–Ord local statistics are given as Equations (1) and (2):

$$G(d) = \frac{\sum_{i=1}^{n}\sum_{j=1}^{n} w_{ij}x_i x_j}{\sum_{i=1}^{n}\sum_{j=1}^{n} x_i x_j} \quad (1)$$

$$G_i^*(d) = \frac{\sum_{j=1}^{n} w_{ij}x_i}{\sum_{j=1}^{n} x_i} \quad (2)$$

where $n$ is the number of spatial units, $x_i$ is the attribute value for feature $i$, $x_j$ is the attribute value for feature $j$, and $w_{ij}$ is the spatial weight between feature $i$ and $j$.

### 2.2.2. Influencing Indicators System for Carbon Peak

Based on the existing literature [17,18,29], this paper selects influencing factors closely related to carbon emissions and carbon peak and constructs the carbon peak trend evaluation index system. As presented in Table 1, the index system encompasses the following 12 indicators: per capita carbon emissions, population size, economic level, and industrial structure (regional GDP, per capita GDP, government budget, residents' savings depository balance, second industry, territorial industry), land use (land), and energy consumption and environmental protection (energy consumption, PM2.5). From a theoretical standpoint, the variables do not strictly adhere to an absolute normal distribution. However, considering that the kurtosis absolute value is less than 10 and the skewness absolute value is less than 3, and when examining the normal distribution plot, it can be reasonably characterized as broadly conforming to a normal distribution.

**Table 1.** Carbon peak evaluation index system.

| NO. | Indicator | Description | Mean | Std. Dev. | Max | Min | Kurtosis | Skewness |
|---|---|---|---|---|---|---|---|---|
| X1 | Per capita carbon emissions | The amount of carbon dioxide emissions produced by an individual on average (metric tons per person) | 0.09 | 0.05 | 0.41 | 0.01 | 7.38 | 2.13 |
| X2 | Population size | The total count of individuals living in a given area or region. (Number of individuals) | 74.49 | 31.42 | 175.63 | 21.91 | 0.66 | 0.92 |

**Table 1.** *Cont.*

| NO. | Indicator | Description | Mean | Std. Dev. | Max | Min | Kurtosis | Skewness |
|---|---|---|---|---|---|---|---|---|
| X3 | Construction land | The percentage of land area dedicated to construction or urban development. (%) | 0.19 | 0.14 | 0.95 | 0.06 | 3.18 | 3.32 |
| X4 | Government general budget | The total amount of money spent by the local government from the general budget. (CNY 100 million) | 53.43 | 32.95 | 23.10 | 18.65 | 9.83 | 2.70 |
| X5 | Residents' savings deposit balance | The total balance of savings deposits held by urban and rural residents, measured in CNY 10 thousand. (CNY 100 million) | 340.37 | 227.36 | 1173.12 | 90.74 | 2.30 | 1.26 |
| X6 | Regional GDP | The total value of goods and services produced within a specific region, measured in the local currency. (CNY 100 million) | 441.91 | 331.63 | 2765.69 | 84.10 | 6.93 | 3.10 |
| X7 | Per capita GDP | The average value of goods and services produced per person within a given region, measured in the local currency. (Unit of currency per person) | 7.28 | 5.55 | 34.62 | 1.79 | 5.68 | 2.13 |
| X8 | Secondary industry | The percentage contribution of the secondary industry (manufacturing, construction) to the overall economic output. (%) | 0.45 | 0.10 | 0.66 | 0.08 | 1.42 | −0.91 |
| X9 | Tertiary industry | The percentage contribution of the tertiary industry (services, commerce) to the overall economic output. (%) | 0.50 | 0.12 | 0.89 | 0.28 | 1.14 | 1.09 |
| X10 | Energy consumption | The quantity of standard coal consumed per CNY 10 thousand of GDP (tons of standard coal per CNY 10 thousand GDP) | 1.58 | 0.80 | 6.06 | 0.06 | 9.20 | 2.33 |
| X11 | PM2.5 emissions | The overall quantity of particulate matter with a diameter of 2.5 μm or less released into the atmosphere in a year (10 thousand metric tons) | 5.69 | 2.68 | 13.65 | 0.11 | 0.151 | −0.80 |
| X12 | Environmental protection penalty | The total count of legal cases or instances where penalties have been imposed for violating environmental protection regulations. (Count) | 17.27 | 32.37 | 198.00 | 0.00 | 9.28 | 2.80 |

2.2.3. Identifying Key Factors Driving Carbon Peak: PCA

The carbon emission level is a complex and extensive system with multiple interacting factors and significant differences exist among its elements. Principal component analysis (PCA) [30] aims to identify a few new variables (principal components) that are independent of each other through variable transformation, with minimal information loss. These new variables represent a linear combination of the original variables, reflecting the information contained in the original variable index to the greatest extent while being independent of each other. In this study, principal component analysis was conducted for the 12 indicators, and the principal components with a cumulative contribution rate of 85% and above were extracted as the main factors influencing carbon peak. Given that a large number of variables make any worthwhile judgment and interpretation impossible, the method of Quartimax rotation in PCA that yields the most interpretable results was applied to the matrix. The composite score of the main factor was then calculated as follows:

$$\begin{cases} z_1 = l_{11}x_1 + l_{21}x_2 + \ldots + l_{p1}x_p \\ z_2 = l_{12}x_1 + l_{22}x_2 + \ldots + l_{p2}x_p \\ z_m = l_{1m}x_1 + l_{2m}x_2 + \ldots + l_{pm}x_p \end{cases} \quad (3)$$

In this context, $z_m$ represents the score of the $m$-th principal component, $x_p$ denotes the standardized value of the $p$-th original variable, and $l_{pm}$ stands for the loading coefficient of the $p$-th original variable in the $m$-th principal component.

2.2.4. Segmenting Carbon Peak Area Systems: K-Means

The principal components obtained from principal component analysis are saved as variables, and the principal element score is calculated using Formula (3). Subsequently, the K-means clustering algorithm is applied to cluster the dominant factors, with three to eight different clustering schemes attempted. The cluster schemes with shorter class spacing are selected, and the results are visualized in ArcGIS to analyze the spatial distribution of the carbon peak area.

The Kruskal–Wallis test (K–W test) is employed to examine whether there are significant differences in the overall distributions of multiple independent samples. In this context, it is used to assess the various components across different clustering areas. The significance level, denoted as $\alpha$, is set to 0.05. If the $p$-value from the test is less than 0.05, the null hypothesis is rejected, indicating that there is a significant difference in the visitation rates of tourists from different household life cycles at the 0.05 significance level. Conversely, if the $p$-value is greater than 0.05, it suggests that there is no significant difference.

The clustering process involves the random selection of K samples as initial cluster centers, calculating the distance or similarity between samples using formula (2), and then allocating the samples to homogenous clusters based on the principle of the nearest center distance.

$$d\left(X'_{mj}, X'_{nj}\right) = \sum_{K=1}^{q}\left(X'_{mj} - X'_{nj}\right)^2 \quad (4)$$

In this context, $K$ represents the number of categories, $q$ denotes the dimensions of cluster indicators, and $X'_{mj}$ and $X'_{nj}$ refer to the standard values for the $j$ indicators of units in group m and group n, respectively.

## 3. Results

### *3.1. Spatial Pattern of Carbon Emissions at the County Level*

In this study, we employed the natural breaks method to examine the spatial distribution of the carbon emissions at the county level in Shandong for the year 2017, as depicted in Figure 1a. The findings revealed that there is an east–west pattern of carbon emissions, with the highest emissions observed in east coast cities. Carbon emissions in each city exhibit a looped pattern, featuring lower emissions in the inner city, substantial increases in the surrounding counties, and subsequent declines in the outermost counties. Moreover, the global spatial autocorrelation results demonstrated that the Getis–Ord General G value for carbon emissions was greater than 0 and passed the significance test, signifying the presence of spatial autocorrelation in this study. Then, using Getis–Ord GI* analyses, we found that carbon emissions exhibited a notable spatial clustering pattern. As illustrated in Figure 1b, the hot spots are situated in the northeast, specifically in Dongying, Qingdao, and Weihai. These counties rely on coastal cities to leverage port trade and have developed industries with high carbon emissions, notably in the petrochemical sector. On the other hand, the cold spots are dispersed in the outermost counties of inland cities in the west and center, primarily encompassing rural areas with comparatively lower socio-economic levels.

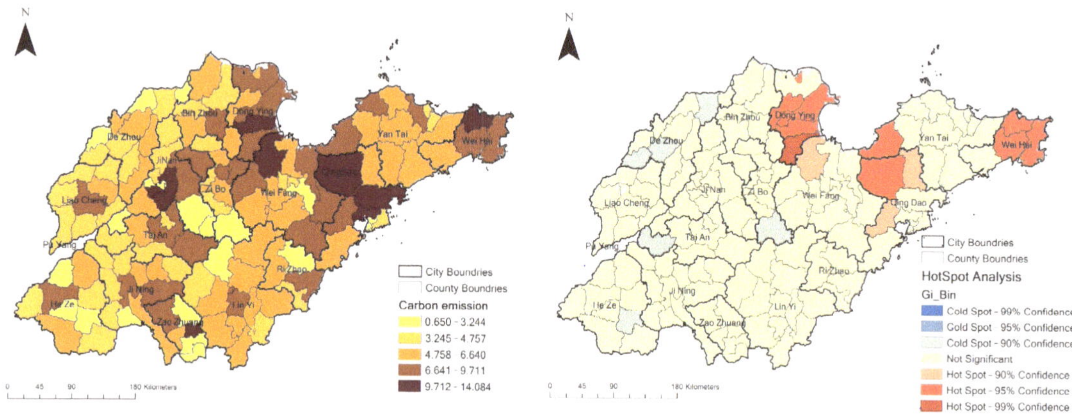


(a) (b)

**Figure 1.** Spatial pattern of carbon emissions in 2017 (**a**) Spatial distribution of carbon emissions (**b**) cold and hot spot of carbon emissions.

### *3.2. Driving Factors of Carbon Peak at the County Level*

The 12 variables presented in Table 1 after nine iterations when a Quartimax rotation method was selected. The Kaiser–Meyer–Olkin (KMO) measure resulted in a value of 0.696, and the Bartlett spherical assay $p$-value was 0.000, both meeting the criteria for factor analysis (KMO > 0.5, $p < 0.05$). Subsequently, five common factors were extracted based on the principle of eigenvalue > 1 and cumulative interpretation of over 80% (Table 2).

**Table 2.** Principal component weight results.

| Component | Eigenvalues | Percentage of Variance (%) | Cumulative Variance Explanation Rate (%) |
|---|---|---|---|
| Component 1 | 3.38 | 28.168 | 28.168 |
| Component 2 | 2.622 | 21.851 | 50.018 |
| Component 3 | 1.856 | 15.463 | 65.481 |
| Component 4 | 1.284 | 10.699 | 76.18 |
| Component 5 | 1.189 | 7.454 | 83.635 |

To improve the interpretability of the common factor on the carbon peak at the county level, we applied rotation to the load matrix, resulting in the rotated component matrix presented in Table 3. In general, the components derived from the PCA encompass urbanization, industrial economy, energy consumption, and other classical factors that influence carbon emissions during socio-economic development. Additionally, these factors include characteristic indicators that reveal the trend of low-carbon transformation in the region such as ecological conservation and green innovation.

**Table 3.** Rotated component matrix.

| NO. | Indicator | Component 1 | Component 2 | Component 3 | Component 4 | Component 5 |
|---|---|---|---|---|---|---|
| X1 | Per capita carbon emissions | −0.318 | 0.185 | 0.703 | 0.398 | −0.370 |
| X2 | Population size | 0.052 | 0.594 | −0.725 | −0.042 | 0.133 |
| X3 | Construction land | 0.784 | 0.096 | 0.274 | −0.298 | 0.091 |
| X4 | Government general budget | 0.178 | 0.832 | −0.002 | −0.174 | −0.100 |
| X5 | Residents' savings deposit balance | −0.027 | 0.801 | −0.087 | 0.211 | 0.171 |
| X6 | Regional GDP | 0.012 | 0.730 | 0.159 | −0.545 | −0.129 |
| X7 | Per capita GDP | 0.311 | 0.478 | 0.682 | −0.182 | −0.173 |
| X8 | Secondary industry | −0.874 | 0.046 | 0.103 | −0.080 | 0.166 |
| X9 | Tertiary industry | 0.880 | 0.218 | 0.025 | 0.010 | −0.058 |
| X10 | Energy consumption | −0.044 | −0.070 | 0.057 | 0.935 | −0.033 |
| X11 | PM2.5 emissions | −0.106 | 0.018 | −0.150 | −0.023 | 0.952 |
| X12 | Environmental protection penalty | −0.510 | 0.264 | −0.705 | 0.140 | −0.025 |

Component 1 can be named "Green Transformation Component". It is characterized by its high correlations with certain indicators, notably X3 construction land (0.784), X9 tertiary industry (0.880), and X8 secondary industry (−0.874), while X1 per capita carbon emissions (−0.318) exhibits a relatively low correlation. These indicators collectively signify a shift in counties away from traditional industrialization pathways towards more energy-efficient and low-carbon service sectors. Based on the component scores (Figure 2a), a notable loop effect was observed in each city, with the highest scores found in the central counties and gradually decreasing towards the outer areas. This observation highlights that the impact of green transformation was most pronounced in the urban core areas.

Component 2 can be named the "Urbanization Component". It is characterized by its high correlations with indicators, notably X2 population size (0.594), X4 government general budget (0.832), X5 residents' savings deposit balance (0.801), and X6 regional GDP (0.730). These correlations unveil patterns of urbanization closely associated with the size of regional population concentrations and levels of economic affluence. An examination of the component scores (Figure 2b) reveals that the eastern urban counties exhibit higher levels of urbanization compared to their western counterparts. Within each city, most of these counties comprise newly developed urban areas and enjoy certain advantages in terms of overall population and socio-economic development.

Component 3 can be named the "Industrial Construction Component". It exerts a discernible positive influence on indicators such as X1 per capita carbon emissions (0.703) and X7 per capita GDP (0.682). Moreover, the indicator X8 secondary industry (0.103) demonstrates a relatively high positive correlation. Conversely, indicators such as X2 population size (−0.725) and X12 environmental protection penalty (−0.705) exhibit negative impacts. These patterns unveil an industry-centric development model within the region, accompanied by associated carbon emissions and environmental pollution challenges. Based on the component scores (Figure 2c), it becomes evident that the level of industrial development in the counties of eastern and northern cities generally surpasses that in the southern and western regions. This discrepancy is particularly pronounced in the majority of counties in northern and eastern cities, such as Dongying, Yantai, and Weihai, as well as in parts of Jinan and Zibo in central cities.

Component 4 can be named the "Energy Consumption Component" with X10 energy consumption (0.935) being the only significant positive factor. Additionally, X3 construction land (−0.298), X6 regional GDP (−0.545), and X7 per capita GDP (−0.182) serve as significant negative indicators. These findings highlight the substantial challenges associated with achieving sustainable development due to serious structural and efficiency problems in the utilization of energy resources in the examined areas. When examining the component scores (Figure 2d), it becomes apparent that the overall spatial pattern of energy consumption and industrial construction components displays a degree of convergence, implying a reliance of industrial construction on energy consumption. However, the energy consumption and industrial construction in Jinan, Weihai, and Qingdao do not exhibit consistency, suggesting that the counties in these cities are largely free from high energy consumption.

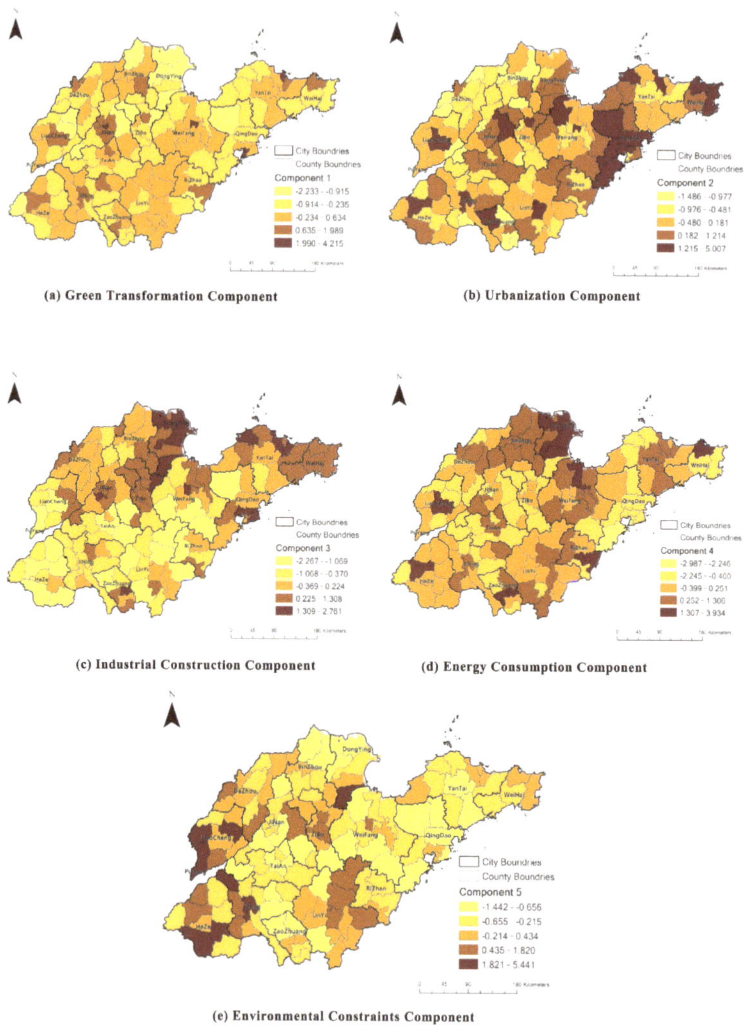


**Figure 2.** Spatial pattern of component of carbon peak at the county level.

Component 5 can be named the "Environmental Constraints Component" characterized mainly by X11 PM2.5 emissions (0.952) with dominant influence, and the negative indicator X1 per capita carbon emissions (−0.370) with relatively high coefficients. These findings indicate a close relationship between the regional carbon emissions problem and the ecological environment. Based on the component scores (Figure 2e), the environmental restriction level is highest in the west and south, particularly in certain counties of Liaocheng, Heze, and Jining. Notably, these regions exhibit lower levels of socio-economic development, indicating that the industrial development of these counties is more challenging and has a more pronounced ecological impact.

In general, PCA extracted 5 components, including urbanization, industrial construction, and energy consumption, which are classical influencing factors that reflect the long-term social and economic development and accumulation in the region [15,17,19]. Additionally, the components of green transformation and environmental constraints depict the region's capacity to transform and develop in line with China's ecological civilization principles, especially following the introduction of carbon-neutral strategies [31,32].

### *3.3. Carbon Peak path at the County Level*

The carbon emission peak characteristics of county-level regions in Shandong Province can be categorized into five cluster areas (Table 4, Figure 3). The spatial pattern of the carbon peak is illustrated in Figure 4.

**Table 4.** Identification of clusters.

| Cluster Area | Component 1 Green Transformation component | Component 2 Urbanization Component | Component 3 Industrial Construction Component | Component 4 Energy Consumption Component | Component 5 Environmental Constraints Component |
|---|---|---|---|---|---|
| Agricultural transformation Pending area | −0.19 | −0.52 | −0.23 | 0.03 | −0.25 |
| Low-carbon lagging area | −0.03 | 0.27 | −0.35 | 0.06 | 2.36 |
| Industrial transformation area | −0.02 | 0.71 | 1.36 | 2.48 | −0.23 |
| Low-carbon potential area | −0.43 | 1.32 | 0.13 | −0.63 | −0.36 |
| Low-carbon demonstration area | 2.61 | −0.27 | 0.69 | −0.96 | −0.27 |

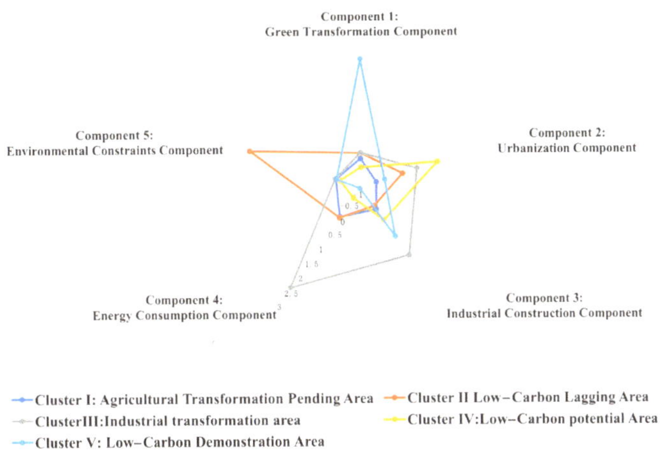


**Figure 3.** Radar chart of the average scores of components for cluster areas.

Cluster I, referred to as the "Agricultural Transformation Pending Area", comprises 79 counties, accounting for 58.1 percent of the total. These areas are primarily concentrated in eastern Shandong Province, with the majority located in inland regions. At the city level, they are mainly situated in peripheral urban and rural fringe areas. The scores of these areas are relatively balanced, but they all remain at a low level, particularly exhibiting the lowest scores for urbanization component (Component 2), which further highlights the underdeveloped status of regional development. Additionally, the subpar scores for industrial construction component (Component 3) and the green transformation component (Component 1) indicate structural obstacles associated with a single economic structure and a weak industrial base.

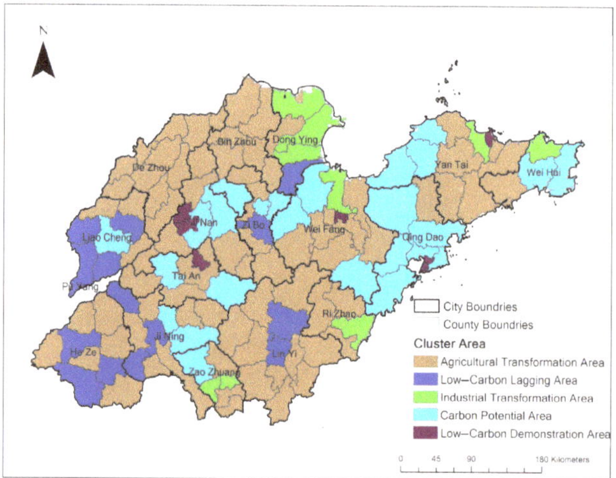


**Figure 4.** The spatial pattern of carbon peak at the county level in Shandong Province.

Cluster II, referred to as the "Low-Carbon Lagging Area", comprises 15 counties, accounting for 11 percent of the total. These areas are primarily located in the western part of Shandong Province, with some scattered in the central area. Although the urbanization component (Component 2) score falls within the middle range, indicating an average level of socio-economic development in the affected areas, the notably low score for the industrial construction component (Component 3), along with high scores for energy consumption component (Component 4) and environmental constraints component (Component 5), indicates that the traditional industrial structure is the primary factor influencing regional carbon emissions and environmental protection.

Cluster III, referred to as the "Industrial Transformation Area", comprises 11 counties, accounting for 8.1 percent of the total. These areas are typically situated in coastal port cities in eastern Shandong Province. The scores for industrial construction component (Component 3) and energy consumption component (Component 4) in this category are significantly higher than in other types, indicating that the leading industrialization process has contributed to a good level of economic development in the region. However, it is crucial to highlight that the elevated score for the environmental constraints component (Component 5) also signifies their significant reliance on the secondary sector, where the scale and intensity of industrial carbon emissions are significantly higher than in equivalent units, leading to a sharp contradiction between environmental pollution and ecological protection.

Cluster IV, referred to as the "Low-Carbon Potential Area", consists of 22 counties, accounting for 16.2 percent of the total. This category is primarily concentrated in the outskirts of Jinan and Qingdao, the most developed cities in Shandong Province, with some scattered in the central areas of the main cities in the east and south. The urbanization component (Component 2) scores in these areas are the highest, aligning with their long-term development factors, such as population, capital, and land agglomerations, reflecting their favorable social and economic conditions. Although the low scores for environmental constraints component (Component 5) and energy consumption component (Component 4) partly indicate the trend of green transformation through regional industrial structure upgrading and efficient energy utilization, the very low score for green transformation component (Component 1) indicates that regional development is still in the transitional stage of structural optimization.

Cluster V, referred to as the "Low-Carbon Demonstration Area", consists of 9 counties, accounting for 6.6 percent. These areas are all located in the urban centers of leading cities in the province, including Jinan, Qingdao, Yantai, and Weifang. Compared to other types, this cluster shows higher scores for industrial construction component (Component 3) and the green transformation component (Component 1), while exhibiting a significant negative influence of energy consumption component (Component 4) and environmental constraints component (Component 5). These findings suggest that the region has successfully departed from the traditional path of urbanization and industrialization and currently finds itself in the latter stages of the "inverted U-shaped" relationship between economic development and carbon emissions. As a result, this cluster area has achieved a remarkable level of economic development with low carbon emissions. The low-carbon model in this region is evidently effective, as the transformation of the industrial structure has largely been accomplished, and a clear decoupling of carbon emissions from economic growth is observed.

In general, areas with a high level of low-carbon development, such as the low-carbon demonstration area and low-carbon potential area, show the potential to reach peak emissions in the near future. These areas should capitalize on their demonstrating functions and play a pivotal role in promoting low-carbon development in their surrounding regions [33]. On the other hand, regions mainly driven by industrial and agricultural sectors, facing challenges in achieving peak emissions, share common difficulties stemming from traditional urbanization and industrial development models. These challenges comprise resource dependency, energy waste, and backward industries [14,34].

## 4. Discussion

### *4.1. Typical Spatial Combination Model of Carbon Peak Based on Urban System Division of Labor*

At the county level, the spatial pattern of carbon peak exhibits distinct typical spatial combination models (Figure 5). This study provides valuable insights for further delineating the characteristics of inner-city types, thus supplementing existing research conducted at the urban and provincial levels [27,35]. When viewed from the city-level typology perspective, it aligns with the division of labor within the urban system, underscoring the significant influence of the urban spatial development model on carbon dioxide emissions and the need for a comprehensive understanding of low-carbon sustainable development [36].

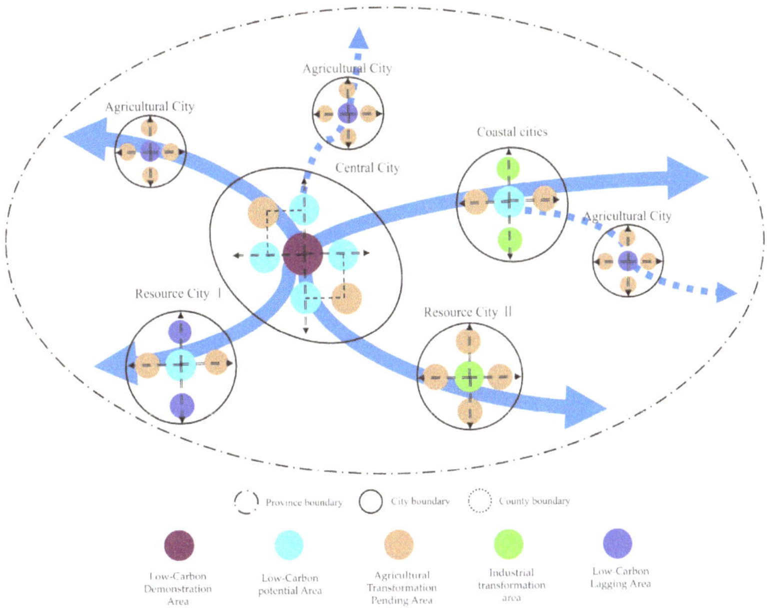


**Figure 5.** Typical spatial combination model of carbon peak based on urban system division of labor.

In its long-term development trajectory, Shandong Province has established a dual-core development pattern with Jinan and Qingdao as the central cities. Jinan is striving to position itself as a financial, logistics, science, technology, and innovation center, focusing on the headquarters economy, service economy, high-end equipment manufacturing, and high-tech industries. On the other hand, Qingdao is leveraging its advantages in openness, aligning with the national Belt and Road strategy, and prioritizing the development of modern service industries, advanced manufacturing, and emerging marine sectors, with the aim of becoming a globally competitive coastal city. As a result, both cities have effectively decoupled social and economic development from carbon emissions, establishing themselves as the "low-carbon demonstration zones" at their core, encircled by "low-carbon spheres of influence", with the "Agricultural Transformation Pending Area" situated at their periphery.

In addition to the construction of dual-core development pattern, the spatial pattern of carbon peak in Shandong Province is also influenced by geographical conditions. Shandong Peninsula is located in the transition area between land and sea. The region has different development conditions and resource endowments, which leads to different development paths in the region. For instance, certain coastal cities have leveraged port trade advantages to progressively cultivate leading industries with high carbon emissions, notably in the petrochemical sector [37]. In the northern region, Dongying is primarily reliant on the Shengli Oilfield, which stands as China's most significant domestic crude oil source. Since the 1960s, it has progressively developed related industrial clusters [38]. Furthermore, coastal cities such as Yantai, Qingdao, and Rizhao have capitalized on China's reform and opening-up policies, along with the forces of economic globalization. These cities have emerged as pivotal locations for national overseas crude oil imports and have established multiple storage and transportation facilities and processing zones, centered around their core trade functions [39]. Their distinctive spatial combination model revolves around the concept of "Low Carbon Potential Area", which is surrounded by the "Industrial Transformation Area" and the "Agricultural Transformation Pending Area". While some counties have made notable economic development progress, and a few exhibit potential for a transition toward a low-carbon economy, the majority of counties face significant challenges associated with industrial transformation.

Shandong Province encompasses several resource cities in its central and western regions, including Taian, Linyi, and Zibo, which boast abundant mineral and agricultural resources. These cities play pivotal roles in industries such as coal, steel, and agricultural products, thereby contributing to the local economy's growth. While certain cities have achieved notable progress in low-carbon development, resulting in a framework with "Low Carbon Potential Area" at its core, surrounded by "Low Carbon Lagging Area" and "Agricultural Transformation Pending Area". On the other hand, some cities still adhere to conventional dynamics of "Industrial Transfer Pending Area" at their core, with an array of "Agricultural Transfer Pending Area" zones surrounding them. In general, these cities grapple with multifaceted challenges in their pursuit of low-carbon development, primarily attributable to their economic structure, energy reliance, and environmental pressures.

The western region of Shandong Province lies within the Huang-Huai Hai impact plain, renowned for its favorable agricultural conditions and historical significance as a crucial grain production hub in China. Within these agricultural cities, the prevailing feature consistently consists of the "Low-Carbon Lagging Areas", encircled by a concentrated and contiguous expanse of the "Agricultural Transformation Pending Area". This agricultural sector is characterized by density and continuity, yet it also grapples with the challenge of transitioning agricultural production patterns.

In general, this typical model, based on various divisions of labor, provides valuable insights into the distinct characteristics and potential of different counties within the city. Such an approach is indispensable for comprehending carbon emission patterns and effectively achieving carbon reduction targets. Particularly, when examining carbon emissions through the lens of urban-rural duality, the introduction of this typical model serves as a crucial complementary tool [39,40]. Traditional urban and rural dual structures may not adequately account for the functional orientation and development of certain counties. By delving into the specific attributes and potential of different counties, we can more accurately identify the primary sources and impacts of carbon emissions, thus enabling the formulation of targeted strategies effective carbon emission reduction.

### 4.2. Exploring Differentiated Paths for Peak Optimization and Policy Implications

By considering the differences in resource endowments and developmental stages, we can assess the influencing factors of carbon emissions in different counties and their evolving trends. This analysis provides valuable support for the classified design and implementation of carbon peak paths. By understanding these factors and their interplay, policymakers can devise targeted strategies to achieve carbon peak goals efficiently and effectively in different counties.

The low-carbon demonstration areas are primarily situated in the central regions of central cities. The low-carbon model implemented in this area is operating effectively and should serve as a catalyst for promoting low-carbon development in the surrounding counties. These counties should prioritize decoupling carbon emissions from economic growth in order to achieve regional emissions peak targets. This can be accomplished by fostering green industries, adopting clean energy sources, disseminating energy-saving and emission-reduction technologies, strengthening carbon market mechanisms, and providing incentives for businesses to take actions that reduce emissions.

The low-carbon potential areas have substantial potential for low-carbon development but have yet to achieve its carbon peak targets. These counties are typically located on the outskirts of a central city or at the core of a coastal or resource city. It means that they are influenced by the low-carbon demonstration area and contribute to the radiation effect for low-carbon development in the surrounding region. These counties need innovative solutions to address energy demand during rapid urbanization and industrialization. This involves transforming high-energy consumption and high-emission development patterns, encouraging green production practices among businesses, and establishing green supply chains.

The low-carbon lagging areas are typically situated on the outskirts of resource cities or at the core of agricultural cities. These areas will continue to prioritize social and economic development in the foreseeable future. However, as these regions undergo industrial upgrading, their existing resource-intensive development model is likely to intensify the conflict between economic growth and carbon emissions. This, in turn, will necessitate increased efforts for carbon reduction and achieving emissions peak targets. To effectively tackle this challenge, it is advisable for these regions to capitalize on their latecomer advantages and expedite their peak process through the implementation of innovative policies and market-driven mechanisms that foster low-carbon transformation and optimize industrial economies, thereby circumventing the conventional industrialization path of "pollute first, treat later". Additionally, these regions should address the complex task of ensuring economic stability and employment, taking cautionary measures to prevent potential issues such as economic downturns and population outflows while concurrently pursuing their low-carbon development objectives.

The industrial transformation areas are typically located on the outskirts of coastal cities or at the core of resource cities. In these areas, there is often a prevailing development model characterized by "high emissions for high economic returns". This model tends to prioritize immediate economic growth while overlooking the environmental consequences linked to high energy consumption. This approach may result in short-term financial gains, but it is unsustainable in the long term and can impede progress toward carbon emissions reduction goals and peak targets. Furthermore, it exposes these regions to vulnerabilities and risks related to sustainable development, including heavy reliance on specific industries and environmental pollution. To address these challenges, these regions should shift their focus in economic development and low-carbon governance.

This shift involves reducing dependence on particular industries and improving energy efficiency. It requires active government intervention through policy support and guidance. Outdated, low-profitability, and highly pollutant industries should be phased out, while emerging, low-carbon industries should be nurtured and expanded. This comprehensive approach will contribute to improving the region's natural environment and building a more sustainable economic foundation for urban development.

As for the agricultural transformation pending area are typically located on the outskirts of various cities. These areas often face unique challenges due to their special status in the context of food security strategies, which may impose limitations on urban expansion and industrial development due to strict farmland protection policies. This complexity adds to the challenge of reversing the current situation of backward development. Therefore, it is advisable for these regions to focus on their regional endowments. While promoting orderly urbanization and industrialization, they should simultaneously strengthen the protection of agriculture and natural ecological environments around cities. Utilizing the "carbon sink" function and ecological compensation benefits of green landscapes and natural environments can create new driving forces for regional development.

## 5. Conclusions

County-level areas, as pivotal units responsible for coordinating diverse initiatives and driving low-carbon development, play a fundamental role in the realization of China's carbon peak and carbon neutrality objectives. Additionally, they serve as a vital complement to ongoing research conducted at the city level. This paper leverages disparities in resource endowments and developmental stages to analyze the magnitude, composition, and evolving patterns of carbon emissions across various regions. The goal is to provide support for the classification design and implementation of peak emission paths in various areas.

Firstly, the driving factors of carbon peak encompass the green transformation component, urbanization component, industrial construction component, energy consumption component, and environmental constraints component. These factors span a broad spectrum of social, economic, and environmental dimensions, offering a comprehensive understanding of the contributing elements to the surge in carbon surge. This analysis enhances readers' grasp of the intricacies of the issue and reflects a region's development over different periods and spatial contexts. Urbanization and industrial construction illustrate the trajectory of economic and social progress, while energy consumption relates to the regional energy structure and technological advancement. The transition towards green practices and environmental constraints are linked to future directions and potential limitations. This consideration of the spatiotemporal dimension underscores the gradual evolution of carbon emission patterns, necessitating the assessment of effects and influences across multiple stages [41,42].

Secondly, the carbon emission peak characteristics of county-level regions in Shandong Province can be classified into five cluster areas. Based on this classification, different paths and strategies for low-carbon development are delineated as follows. The "Low-Carbon Demonstration Area" and "Low-Carbon Potential Area" are strategically designed to pioneer the promotion of environmentally friendly development in neighboring regions and to facilitate a systematic transition towards achieving the carbon emission peak. In the case of "Low-Carbon Lagging Area", it is crucial to place paramount attention on social and economic development. Accelerating the achievement of emission peak targets achievement the advantages of backward development, deploying innovative policies, and market-oriented mechanisms to facilitate the transition towards low-carbon practices and optimizing industrial economies. Ensuring economic stability and safeguarding employment are imperative steps to mitigate potential challenges such as economic downturns and population decline. Additionally, the "Agricultural Transformation Pending Area" and "Industrial Transformation Area" belong to the category of regions facing significant challenges in achieving the carbon emission peak. These challenges are often rooted in conventional urbanization and industrialization, and they encompass issues such as resource dependency, energy inefficiency, and a lagging pace of industrial development.

Thirdly, typical spatial combination models of carbon peak at the county level, based on the division of labor within the urban system, were identified. The division of labor within the urban system encompasses central cities, eastern coastal cities, central and western resource-based cities, and agricultural cities. The carbon mixing patterns exhibited by these cities demonstrate the influence of urban positioning on carbon trajectories. In central cities, coastal cities, and certain developed resource-based cities, a typical spatial combination model emerges: the core of the central city serves as the focal point for either low-carbon demonstration or potential areas, encompassed by low-carbon lagging areas or industrial transformation areas undergoing development. This distributional model can be interpreted through the lens of Hirschman's trickle-down effect in regional economics [43]. Urban centers tend to function as "polarized" zones characterized by low energy consumption, minimal emissions, and high value-added industries. Consequently, advanced development factors like green production, innovative technologies, and low-carbon concepts diffuse to influence less developed regions. This developmental trajectory aims to shift from isolated and imbalanced growth patterns towards interconnected and harmonized systems. In terms of the under-development and traditional agricultural cities, the core comprises the "Low-Carbon Lagging Area" and the "Industrial Transformation Area", surrounded by the "Agricultural Transformation Pending Area". In these cities, the transition to a low-carbon economy often coincides with high energy consumption and emissions due to their reliance on energy sources such as coal, oil, and gas. Traditional economic frameworks constrict cities' sustainable development and impede the realization of low-carbon economic growth. Consequently, there's a surge in environmental pressures, leading cities to grapple with severe environmental issues such as air pollution and water scarcity. Addressing these challenges necessitates increased investment in green technologies and low-carbon innovation in the future. This investment is crucial for bridging the gap in technological upgrading and advancing the trajectory of low-carbon development.

The identification of typical models at the micro level, which concentrates on low-carbon development at the county level while considering the distinct attributes and potential of diverse regions within cities, serves as a robust complement to prior investigations conducted at the city and provincial levels [17,27]. This underscores the importance of understanding the fundamental patterns of variation within cities in order to achieve the overarching goal of carbon peak. Precise optimization strategies and trajectories must be devised by comprehensively comprehending these variations.

This study has limitations. Firstly, it relies on cross-sectional data, which cannot effectively capture the evolving driving factors and spatial pattern of carbon peak. Longitudinal studies conducted over time would be more influential in uncovering the dynamic trends in carbon emissions, including changes in carbon-intensity factors. Second, while conducting studies within provincial administrative units has its advantages for management and policy formulation, it may overlook cities located on the periphery of the administrative region. This could result in missing significant spatial interactions with neighboring cities or districts in adjacent provinces. Conducting larger-scale studies can mitigate the constraints imposed by boundary effects and provide a more comprehensive understanding. Thirdly, certain interpretations of carbon emissions remain at a qualitative level, such as the influence of trade on coastal carbon emissions. Conducting quantitative studies in future research endeavors could enhance the precision of these effects, further substantiating our findings. Lastly, the traditional k-means method does not account for the spatial dependence present within the geographical clusters. In subsequent research, we intend to integrate advanced spatial clustering methods to holistically address spatial dependence.

**Author Contributions:** Conceptualization, X.H. and L.Z.; methodology, N.Q.; formal analysis, X.H. and N.Q.; investigation, P.Q.; software, B.S. and J.W.; data curation, P.Q. and J.W.; writing—original draft, X.H. and N.Q.; writing—review and editing, P.Q. and L.Z.; visualization, N.Q. and B.S.; supervision, X.H. and L.Z. All authors have read and agreed to the published version of the manuscript.

**Funding:** This research was funded by the Natural Science Foundation of Shandong Province grant number ZR2023QE242.

**Institutional Review Board Statement:** Not applicable.

**Informed Consent Statement:** Not applicable.

**Data Availability Statement:** The data cannot be made available due to confidentiality reasons.

**Conflicts of Interest:** The authors declare no conflict of interest.

## References

1. Rosa, E.A.; Dietz, T. Human drivers of national greenhouse-gas emissions. *Nat. Clim. Chang.* **2012**, *2*, 581–586. [CrossRef]
2. Atici, C. Carbon emissions in Central and Eastern Europe: Environmental Kuznets curve and implications for sustainable development. *Sustain. Dev.* **2009**, *17*, 155–160. [CrossRef]
3. Lv, T.; Hu, H.; Zhang, X.; Wang, L.; Fu, S. Impact of multidimensional urbanization on carbon emissions in an ecological civilization experimental area of China. *Phys. Chem. Earth Parts A/B/C* **2022**, *126*, 103120. [CrossRef]
4. Liu, K.; Tao, Y.; Wu, Y.; Wang, C. How does ecological civilization construction affect carbon emission intensity? Evidence from Chinse provinces' panel data. *Chin. J. Popul. Resour. Environ.* **2020**, *18*, 97–102. [CrossRef]
5. Dudley, B. BP statistical review of world energy 2019. In *British Petroleum Statistical Review of World Energy*; BPLC, Ed.; Pureprint Group Limited: Uckfield, UK, 2019.
6. IEA. Enhancing China's ETS for Carbon Neutrality: Focus on Power Sector, IEA. Available online: https://www.iea.org/reports/enhancing-chinas-ets-for-carbon-neutrality-focus-on-power-sector (accessed on 8 August 2023).
7. Yu, G.; Hao, T.; Zhu, J. Discussion on action strategies of China's carbon peak and carbon neutrality. *Bull. Chin. Acad. Sci. (Chin. Version)* **2022**, *37*, 423–434.
8. Du, L.; Wei, C.; Cai, S. Economic development and carbon dioxide emissions in China: Provincial panel data analysis. *China Econ. Rev.* **2012**, *23*, 371–384. [CrossRef]
9. Hao, Y.; Chen, H.; Wei, Y.; Li, Y. The influence of climate change on $CO_2$ (carbon dioxide) emissions: An empirical estimation based on Chinese provincial panel data. *J. Clean. Prod.* **2016**, *131*, 667–677. [CrossRef]
10. Boyd, G.A.; Hanson, D.A.; Sterner, T. Decomposition of changes in energy intensity: A comparison of the Divisia index and other methods. *Energy Econ.* **1988**, *10*, 309–312. [CrossRef]
11. Eggleston, H.S.; Buendia, L.; Miwa, K.; Ngara, T.; Tanabe, K. 2006 IPCC Guidelines for National Greenhouse Gas Inventories. 2006. Available online: https://www.osti.gov/etdeweb/biblio/20880391 (accessed on 8 August 2023).
12. Park, S. Decomposition of industrial energy consumption: An alternative method. *Energy Econ.* **1992**, *14*, 265–270. [CrossRef]
13. Kennedy, C.; Steinberger, J.; Gasson, B.; Hansen, Y.; Hillman, T.; Havránek, M.; Pataki, D.; Phdungsilp, A.; Ramaswami, A.; Mendez, G.V. Methodology for inventorying greenhouse gas emissions from global cities. *Energy Policy* **2010**, *38*, 4828–4837. [CrossRef]
14. Kais, S.; Sami, H. An econometric study of the impact of economic growth and energy use on carbon emissions: Panel data evidence from fifty eight countries. *Renew. Sustain. Energy Rev.* **2016**, *59*, 1101–1110. [CrossRef]
15. Xu, X.; Yang, G.; Tan, Y.; Zhuang, Q.; Tang, X.; Zhao, K.; Wang, S. Factors influencing industrial carbon emissions and strategies for carbon mitigation in the Yangtze River Delta of China. *J. Clean. Prod.* **2017**, *142*, 3607–3616. [CrossRef]
16. Yi, Y.; Qi, J.; Chen, D. Impact of population agglomeration in big cities on carbon emissions. *Environ. Sci. Pollut. Res.* **2022**, *29*, 86692–86706. [CrossRef] [PubMed]
17. Su, W.; Liu, Y.; Wang, S.; Zhao, Y.; Su, Y.; Li, S. Regional inequality, spatial spillover effects, and the factors influencing city-level energy-related carbon emissions in China. *J. Geogr. Sci.* **2018**, *28*, 495–513. [CrossRef]
18. Gao, P.; Wang, Y.; Zou, Y.; Su, X.; Che, X.; Yang, X. Green technology innovation and carbon emissions nexus in China: Does industrial structure upgrading matter? *Front. Psychol.* **2022**, *13*, 951172. [CrossRef] [PubMed]
19. Liu, Z.; Guan, D.; Wei, W.; Davis, S.J.; Ciais, P.; Bai, J.; Peng, S.; Zhang, Q.; Hubacek, K.; Marland, G. Reduced carbon emission estimates from fossil fuel combustion and cement production in China. *Nature* **2015**, *524*, 335–338. [CrossRef]
20. Chen, H.; Yi, J.; Chen, A.; Peng, D.; Yang, J. Green technology innovation and $CO_2$ emission in China: Evidence from a spatial-temporal analysis and a nonlinear spatial durbin model. *Energy Policy* **2023**, *172*, 113338. [CrossRef]
21. McCormick, K.; Anderberg, S.; Coenen, L.; Neij, L. Advancing sustainable urban transformation. *J. Clean. Prod.* **2013**, *50*, 1–11. [CrossRef]
22. Moloney, S.; Horne, R. Low carbon urban transitioning: From local experimentation to urban transformation? *Sustainability* **2015**, *7*, 2437–2453. [CrossRef]
23. Wu, R.; Tan, Z.; Lin, B. Does carbon emission trading scheme really improve the $CO_2$ emission efficiency? Evidence from China's iron and steel industry. *Energy* **2023**, *277*, 127743. [CrossRef]
24. Liu, Z.; Sun, H. Assessing the impact of emissions trading scheme on low-carbon technological innovation: Evidence from China. *Environ. Impact Asses* **2021**, *89*, 106589. [CrossRef]
25. Mi, Z.; Zheng, J.; Meng, J.; Zheng, H.; Li, X.; Coffman, D.; Woltjer, J.; Wang, S.; Guan, D. Carbon emissions of cities from a consumption-based perspective. *Appl. Energy* **2019**, *235*, 509–518. [CrossRef]

26. Zhu, E.; Qi, Q.; Chen, L.; Wu, X. The spatial-temporal patterns and multiple driving mechanisms of carbon emissions in the process of urbanization: A case study in Zhejiang, China. *J. Clean. Prod.* **2022**, *358*, 131954. [CrossRef]
27. Tian, S.; Xu, Y.; Wang, Q.; Zhang, Y.; Yuan, X.; Ma, Q.; Chen, L.; Ma, H.; Liu, J.; Liu, C. Research on peak prediction of urban differentiated carbon emissions--a case study of Shandong Province, China. *J. Clean. Prod.* **2022**, *374*, 134050. [CrossRef]
28. Chen, J.; Liu, J.; Qi, J.; Gao, M.; Cheng, S.; Li, K.; Xu, C. City-and county-level spatio-temporal energy consumption and efficiency datasets for China from 1997 to 2017. *Sci. Data* **2022**, *9*, 101. [CrossRef] [PubMed]
29. Yang, X.; Sima, Y.; Lv, Y.; Li, M. Research on Influencing Factors of Residential Building Carbon Emissions and Carbon Peak: A Case of Henan Province in China. *Sustainability* **2023**, *15*, 10243. [CrossRef]
30. Bro, R.; Smilde, A.K. Principal component analysis. *Anal. Methods-UK* **2014**, *6*, 2812–2831. [CrossRef]
31. Ge, W.; Xu, Y.; Razzaq, A.; Liu, G.; Su, X.; Yang, X.; Ran, Q. What drives the green transformation of enterprises? A case of carbon emissions trading pilot policy in China. *Environ. Sci. Pollut. Res.* **2023**, *30*, 56743–56758. [CrossRef]
32. Hou, J.; Teo, T.S.; Zhou, F.; Lim, M.K.; Chen, H. Does industrial green transformation successfully facilitate a decrease in carbon intensity in China? An environmental regulation perspective. *J. Clean. Prod.* **2018**, *184*, 1060–1071. [CrossRef]
33. Du, H.; Chen, Z.; Mao, G.; Li, R.Y.M.; Chai, L. A spatio-temporal analysis of low carbon development in China's 30 provinces: A perspective on the maximum flux principle. *Ecol. Indic.* **2018**, *90*, 54–64. [CrossRef]
34. Chen, L.; Xu, L.; Cai, Y.; Yang, Z. Spatiotemporal patterns of industrial carbon emissions at the city level. *Resour. Conserv. Recycl.* **2021**, *169*, 105499. [CrossRef]
35. Song, Q.; Zhou, N.; Liu, T.; Siehr, S.A.; Qi, Y. Investigation of a "coupling model" of coordination between low-carbon development and urbanization in China. *Energy Policy* **2018**, *121*, 346–354. [CrossRef]
36. Liu, S.; Shen, J.; Liu, G.; Wu, Y.; Shi, K. Exploring the effect of urban spatial development pattern on carbon dioxide emissions in China: A socioeconomic density distribution approach based on remotely sensed nighttime light data. *Comput. Environ. Urban Syst.* **2022**, *96*, 101847. [CrossRef]
37. Cheshmehzangi, A.; Tang, T. Shandong Byland City Cluster: From an Agricultural Hub to Three Economic Circles and Industrial Clusters. In *China's City Cluster Development in the Race to Carbon Neutrality*; Springer: Berlin/Heidelberg, Germany, 2022; pp. 107–118.
38. Ma, S.; Zeng, G. Effects of network closure on cooperative innovation: Evidence from Dongying's petroleum equipment industry in China. *Chin. Geogr. Sci.* **2019**, *29*, 517–527. [CrossRef]
39. Peng, Y.; Zhang, X.; Li, Z.; Du, D. Evolution of freight spatial pattern of China's coastal ports facing the Polar Silk Road. *Res. Transp. Bus. Manag.* **2022**, *45*, 100900. [CrossRef]
40. Xu, C. Towards balanced low-carbon development: Driver and complex network of urban-rural energy-carbon performance gap in China. *Appl. Energy* **2023**, *333*, 120663. [CrossRef]
41. Jia, Z.; Wen, S.; Liu, Y. China's urban-rural inequality caused by carbon neutrality: A perspective from carbon footprint and decomposed social welfare. *Energy Econ.* **2022**, *113*, 106193. [CrossRef]
42. Xu, S.; He, Z.; Long, R.; Chen, H. Factors that influence carbon emissions due to energy consumption based on different stages and sectors in China. *J. Clean. Prod.* **2016**, *115*, 139–148. [CrossRef]
43. Aghion, P.; Bolton, P. A theory of trickle-down growth and development. *Rev. Econ. Stud.* **1997**, *64*, 151–172. [CrossRef]

sustainability MDPI

*Review*

# Research on Carbon Emission Structure and Model in Low-Carbon Rural Areas: Bibliometric Analysis

**Zejun Yu [1], Yao Wang [2,*], Bin Zhao [3], Zhixin Li [2,*] and Qingli Hao [1]**

1 School of Architecture, Tianjin University, Tianjin 300072, China; yuzejun1994@tju.edu.cn (Z.Y.); qingli_h@tju.edu.cn (Q.H.)
2 School of Architecture, Tsinghua University, Beijing 100190, China
3 School of Public Administration, Heilongjiang University of Science and Technology, Harbin 150006, China; zhbin202206@163.com
* Correspondence: wangyao22@mails.tsinghua.edu.cn (Y.W.); lizhixin22@mails.tsinghua.edu.cn (Z.L.)

**Abstract:** As the energy crisis and environmental problems are exacerbated, rural carbon emissions have gradually attracted increasing attention. Meanwhile, rural areas have the important function of ecological carbon sequestration, so the research field of carbon emissions has gradually expanded from urban to rural areas in recent years. To systematically sort out the research progress regarding low-carbon rural areas, a total of 583 papers published during 2013–2023 were acquired through the bibliometric analysis method from the "professional field of science of building technology" in the core database of Web of Science. Then, the research stage, author cooperation, institutional cooperation, national distribution, and keyword co-occurrence of "low-carbon rural areas" were analyzed via data visualization analysis software VOSviewer, and the current hot issues, such as carbon emission calculation list and elements, carbon emission measurement method/model, and energy saving and emission reduction paths, were further summarized and reviewed. The statistical analysis results show that: (1) in most of the current research on rural carbon emissions, geographical characteristics and rural carbon emissions have been rarely analyzed; (2) there is a lack of interdisciplinary research, e.g., less cooperation with geographic information or other disciplines; (3) the differences and characteristics of rural carbon emission factors in different economies and regions have rarely been analyzed. Based on the analysis, this research pointed out that future research can be comprehensively carried out in formulating rural carbon emission inventory, establishing a multi-scale and multi-regional statistical method and so on. This research also provides a comprehensive analysis and summary of the existing research on the structure and model of carbon emissions in low-carbon rural areas and also presents the problems that need to be paid attention to in the future research, pointing out the research direction for relevant scholars.

**Keywords:** low-carbon rural area; carbon emission structure; bibliometrics; VOSviewer

**Citation:** Yu, Z.; Wang, Y.; Zhao, B.; Li, Z.; Hao, Q. Research on Carbon Emission Structure and Model in Low-Carbon Rural Areas: Bibliometric Analysis. *Sustainability* **2023**, *15*, 12353. https://doi.org/10.3390/su151612353

Academic Editors: Elena Lucchi, Tianyi Chen and Wen Zhang

Received: 10 July 2023
Revised: 4 August 2023
Accepted: 9 August 2023
Published: 14 August 2023



## 1. Introduction

With global climate change, carbon emissions have become an increasing concern in recent years. According to the report of the International Energy Agency (IEA), global energy carbon emissions reached a record of 36.8 billion tons in 2022 [1]. Rural areas are important areas for human carbon emissions. According to the data released by the World Bank on 21 January 2023, the global rural population touched 4526.1 million in 2022, representing more than half of the world's total population [2]. Therefore, it is particularly important and necessary to study rural carbon emissions. In addition, relevant research shows that the research on carbon emission management is relatively extensive in rural areas compared to urban areas, and thus rural areas have great potential for energy conservation and emission reduction. Hence, it is necessary to explore and summarize the literature data related to rural carbon emissions in recent years.

In recent years, bibliometrics, as an important retrospective method for related research, has been widely used in various research fields. For example, Marina Figueiredo Muller used the multi-criteria decision analysis method to systematically review the literature on sustainability factors, interoperability issues, and life cycle stages of buildings. Through the screening and evaluation of a large number of documents, the author summarized the correlation between life cycle stage and sustainability, and provided the influence matrix between two fields [3]. Based on the method of artificial intelligence, Jack Ngarambe reviewed the current research aiming to improve the thermal comfort of indoor space, summarized the thermal comfort prediction models focusing on various machine learning (ML) algorithms and their deployment in building control systems to achieve the purpose of energy saving, discussed the gaps in the existing literature, and emphasized the potential research direction in the future [4]. Shameri, M.A. systematically reviewed the previous research on building double-skin facade system (DSFS), and summarized the development and future research direction of DSFS [5]. Abdelaziz, E.A. carried out a comprehensive literature review on industrial energy conservation through management, technology, and policy, and found that energy-saving technology can save a lot of electricity, emissions, and water and electricity charges [6]. Thevega, T. summarized building fireproof materials, and put forward the engineering process and tools for evaluating the applicability of coating materials to different building types [7]. Yin, X.F. explored the latest technology of OSC BIM through the bibliometric–qualitative review method, and determined the research trend and the knowledge gap that could be solved in the future research of OSC BIM. This research improved AEC practice by synthesizing the latest technology of OSC BIM and revealing the research demand in this field, thus contributing to the knowledge system [8]. It can be seen that bibliometric technology has been widely used in a large number of research fields, and its accuracy and utility have been verified. In this research, therefore, this method was adopted to systematically analyze the field of rural carbon emissions.

It is difficult to fully and clearly reveal the cooperation network, knowledge structure, and evolution trend of carbon emission structure and model research in low-carbon rural areas through the traditional literature review formed by manually reading a large amount of literature data. Therefore, VOSviewer software (Version 1.6.18) was used for analysis in this research. This investigation aims to systematically generalize and summarize the overall situation of rural carbon emissions-related research in the core collection of Web of Science in the period 2013–2022, including the comparative analysis of the number of articles published by authors, the number of articles published by institutions, the distribution of countries and regions, and the co-occurrence of keywords, and to predict the research direction and trend of rural carbon emissions. The rapid increase in the number of documents published by researchers in this field has posed a new challenge to document data mining. Therefore, it is necessary for this research to systematically review this research field by bibliometrics. In this research, the research data on the carbon emission structure and model of low-carbon rural areas were analyzed from the following five aspects: (1) knowledge map analysis of literature data; (2) analysis of research progress in several hot areas of low-carbon rural areas; (3) discussion and analysis of the research results, and identification and summarization of the research topics, research gaps, and research trends on rural carbon emissions; (4) forecasting of the research prospects on rural carbon emissions; (5) presentation of conclusions and research limitations.

## 2. Methodology

### 2.1. *Bibliometric Analysis Method*

Bibliometrics, which refers to a set of research methods based on literature characteristics and systems, provides an effective tool for quantitative analysis in specific research fields. Bibliometrics uses mathematical and statistical methods to explore various features of documents, including distributed architecture, collaborative relationships, and different modes. Bibliometric analysis, as one of the most effective and extensive quantitative methods, can measure the contributions of different aspects in special research topics. According to Zhou et al., [9] bibliometric analysis not only provides an effective tool to change from the micro level to the macro level but also helps scholars identify future research trends and hot topics.

*2.2. Literature Screening*

The research results published in the field of rural carbon emissions on the Web of Science (WoS) during 2013–2023 were summarized by using the overall research method (date of retrieval: 19 March 2023).

In this research, the core collection of WoS served as the literature data source, and retrieval was performed using topic fields under an advanced search mode: (((TS = ("Rural" OR "rural area" OR "village" OR "countryside" OR "country" OR "rural areas" OR "villages" OR "town ")) OR TS = ("village" OR "villages" OR "village distribution" OR "rural settlement" OR "traditional village" OR "settlements" OR "community" OR "rural village")) OR TS = ("farmhouse" OR "rural house" OR "house in village" OR "rural housing" OR "agricultural housing" OR "rural houses" OR "house in the countryside" OR "rural building")) AND TS = ("carbon emission" OR "carbon emissions" OR "carbon dioxide emissions" OR "carbon dioxide emission" OR "$CO_2$ emission" OR " $CO_2$ emissions" OR "carbon" OR "carbon discharge") on 19 March 2023. Specifically, the research direction was limited to construction building technology. The time span was set as 2013–2023. A total of 584 literature documents were obtained, among which 1 invalid document was excluded through data cleaning, deduplication, and preprocessing, and finally, 583 articles were acquired. The concrete steps are shown in Figure 1.

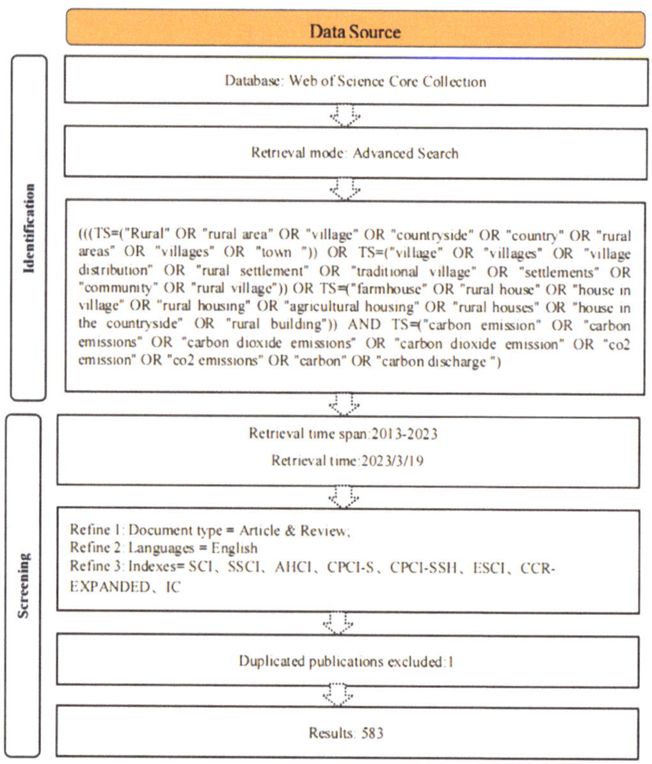


**Figure 1.** The procedure of the bibliometric analysis on the carbon emission structure and model in low-carbon rural areas.

### 2.3. Analysis Tool: VOSviewer

VOSviewer is a social network analysis (SNA) tool used to analyze and visualize complex networks. The software can be used to explore different types of knowledge networks, such as keyword co-occurrence, and cooperation network between authors and institutions. By mining potential topics and keywords, users can deeply understand the development trends and hot issues in the relevant literature fields [10]. In 2007, Leydesdorff, L. et al. used VOSviewer to analyze the data of journal papers in SCI literature library, and used the index—"node betweenness centrality"—to evaluate the interdisciplinary degree and interdisciplinary cooperation between journals [11]. In 2012, Waltman, L., et al. proposed a document classification method based on VOSviewer, establishing a journal classification system by virtue of the citation relationships between journals and displaying the interdisciplinary situation in the form of clustering heat maps [12]. In 2017, Glänzel, W. et al. analyzed the literature data in the field of astronomy by using VOSviewer combined with the LDA topic model, and improved the accuracy of literature classification and topic identification through the "core literature" strategy [13].

## 3. Knowledge Map Analysis via VOSviewer

Based on the bibliometrics theory, the current research situation with regard to rural carbon emissions was mainly combed from keywords, countries, institutions, authors, and time axes in this research.

### 3.1. Keyword Co-Occurrence Analysis

Keywords are highly summarized and generalized by researchers, which are mainly used to highlight the focus of their research content and help readers quickly understand their main research background. In this research, therefore, the keywords of authors in this research field were explored and analyzed using VOSviewer to exhibit high-frequency keywords and their co-occurrence relationship. Among the 1988 keywords, 70 ones were cited four or more times, specifically as shown in Figure 2.

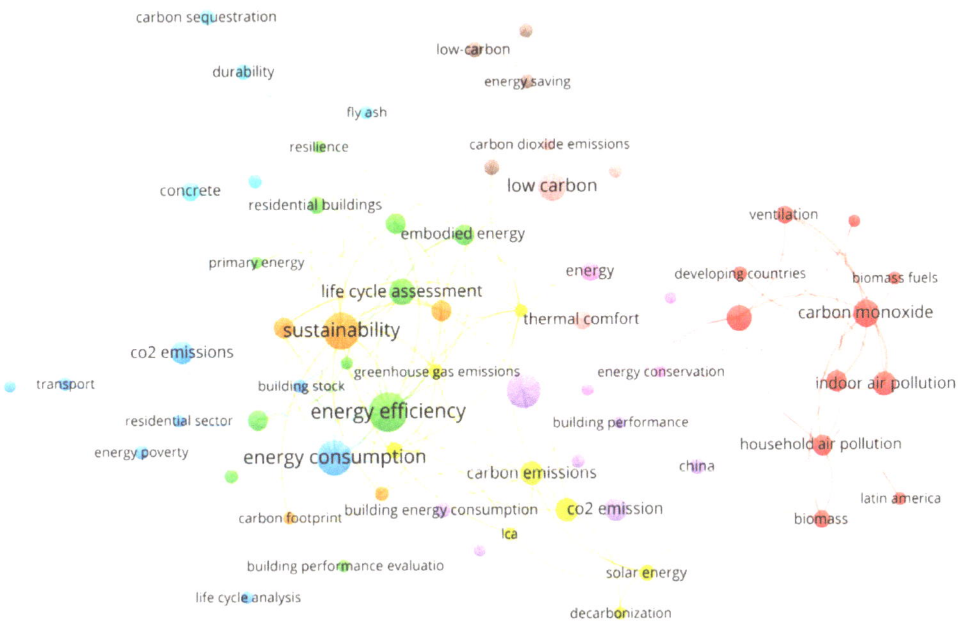


**Figure 2.** Keyword co-occurrence network analysis.

It can be seen from the figure that the core keyword in the literature data is energy efficiency [14,15]. Meanwhile, the research results also reveal that other several key research topics arousing widespread concern may be related to different research dimensions, such as sustainability (namely, sustainable development indexes, sustainable development) [15,16], energy consumption (namely, solar energy utilization and sustainable consumption) [17–19], climate change (namely, climate risk management) [20,21], and low carbon (namely, low-cost housing) [22,23]. The co-occurrence relationship and co-occurrence times between high-frequency keywords (the thicker the ligature, the more the co-occurrence times there were) are helpful in understanding and reflecting the research topics of rural carbon emissions in the past.

This reflects that in the past research on rural carbon emissions, energy efficiency was a core concern. Topics such as sustainable development, energy consumption, climate change, and low carbon are diverse in different research dimensions. By analyzing the co-occurrence relationship and co-occurrence times between high-frequency keywords, the theme and relationship of rural carbon emission research in the past can be understood more deeply. These results provide important reference points and guidance for the further discussion and management of rural carbon emissions. In the future research, the comprehensive solutions to sustainable development, energy consumption, climate change, and low carbon can be further explored on the basis of energy efficiency, so as to achieve the goal of reducing rural carbon emissions and achieving sustainable development.

### 3.2. Analysis of Research Stages

Using the time dimension to search for the whole research context of low-carbon rural areas can better understand the concerns of the entire academic community about low-carbon rural areas or the trend and direction of rural carbon emission research. In this research, the research topic trends in this field were analyzed through authors' keywords in the literature data. The time span was set to 2003–2022, the minimum frequency of words to six, and annual number of words to four.

Generally, the keywords in a document are highly related to the content of the document, through which the main aspects of the research topic in this research field can be judged [24]. Although high-frequency keywords are displayed in the co-occurrence of keywords, the evolution trend of rural carbon emissions-related research topics for many years can be further understood by analyzing the keyword topic trend. Figure 3 shows research scholars' focus on the research topics regarding rural carbon emissions in different years. These topics are related to the research field of rural carbon emissions in many aspects. For example, in 2017, energy efficiency was the most discussed topic and also an important research field of rural carbon emissions. In 2018, meanwhile, climate change was the main research topic, which was one of the key factors leading to rural carbon emissions. In 2019, sustainability ranked first, which also formed another key area of rural carbon emissions. In this analysis, solar energy was found to be a hot topic in 2021. Thus, it can be seen that the current research status has gradually shifted from "reducing sources" to "opening sinks"; that is, reducing carbon sources and increasing carbon sink types to improve the optimization efficiency of carbon emissions. It is revealed that the current research on reducing carbon emission sources has entered a bottleneck, and new technologies and methodologies are needed to break such shackles.

To better analyze the evolution process of this field, we divided the low-carbon development into different stages combining the time and content of the relevant literature. Specifically, it was generally divided into three stages—budding stage, developmental stage, and deepening stage—according to the timeline.

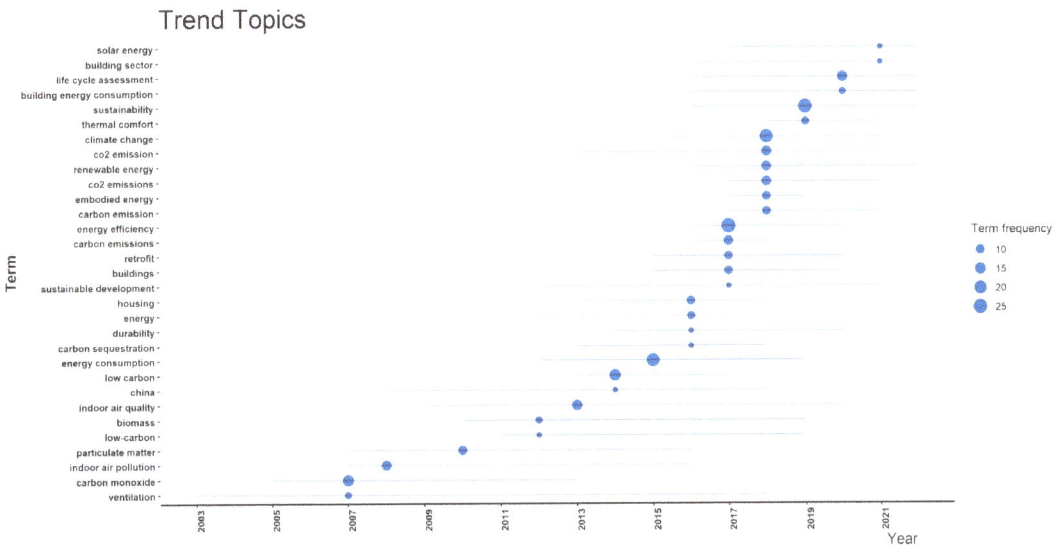


**Figure 3.** Keyword topic trend chart.

#### 3.2.1. Budding Stage

The years 2003–2010 comprise the budding stage of low-carbon rural research, during which the application of the low-carbon concept in rural production and life was preliminarily explored with the keywords of low carbon, indoor air quality [25,26], and carbon monomer [27,28]. The concepts of low carbon and energy saving and emission reduction were popular in rural areas, and the research at this stage was mainly qualitative research. In addition, more attention was paid to the low carbon concept and low carbon behavior and lifestyle. The research areas were mainly traditional developed countries, such as Sweden, Germany, Denmark, Finland, and Britain. For example, Burford, N. and Neil, T. discussed four strategies for the new sustainable development of low-energy houses in traditional villages in rural Scotland. Anchoring, Court, Street, and Croft summarized these principles into four new models to design more sustainable housing in rural. Burford, N. discussed the role of landscape aesthetics and sustainability in the prototype development of new rural housing in response to local landscape quality, mixed land use, urban density, and local renewable energy production. These results provide a semi-quantifiable prototype of spatial development, and these requirements are integrated into the general conceptual framework of rural sustainable life, providing an alternative method for solving Scottish policy legislation [29]. Burford, N. explored the new spatial form of self-sufficient housing in rural land owned by Scottish National Trust in Coton, Perth, predicted that the demand for new housing in nearby rural areas would increase by 33%, put forward suggestions on substitutive theoretical model and spatial concept for affordable and low-energy regional responsive housing development, and solved the wider requirements of the Scottish environment in terms of sustainability, low energy use, low-carbon material resources, and competitive market [30].

#### 3.2.2. Developmental Stage

The years 2011–2018 comprise the developmental stage, in which the technical means and implementation approaches of low-carbon rural areas were further perfected, trying to introduce low-carbon planning into different rural types with research keywords of energy consumption, carbon sequestration [31,32], and energy efficiency. Quantitative research gradually replaced qualitative research. Urban carbon emission research and methods were gradually introduced into rural areas, and the measurement and management of rural carbon emissions became the norm. At this stage, traditional developed countries were still the main research focus, but many developing countries were also concerned about carbon emissions at this stage. For example, Ge, J. used the data of carbon emissions and carbon sequestration of villages and the management functions of grass-roots governments to select four factors influencing the carbon emissions of these villages, including natural ecology, economic industry, residential development, and infrastructure, prepared a carbon emission inventory for these villages, and collected the data on emission activity level and emission factors. Finally, a method for evaluating village carbon emissions was developed. Meanwhile, four landforms (mountains, hills, plains, and islands) and eight villages with different industrial types were selected for case study and evaluation of carbon emissions. According to the evaluation results, the carbon emissions of villages were divided into four types [33]. Wu, G. focused on the regional household energy consumption and related carbon dioxide ($CO_2$) emissions, and found that family size, income, and education level are the main factors that affect carbon dioxide emissions [34]. Li, X.C. investigated the energy consumption of rural households from the aspects of energy (coal, biogas, straw gas, liquefied petroleum gas, electricity, and firewood), utilization, selection, cost, and system. The results show that coal is still the main energy source for heating in winter, accounting for the largest part of household energy consumption [35]. With the development of renewable energy, cooking with coal, liquefied petroleum gas, and firewood has been replaced by renewable energy such as biogas and straw gas, which have higher calorific value and lower prices. The research also shows that the energy consumption structure of rural households is changing from the traditional inefficient biomass energy to the comprehensive consumption of traditional energy and renewable energy. Renewable energy not only solves the problem of rural energy shortages but also helps to save fossil resources and protect the environment.

Siudek, Aleksandra mentioned that the use of renewable energy (RE) in newly built and renovated single-family houses in rural Poland will improve energy efficiency and reduce carbon dioxide emissions. Replacing coal with environmentally friendly renewable energy can reduce carbon dioxide emissions by approximately 90% every year [36].

#### 3.2.3. Deepening Stage

From 2019 to now is the deepening stage, in which the main research direction of low-carbon rural areas was more inclined to "reducing sources and opening sinks" and various means were adopted to reduce carbon emissions. According to the keyword analysis of the retrieved literature, it could be concluded that the main current research direction of low-carbon rural areas lies in "reducing sources and opening sinks", in which reducing sources refers to reducing carbon emissions and opening sinks means expanding the access channels and energy efficiency of new and renewable energy. With the advancement of research, the effect of reducing rural carbon emissions is becoming lower and lower, so the research direction and proportion are gradually inclined toward the development and use of clean energy. To solve the energy shortage in rural areas and improve energy efficiency, Chen, X.P. proposed a hybrid distributed power generation (DG) system including wind energy, hydrogen energy, and fuel cells to supplement the main power grid. Mura, P. proposed a demonstration project in a village in Sardinia, which consists of photovoltaic collectors connected to the network to provide electricity for households and hydrogen for transportation [37]. Yang, T.R. discussed the influence of residential community types on the building energy consumption and solar energy trade-off, aiming to achieve the best conditions of zero non-renewable energy communities, identified six typical Chongming community types, i.e., farmhouse, linear village, core village, townhouse, flat slab house, and high-rise building, and estimated the influence of increasing solar panel installation through Monte Carlo simulation [38]. Skowronski, P. discussed the economic and market potential of solar energy to prepare hot water in rural areas. According to estimates, the solar energy system could meet 30–45% of the energy demand of warm-water power generation in rural areas at a reasonable cost, and correspondingly, reduce $CO_2$ emissions.

The realization rate of the economic potential of solar water heaters depends on the subsidy for equipment installation [39]. There is a large space for research on low-carbon rural areas. At the same time, the rural energy structure and the comparative regional differences also remain to be further explored.

### *3.3. Research Countries, Authors, and Groups*

Through the comprehensive analysis of authors and their literature fields, the current key academic focus in rural carbon emissions can be figured out, which can better clarify the research gaps and new research objectives.

#### 3.3.1. Analysis of Countries

By analyzing the number of papers published in different countries, the current research investments and the region and scope of the current research results can be revealed, which can facilitate the better analysis of application prospects and future trends.

The world distribution map of papers released in different countries is as shown in Figure 4. Although rural carbon emissions have always been a global concern, the research intensity may vary from region to region. Therefore, it is very important to prove the global distribution of rural carbon emissions research by analyzing papers published in different countries and regions. A total of 89 countries participated in the research, among which over nine papers were published in more than 22 countries, and China ranked first in the number of papers published (170). The time-dependent changes in carbon emissions from urban and rural houses were investigated from the aspect of terminal use; in particular, the comprehensive emission factor of electricity and heating power in time series was considered. Then, it was found that the carbon emissions from private transportation grew fastest, and the carbon emissions produced by space heating and cooling were approximately 40% and 30% [40].

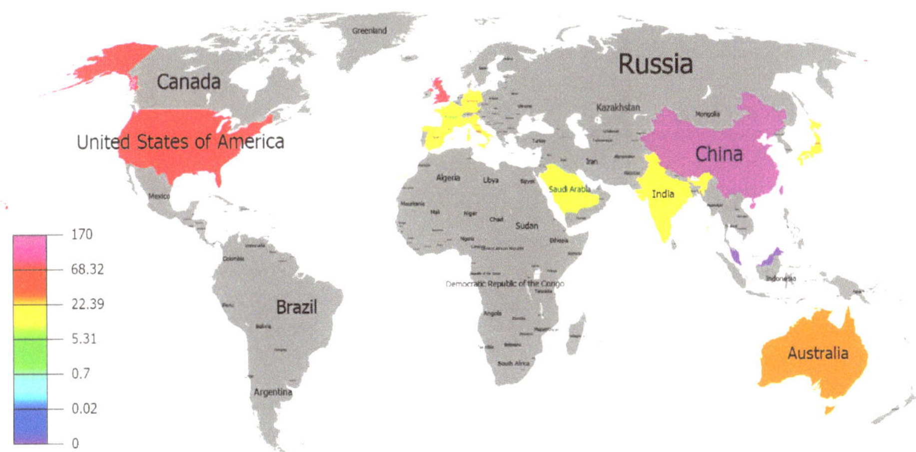


**Figure 4.** Distribution map of papers published in different countries.

Secondly, the United States (77 papers) pointed out that the construction industry was the emitter of the majority of thr greenhouse gas carbon dioxide ($CO_2$) partly because of the operation scale and a great variety of equipment [41]. Kaya, Y. proposed in 1993 that building carbon emissions can be calculated using the Kaya formula used to measure the effects of economic growth, population growth, energy efficiency, and energy structure on greenhouse gas emissions, specifically $CO_2$ emissions = population × per capita energy consumption × energy intensity × $CO_2$ emission coefficient. Usually, the calculation boundary of building carbon emissions includes the manufacturing, transportation, and installation of construction materials, energy consumption during the use of buildings, and maintenance and demolition of buildings. The calculation method of building carbon emissions based on the Kaya formula can provide guidance and support for the sustainable development of the construction industry. In addition, this method can identify and quantify the source and influencing factors of building carbon emissions so as to formulate the corresponding emission reduction measures and reduce building carbon emissions. Meanwhile, this method can provide a scientific basis for the environmental protection policy making of the construction industry and boost its sustainable development and green transformation [42].

Thirdly, the United Kingdom (69 articles) described the application of the data-driven localized geographic information system (GIS) in Oxford, England, to model, draw, and manage household energy use and carbon emissions at the residential and community levels [43]. It has been discovered that the neighborhood energy model helps to summarize the demand for cost-effective energy action measures. It is estimated that the energy consumption will be reduced by 44% and the energy cost will be reduced by 38% after the comprehensive transformation of surveying and mapping areas. If actions are taken, energy use can be reduced and the fuel shortage in neighboring areas can be mitigated [44]. Based on the interconnection between different countries, it is shown that in recent years, with the increasingly severe global climate change and the gradual improvement of carbon emission reduction targets, developed countries have carried out a lot of research work in the field of carbon emissions [36]. With the implementation of relevant policies and the advancement of technological progress, however, some developed countries seem to be paying less attention to carbon emissions, which may be ascribed to some progress in carbon emission reduction in such countries.

In addition, the research in the fields of green energy, climate adaptation, and carbon capture in developed countries has also received more attention. Secondly, China and other developing countries continue to exert their efforts in the field of rural low-carbon research. With the rapid economic development and accelerated urbanization in China, carbon emission has become a problem to be solved urgently. Rural areas are also an important source of carbon emissions. Therefore, it is very significant for China to make continuous efforts in the field of rural low-carbon research. In recent years, China has achieved some progress in the field of rural low-carbon research, such as developing green agriculture, improving rural ecological environment protection, and popularizing clean energy. In addition, China is also exploring policies and technologies for carbon emission reduction in rural areas, including carbon trading, carbon sequestration, and biomass energy.

As revealed by the literature analysis, the emphasis on rural carbon emissions varies with the rural developmental stage and resource endowment in different countries. Developed countries are characterized by a high degree of rural mechanization and a large number of equipment activities. More emphasis is on scientific research and development to reduce carbon emissions from construction and agricultural equipment operation. Developing countries should explore more policies and technologies for carbon emission reduction in rural areas, including carbon trading, carbon sequestration, and biomass energy, and increase the generation of clean energy. Generally speaking, carbon emission is a global environmental problem, which breaks through the past pure concern about the emission reduction responsibility of developed countries and highlights the important role played by developing countries in emission reduction and climate change. It calls on the international community to implement a global carbon emission reduction plan to ensure that all countries work together and cooperate to reduce carbon emissions. Such global cooperation and efforts will ensure the fairness and effectiveness of carbon emission reduction measures and lay the foundation for achieving global climate goals. At the same time, this also means that carbon emission reduction is not only the responsibility of developed countries but also needs to be shared by all countries in the world. Developing countries should strengthen their own research and exploration, seek emission reduction schemes suitable for their own national conditions, and promote the implementation of these schemes within the framework of sustainable development. This will help to balance global emission reduction efforts and promote sustainable economic growth and social development in developing countries.

3.3.2. Analysis of Research Authors

By analyzing the authors of the 583 articles on rural carbon emissions, the authors publishing most papers were determined, including their cooperative relationships and cooperation network. In a new research field, the number of papers published by an author could represent their research contribution and professional level in this field. In this research, the minimum number of papers published was set to two, and 111 authors met the standard. The authors publishing the largest number of papers were Gupta, Rajat, and Zhang, Y. who each published five articles on rural carbon emissions, followed by Shan M., Smith, K.R., and Yang X.D. who each published four papers.

In addition, it could be clearly found from Figure 5 that 111 authors could be divided into 33 cooperation networks, and a few authors were independent researchers. Each cooperative group was named after the author publishing the largest number of papers in a cooperation network. For instance, the most influential cooperation networks were Shan, M., including authors Shan, M., Lu, F., Nie, Y.Z., Yuan, Y.P., Yang, M., Deng, M.S., and Ma, R.J. Such authors mainly summarized the use of renewable energy in buildings, and pointed out some problems and basic laws of renewable energy utilization. Through two demonstration projects in different rural areas, it has been proved that it is feasible to realize "low carbon" or "zero coal" villages by implementing different renewable energy technologies [45]. Meanwhile, it has been pointed out that the environment will be seriously polluted by combusting solid fuels in traditional heating furnaces, and the PM2.5, CO, and $SO_2$ emissions from biomass particle heating furnaces using wood pellets are the lowest [46,47]. The second most influential cooperation network was Gupta R. who implemented the alternative method of reducing domestic energy demands by changing the understanding and behaviors of house owners and implementing physical transformation. The results show that energy transformation is quite effective in reducing residential energy use (75% of households reduce the use of gas and/or electricity) [48,49]. It is estimated that energy consumption will be reduced by 44% and energy cost will be reduced by 38% by implementing individual and combined energy improvement measures. If actions are taken, energy use can be reduced and the fuel shortage in neighboring areas can be relieved.

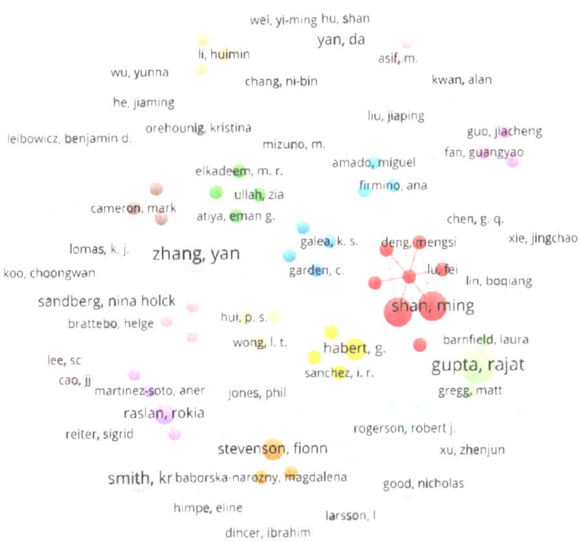


**Figure 5.** Cooperation network of authors.

It could be observed from Figure 5 that the cooperation networks of authors did not overlap, indicating the relative independence of rural carbon emission researchers who may differ in their concerns and research methods. This phenomenon reflects the complexity and diversity of rural carbon emission research fields. The carbon emissions in rural areas are affected by geographical, climatic, ecological, and economic factors, and great regional differences are manifested [44]. Hence, researchers need to study from multiple angles to fully understand and master the knowledge in this field. At the same time, this relative independence also brings opportunities for researchers to cooperate and communicate, through which they can share their research results and experience, thus improving the research level and quality of the whole field. However, this relative independence may also lead to information asymmetry and knowledge barriers among researchers. If there is not enough communication and cooperation between different researchers, they may miss some important research results and discoveries, and may even produce redundant and repetitive research. Therefore, researchers dedicated to the rural carbon emission research should strengthen exchanges and cooperation and establish a closer and more organic cooperation network, which can promote the sharing and dissemination of knowledge, thus improving the research level and quality of the whole field.

#### 3.3.3. Analysis of Research Institutions

The institutional cooperation network included 94 institutions and 153 cooperative ties. As shown in Figure 6, the Harbin Institution of Technology ranked first (12 papers). Taking the cold region as the research object, this research institution analyzed the structure and mode of rural energy consumption and the main problems of rural planning and construction, put forward planning technical measures and construction schemes considering terrain selection, divided functional areas in towns and villages, encouraged the construction of townhouses, built multiple windbreaks, explored energy from rural production and household waste, and developed solar energy, biogas, wind energy, and other energy sources to adapt to the characteristics of cold climate [32,43,50], followed by Tsinghua University (11 papers). Tsinghua University mainly focused on carbon emissions. In addition, the influence of the unbalanced proportion of forest land on urban carbon emissions was also studied [51–53]. Meanwhile, it has been found that many people in cold rural areas still use coal as the main heat source, and traditional cooking and heating by burning solid fuel in rural areas in the winter has also caused serious environmental pollution [46]. It is very important to choose clean heating technology that can replace raw coal, which may be useful to other countries facing similar environmental problems [54]. The Chinese Academy of Sciences (10 papers) paid attention to the carbon emissions generated by household consumption and the impact of overall emission changes on the sustainable development of the Qinghai–Tibet Plateau with the highest altitude in the world [15,55,56], reflecting that researchers from the Chinese Academy of Sciences concern various factors in environmental protection and put efforts into solving practical problems through scientific research. Particularly, they focus on the influence of regional characteristic differences and spatial heterogeneity on carbon emissions. Significant regional differences in carbon emissions result from the differences in geographical location and climate. Therefore, taking regional characteristics into account, carbon emissions can be evaluated more accurately, so as to formulate the corresponding environmental protection policies. In addition, researchers from the Chinese Academy of Sciences also concern the carbon emissions from rural household consumption, reflecting that they lay emphasis on the environmental impact of human activities. With the improvement of people's living standards and the change in the consumption structure, the impact of rural household consumption on carbon emissions has become increasingly prominent. Therefore, investigating the influences of the carbon emission from household consumption and the overall emission changes on the sustainable development of Qinghai–Tibet Plateau can provide a more scientific basis for formulating environmental protection policies. Other institutions, including the Hong Kong Polytechnic University (eight papers) and Xi'an University of Architecture and

Technology (8 papers), analyzed the different possibilities of cost/benefit in reducing the carbon emission level [20,57,58]. In rural areas, biomass (wood and agricultural wastes, such as straws, corn vines and branches, wood firewood) is the most commonly used fuel type, which will emit a variety of air pollutants: particulate matter (PM), CO, NMHC, $CH_4$, high-content black carbon (BC)—a greenhouse effect aerosol, and organic carbon (OC)—a cooling effect aerosol [21]. The University of Sheffield (seven papers) deeply interviewed different British family environments (owner-occupied and leased; additional care, shelter and nursing homes; urban and rural areas) to explore how the elderly living with low-carbon thermal technology expressed thermal comfort. The survey results show that the elderly attach great importance to comfort and radiance, and achieve it in many ways that may run counter to the policy objectives. In the rural residence occupied by the owner, the wood-burning stove is reserved after the floor heating/heat pump is installed, so as to provide visible brilliance and hospitality for the guests. In nursing homes, fake fireplaces provide comfort and brilliance without affecting the concerns about risks. The research shows that the assumed emission reduction brought by implementing low-carbon heating technology may be overestimated because the equipment is supplemented in the family manufacturing practice to provide comfort and social contact [49,59,60]. Colorado State University (six papers) analyzed the total carbon (TC) and radioactive carbon (C-14) of fine particles collected from two cities, four neighboring cities, and six remote locations in the United States. It is found that in summer, the carbon concentration in urban and rural areas is similar. In winter, the carbon content in urban areas is more than twice that in neighboring areas, and the carbon content in urban fossils in both seasons is 4–20 times that in neighboring rural areas. Organic carbon (OC) and elemental carbon (EC) were obtained by TOR analysis, and these data and radioactive carbon data were used to estimate the characteristic fossils in winter and summer and the EC/TC ratio in the same period. These ratios were applied to the carbon data monitored between protected visual environmental network institutions to estimate the carbon proportion of most rural areas in the United States in the same period [61–63].

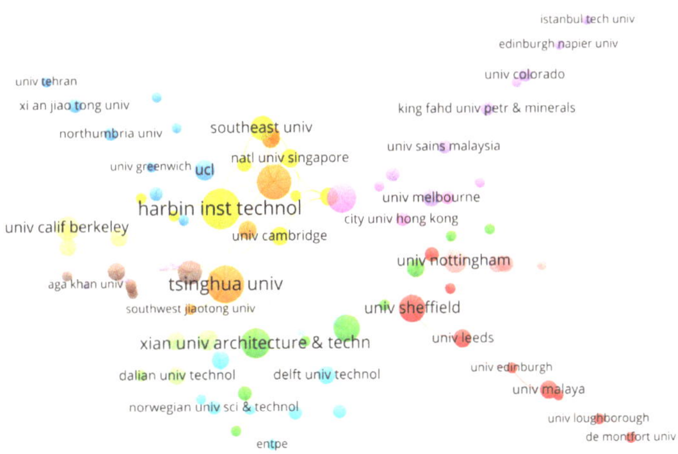


**Figure 6.** Cooperation network of institutions.

This analysis of institutions shows that different institutions have different research interests in rural carbon emissions. The University of Sheffield, Colorado State University, and other institutions paid more attention to the research and development of some technical models. Harbin Institution of Technology and other units focused on the influences of different behaviors and factors on rural carbon emissions, and attention was also paid to computational boundaries and methods. This reflects that in environmental protection, different research institutions turn their eyes to different regions and scenarios, and meanwhile, different calculation methods and parameters need to be considered to more accurately evaluate carbon emissions and formulate the corresponding environmental policies. To evaluate the carbon emissions in rural areas more accurately, therefore, the geographical location, climate conditions, energy structure, and production modes should be comprehensively considered.

Generally speaking, these studies provide valuable information for solving the problems of rural energy consumption and low-carbon planning and construction and can be referenced by other countries facing similar environmental challenges. In future research work, clean heating technology that can replace raw coal can be explored to reduce the air pollution caused by the use of biomass fuel in rural areas and promote the development and utilization of sustainable energy. These research results are of great significance for promoting rural sustainable development and low-carbon transformation and provide a beneficial reference for relevant policy makers, academic circles, and all walks of life.

## 4. Current Research Progress and Hotspots in Low-Carbon Rural Areas

### *4.1. Low-Carbon Rural Structure*

The research on the structure of rural carbon emissions is mainly divided into two categories: carbon source and carbon sink. The inventory of rural carbon sources and carbon sinks is slightly different in different countries and different regions [15,64]. In this research, the carbon source and sink inventory in the current mainstream research on the carbon emissions in low-carbon rural areas were summarized and analyzed [15,15,65,66].

Through the analysis, the existing literature can be basically divided into the following aspects (Table 1): the first-level carbon sources refer to the source indexes of carbon emissions, under which there are two second-level indexes: domestic carbon emissions and production carbon emissions. This division method is based on the consideration that the industrial structure of rural areas in China is dominated by agriculture and labor export, the energy structure by biomass and fossil energy, and the carbon emission source by domestic carbon emission. The second-level index—domestic carbon emissions—contains four classes of third-level indexes: domestic energy carbon emissions, transportation carbon emissions, building operation carbon emissions, and domestic waste carbon emissions. The second-level indexes are adjusted according to the characteristics of the target village, and production carbon emissions mainly include three classes: production energy carbon emissions, farmland carbon emissions, and animal husbandry carbon emissions, in which farmland carbon emissions are subdivided into farmland carbon emissions and pesticide and chemical fertilizer carbon emissions. Each class of third-level indexes also contains some sub-indexes. Such sub-indexes should be accounted for appropriately to reduce errors as much as possible. When calculating the net value of carbon emissions, the current total carbon emission can be obtained by total carbon emission, i.e., carbon source index, minus carbon sink index.

**Table 1.** Decomposition of rural carbon source and carbon emission indexes.

| First-Level Index | Second-Level Index | Third-Level Index | Concrete Content |
|---|---|---|---|
| Carbon source index | Domestic carbon emission | Domestic energy carbon emission | Daily use time and total power |
| | | Building carbon emission | Heating and cooling |
| | | Transportation carbon emission | Vehicle operation |
| | | Household waste carbon emission | Household waste |
| | Production carbon emission | Production energy carbon emission | Planting and harvesting |
| | | Farmland carbon emission | Seed, agricultural film, pesticide, fertilizer, and irrigation |
| | | Animal husbandry carbon emission | Excretion |
| Carbon sink index | Natural carbon sink | Natural forest carbon sink | Area, and forestry category |
| | Artificial carbon sink | Artificial forest carbon sink | Planting area and category |
| | | Planting carbon sink | Planting area and category |

In these sectors, the most important source of rural carbon emissions is the carbon emissions generated by agricultural production and rural residents' life.

#### 4.1.1. Agricultural Carbon Emission

In the agricultural system, agricultural production activities such as livestock breeding and crop planting are the most active. Methane can be produced by rumen microbial fermentation and animal manure treatment in animals [67]. The production and use of chemical fertilizers, pesticides, and agricultural films in the process of crop planting, the power consumption of agricultural machinery fuel and farmland irrigation [68,69], and the exposure and erosion of agricultural soil caused by ploughing will increase rural agricultural carbon emissions; regional differences will be enormously influenced by the developmental stage and agricultural habits [70].

#### 4.1.2. Rural Residents' Life

In the life of residents, carbon emissions will be produced by energy consumption resulting from heating, cooling, and lighting needs in daily life [71]. Additionally, indirect carbon emissions will be caused by producing the commodities and services needed by basic necessities of life of rural residents. With the continuous changes in rural residents' living standard and demands, the rural indirect carbon emission structure becomes more complicated, and the carbon emission source identification in rural areas is taken as the emphasis in many case studies [72].

### *4.2. Calculation Model of Low-Carbon Rural Areas*

The measurement method of low-carbon rural areas, which has always been an important direction of academic research, derives from urban carbon emission accounting, but because the carbon emission inventory is different from that of urban areas, the methods adopted are different, as well. The rural carbon emission effect originates from agricultural production activities and residential energy consumption. Due to the lack of unified accounting standards, the total rural carbon emissions calculated by different calculation ranges, methods, and indexes are quite different.

To understand the current quantitative method for low-carbon rural areas in a more detailed way, the followings were input into the literature retrieval page of the core collection of WoS: (TS = ("low-carbon rural area" OR "low-carbon rural areas" OR "low-carbon" OR "low-carbon rural areas" OR "low-carbon rural")) AND TS = ("carbon emission measurement" OR " measurement of carbon emissions" OR "calculation of building carbon dioxide emission" OR "carbon emissions" OR "carbon emission calculation"), while other options were the same as above. A total of 76 papers were acquired, followed by the keyword analysis, so as to obtain Figure 7. The following conclusions could be drawn from the keyword analysis (Table 2):

The rural carbon emission measurement methods are mainly divided into three main classes. The first class is the carbon dioxide emission accounting method used to support the carbon trading market, which is mainly formed according to the methodology and guideline system released by such institutions as IPCC and is generally accepted by various countries in the world [73]. The second class refers to the carbon emission estimation methods based on the input–output method and life cycle assessment method, where the latter is applied to measure the greenhouse gas emissions in each production step and the whole life cycle of products and the former is more applicable to the estimation of emissions at such levels as the national macro-level and industrial meso-level [74]. The third class is the carbon emission model measurement method based on the factor decomposition method, which analyzes the varying interactions between relevant factors and carbon emissions through modeling and performance estimation [75].

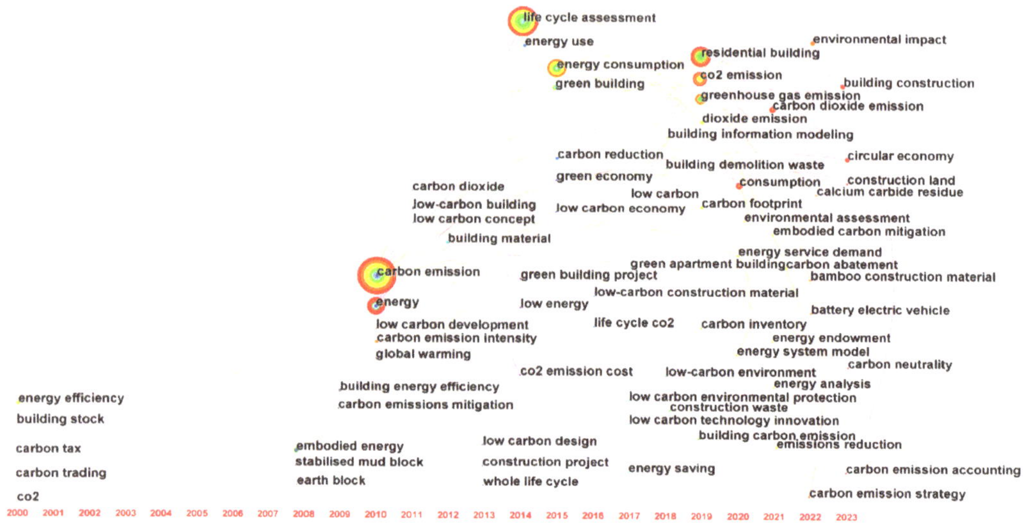


**Figure 7.** Retrieval content and keyword analysis results.

**Table 2.** Carbon emission calculation methods and characteristics.

| Calculation Method | Data Source | Model | Advantage | Disadvantage | Research Method | Reference |
|---|---|---|---|---|---|---|
| Decomposition method | Statistical yearbooks | STIRPAT, grid model, Kaya equation | Data availability | Subjectively defined influencing factors; results unverified | | [76–78] |
| Input–output method | Input–output tables | Relevant carbon emission models | Supply chain emission; data availability | Subjectively defined construction-related industries, without full coverage; failure to cover all production materials | From top to bottom | [79,80] |
| Statistical method | Statistical yearbooks and industrial statistics | Emission factor method | Simplicity, and relatively reliable data | Failure to cover all production materials | | [80,81] |
| Simulation modeling method | Building parameters | Modeling tools integrating energy modules | Macroscopic energy evaluation | Manual insertion of parameters; data are not intuitive, remaining to be explained | | [41,42] |
| Architectural prototype method | Building attributes | Actual building projects | Accuracy, clear boundary, and comparable results | Difficulty in concrete data acquisition | From the bottom up | [43,44] |

Rural areas are vast and sparsely populated compared to rural areas, making it rather difficult to collect data. Hence, the carbon emission coefficient method is currently used to account for carbon emissions from rural agricultural production and direct carbon emissions from residents' lives, and the input–output method to account for the indirect carbon emissions from residents' lives, which has gradually predominated carbon emission accounting [82]. Different measurement methods slightly differ in the difficulty in data collection and estimation precision [83].

### 4.3. *Rural Energy Conservation and Emission Reduction Paths*

An important academic research direction at present lies in improving rural carbon emissions, reducing carbon emissions and increasing carbon sinks by various means, and creating low-carbon rural areas. To comprehend the energy conservation and carbon emission reduction paths of low-carbon rural areas more comprehensively, the following were input into the literature retrieval page in the core collection of WoS: (TS = ("Rural" OR "rural area" OR "village" OR "countryside" OR "country" OR "rural areas" OR "villages" OR "town ")) AND TS = ("energy saving and emission reduction" OR " energy conservation and emission reduction" OR "energy saving" OR "energy conservation" OR "energy conservation and emissions reduction" OR "energy-saving and emission-reducing" OR "energy-saving emission reduction" OR "energy saving and emissions reduction"), the research field was limited to construction building technology, and a total of 89 papers were retrieved, followed by the keyword analysis to obtain Figure 8. The following conclusions could be drawn from the keyword analysis:

**Figure 8.** Keyword analysis.

Through the literature retrieval and further analysis of rural energy conservation and emission reduction pathways, low-carbon rural areas are created mainly from two aspects: source expansion and consumption reduction.

#### 4.3.1. Source Expansion

Rural areas are vast, sparsely populated, and rich in carbon sink resources, which can balance their own carbon emissions and back-feed urban areas, with great potentials, gradually attracting attention. On the one hand, the research focus is on increasing the carbon sink of rural areas and evaluating the carbon reserves and carbon sequestration capacity of herbaceous plants, shrubs, rivers and lakes, wetlands, forest floors, mineral soils, and trees. For example, Xu, W. has proposed and formulated many innovative methods to mitigate the impact of excessive greenhouse gas emissions on global climate change. Sequestration of carbon to terrestrial ecosystems is one of the important clean development mechanisms [84].

On the other hand, rural areas are a good carrier for the utilization of many new energy sources. In recent decades, the global energy demand has been increasing. As climate change is aggravated and environmental awareness is enhanced, seeking for alternative solutions to renewable energy has become the focus of common concern of governments and the international community. In this context, compared with urban areas, rural areas are rich in building roofs, water, and forest resources, serving as a good carrier of new energy, which is non-renewable. Solar energy is considered one of the best alternative energy sources for traditional energy sources. Rational use of solar energy is of great significance for alleviating energy depletion and reducing energy carbon emissions. Compared with wind power generation and hydropower generation technologies, solar power generation is the renewable energy power generation technology with the greatest potential for large-scale development and application, and it has broad market demands and good development prospects. Taking the cold regions as the research object, Xu, D.M. analyzed the structure and mode of rural energy consumption and the main problems of rural planning and construction, put forward the technical planning measures and construction schemes, designed, arranged, and divided the functional areas in towns and villages, encouraged the construction of row houses and multiple windbreaks, explored energy from rural production and domestic garbage, and developed solar energy, biogas, wind energy, and other energy sources to adapt to the characteristics of cold climate [43]. Xu Y, et al. analyzed the development trend of rural solar water heaters, solar cookers, and solar houses, predicted their development potential, and calculated the energy saving, $CO_2$ emission reduction, and their costs. The energy analysis results show that the utilization of solar energy resources in rural areas is increasing rapidly, with huge development potential, and the utilization efficiency needs to be improved. The three methods of solar energy utilization have exhibited remarkable energy saving and emission reduction effects. According to the EU emission trading price, the $CO_2$ emission reduction cost of solar houses and solar furnaces is negative. It is suggested that the government should rationally plan and utilize rural solar energy resources while increasing investments to form scale benefits [85].

### 4.3.2. Consumption Reduction

The related research is mostly based on the application of new carbon reduction technologies in rural areas, while advocating for innovative mechanisms and encouraging rural residents to practice the concept of low-carbon life. Hao, A.M. found that there is an "inverted U" curve relation between digital rural construction and rural carbon emissions. Agricultural planting structure and agricultural technical efficiency are important ways of reducing agricultural carbon emissions in digital village construction. The research findings also reveal that the higher the level of economic development, the stronger the carbon emission reduction effect of digital village construction [86]. Zhou, C.Y. deeply studied the envelope structure of rural houses in severe cold regions and put forward the low-carbon design principles and key technologies for the envelope structures of rural houses: shape control, utilization of natural materials, window design, and the effective utilization of solar energy. These technologies can reduce energy consumption, control greenhouse gas emissions, and relieve environmental pollution [87].

## 5. Discussion

The research field of rural carbon emission is a field of global concern, and all countries are carrying out related research. Generally speaking, the existing research also has some limitations, mainly manifested in the following aspects:

First, the research on carbon emissions in low-carbon rural areas still focuses on a few countries, lacking cross-regional and interdisciplinary cooperation. Due to the differences in developmental stages and energy structure, the experience and methods of relevant countries cannot be directly copied. Low-carbon rural carbon emissions is a global problem to be solved through global cooperation and efforts. Therefore, more countries and institutions are needed to join in the research on carbon emissions in low-carbon rural areas to jointly promote the development of this field.

Second, each country and region should formulate appropriate standards to deepen the research on rural carbon emissions and carbon sinks accounting based on the current conditions and economic development. On the one hand, countries and regions can carry out more in-depth and detailed research on the current measurement of rural carbon emission sources, which is not scientific and comprehensive enough, such as the large differences in the accuracy of data collection, the lagging carbon emission coefficients, the crude calculation methods, and the lack of dynamic adjustments. On the other hand, considering the unevenness of regional development and the balance between carbon emissions and economic development, basic data statistics should be carried out for the less-developed regions, and appropriate and simple carbon emission calculation methods should be established.

Thirdly, the current research on the rural comprehensive energy conservation and emission reduction path lacks comprehensive consideration and comprehensive carbon emission reduction optimization strategies based on carbon emission reduction optimization efficiency ranking, not to mention comprehensive consideration, optimization prediction, and modeling. Hence, it is necessary to divide villages into specific categories according to the influencing factors of different countries for quantitative research, such as geographical features and population structure. Therefore, multiple factors should be comprehensively considered in future research to more comprehensively evaluate rural carbon emissions and propose the corresponding solutions.

Considering the shortcomings of the existing research, future research on low-carbon rural area-related problems can be comprehensively carried out from the following perspectives:

First, a scientific and convenient rural carbon emission inventory should be formulated in accordance with the basic conditions of different regions, which can include: the chemical fertilizers, pesticides, and agricultural films input in rural crop plantation, intestinal fermentation, and fecal discharge during the growth of livestock; residential clothing, food, housing, and travel activities, and the fossil energy consumption throughout production and living activities as sources of rural carbon emissions; and the greenhouse gases to be accounted for, including $CO_2$, $N_2O$, and $CH_4$, and so on.

Secondly, a multi-scale and multi-regional statistical method of carbon emissions, especially the carbon emission coefficient, should be established to determine the carbon emission coefficients of different greenhouse gases, different varieties of crops and livestock, and their planting and breeding methods as well as the carbon sequestration effects of various conservation tillage measures in different regions so that the accounting results of rural carbon effects can be pertinent.

Finally, according to the geographical resource endowment conditions and rural developmental stages of different countries, the principles of rural low-carbon development should be formulated, and rural energy conservation and emission reduction paths specific to different types of villages should be scientifically determined, specifically including: reducing energy consumption, pollution and emission, and increasing carbon sinks, which aims to realize a win–win pattern of rural carbon emission reduction and economic development.

## 6. Conclusions

With the increasing concern around low-carbon rural carbon emissions, a comprehensive view on the low-carbon rural carbon emissions research in the recent decade was proposed in this research. This research was mainly based on the bibliometric analysis results of 583 low-carbon rural carbon emission-related papers published during 2013–2023, and the following results were found:

The analysis of high-frequency keywords shows that keywords such as sustainability, energy consumption, climate change, and low carbon have always been the focus of researchers devoted to the research on low-carbon emissions in low-carbon rural areas. The topic trend analysis of keywords can allow us to further understand the evolution trend of rural carbon emissions research topics for many years and provide effective guidance for the future research on carbon emissions in low-carbon rural areas.

According to the results of the authors' cooperation network analysis, Gupta, R. and Zhang, Y. are the main contributors to the research on carbon emissions in low-carbon rural areas, meaning that they have a major influence in this field. In addition, Shan, M., Smith, K.R., and Yang, X.D. have investigated carbon emissions in rural carbon villages too. Very obviously, scholars from developing economies will conduct more extensive research work on carbon emissions in low-carbon rural areas, since the carbon emissions in low-carbon rural areas have become one of the most critical channels for efficient energy conservation.

According to the analysis results of institutional cooperation networks, it has been discovered that Harbin Institution of Technology, Tsinghua University, the Chinese Academy of Sciences, the Hong Kong Polytechnic University, Xi'an University of Architecture and Technology, the University of Sheffield, Colorado State University, and Oxford Brookes

University are the main contributors to the research on carbon emissions in low-carbon rural areas. This indicates that the work of these institutions in the field of carbon emission research in low-carbon rural areas is very valuable and significant.

In addition, the analysis results of country and region distribution reveal that China, the United States, and the United Kingdom are the main countries making efforts into the research on carbon emissions in low-carbon rural areas. This shows that developed countries invest more into the research on carbon emissions in low-carbon rural areas on the whole and such countries have extensive influence in this field.

The development of low-carbon rural areas is of positive significance for protecting the environment, promoting sustainable development, and narrowing the gap between urban and rural areas. However, obvious differences are observed in the development of low-carbon rural areas with different geographical characteristics, which also provides a broad prospect for related research. With increasing attention being paid to climate change, countries around the world are taking action to mitigate the impact of climate change. Among them, peak carbon dioxide emissions and carbon neutrality have become one of the main ways of achieving emission reduction targets in the world. The traditional carbon emission assessment method only considers the energy consumption of enterprises and households while ignoring the influence of geographical characteristics on carbon emissions. Therefore, when evaluating the carbon emissions of low-carbon rural areas, the influence of geographical characteristics should be considered and adaptation should be made to the developmental stage, so as to better serve rural grassroots to realize energy transformation and upgrading.

Finally, the domestic and foreign literature documents on low-carbon rural optimization under different geographical characteristics were analyzed and summarized, and the related research stages, processes, and frontiers in this field were mainly presented. Meanwhile, the research limitations were also reviewed, and research prospects were noted. In addition, the connotations, research boundary, and field frontier of low-carbon rural areas were also sorted out, which will provide reference for the research on rural geography, urban–rural planning, and environmental science and probably extend the current research to more extensive special research fields. However, this research has certain limitations, since the research data mainly came from the core database of Web of Science. The research sample size was relatively small, and the relevant research content might not be deep enough. The research on low-carbon rural areas under different geographic features should be a research topic of practical guiding significance.

Despite these limitations, this research is of more theoretical significance than practical significance. Hence, it is necessary to form more valuable research content in the follow-up work by combining the specific practice of low-carbon rural optimization.

**Author Contributions:** Conceptualization, Z.Y.; methodology, Y.W.; writing—review and editing, B.Z. and Z.L.; supervision, project administration and funding acquisition, Z.L. and Q.H. All authors have read and agreed to the published version of the manuscript.

**Funding:** This research is funded by National Natural Science Foundation of China (No. 52078265).

**Conflicts of Interest:** The authors declare no conflict of interest.

## References

1. Canton, H. International energy agency—IEA. In *The Europa Directory of International Organizations 2021*; Routledge: Abingdon, UK, 2021; pp. 684–686.
2. Available online: https://data.worldbank.org/indicator/SP.RUR.TOTL.ZS?locations=IN (accessed on 1 August 2023).
3. Muller, M.F.; Esmanioto, F.; Huber, N.; Loures, E.R.; Junior, O.C. A systematic literature review of interoperability in the green Building Information Modeling lifecycle. *J. Clean. Prod.* **2019**, *223*, 397–412. [CrossRef]
4. Ngarambe, J.; Yun, G.Y.; Santamouris, M. The use of artificial intelligence (AI) methods in the prediction of thermal comfort in buildings: Energy implications of AI-based thermal comfort controls. *Energy Build.* **2020**, *211*, 109807. [CrossRef]
5. Shameri, M.A.; Alghoul, M.A.; Sopian, K.; Zain, M.F.M.; Elayeb, O. Perspectives of double skin façade systems in buildings and energy saving. *Renew. Sustain. Energy Rev.* **2011**, *15*, 1468–1475. [CrossRef]

6. Abdelaziz, E.A.; Saidur, R.; Mekhilef, S. A review on energy saving strategies in industrial sector. *Renew. Sustain. Energy Rev.* **2011**, *15*, 150–168. [CrossRef]
7. Thevega, T.; Jayasinghe, J.A.S.C.; Robert, D.; Bandara, C.S.; Kandare, E.; Setunge, S. Fire compliance of construction materials for building claddings: A critical review. *Constr. Build. Mater.* **2022**, *361*, 129582. [CrossRef]
8. Yin, X.; Liu, H.; Chen, Y.; Al-Hussein, M. Building information modelling for off-site construction: Review and future directions. *Autom. Constr.* **2019**, *101*, 72–91. [CrossRef]
9. Zhou, F.; Zhang, T.; Jin, Y.; Ma, Y.; Xian, Z.; Zeng, M.; Yu, G. Worldwide tinnitus research: A bibliometric analysis of the published literature between 2001 and 2020. *Front. Neurol.* **2022**, *13*, 828299. [CrossRef]
10. Van Eck, N.; Waltman, L. Software survey: VOSviewer, a computer program for bibliometric mapping. *Scientometrics* **2010**, *84*, 523–538. [CrossRef]
11. Zhou, Y.; Yan, X. Mapping research trends of building information modeling (BIM) using bibliometric analysis. *Autom. Constr.* **2017**, *80*, 1–13.
12. Yap, M.H.; Wong, J.T.H.; Wong, K.W.; Ngo, T. A bibliometric analysis of building information modeling literature from 2000 to 2015. *Autom. Constr.* **2017**, *84*, 195–208.
13. Chen, Y.; Lin, J. A bibliometric analysis of intelligent building research from 1994 to 2017. *J. Intell. Robot. Syst.* **2019**, *93*, 33–46.
14. Shaw, V.L.; Hunter, A.J.; Mortimer, N.D. Sustainable Energy Development for Rural China. In Proceedings of the ISES World Congress 2007 (Vol. I–Vol. V) Solar Energy and Human Settlement, Beijing, China, 18–21 September 2007; Springer: Berlin/Heidelberg, Germany; pp. 2578–2582.
15. Fan, Y.; Fang, C. Insight into carbon emissions related to residential consumption in Tibetan Plateau—Case study of Qinghai. *Sustain. Cities Soc.* **2020**, *61*, 102310. [CrossRef]
16. Congedo, P.M.; Baglivo, C.; Zacà, I.; D'agostino, D.; Quarta, F.; Cannoletta, A.; Marti, A.; Ostuni, V. Energy retrofit and environmental sustainability improvement of a historical farmhouse in Southern Italy. *Energy Procedia* **2017**, *133*, 367–381. [CrossRef]
17. Sunaga, N.; Onodera, H.; Kumakura, E.; Nakano, I.; Roh, H. Solar town Fuchu–Plan and performance. *Procedia Eng.* **2017**, *180*, 1433–1442. [CrossRef]
18. Zhu, X.; Liu, J.; Yang, L.; Hu, R. Energy performance of a new Yaodong dwelling, in the Loess Plateau of China. *Energy Build.* **2014**, *70*, 159–166. [CrossRef]
19. He, J.; Yue, Q.; Li, Y.; Zhao, F.; Wang, H. Driving force analysis of carbon emissions in China's building industry: 2000–2015. *Sustain. Cities Soc.* **2020**, *60*, 102268. [CrossRef]
20. Ali, G.; Abbas, S.; Pan, Y.; Chen, Z.; Hussain, J.; Sajjad, M.; Ashraf, A. Urban environment dynamics and low carbon society: Multi-criteria decision analysis modeling for policy makers. *Sustain. Cities Soc.* **2019**, *51*, 101763. [CrossRef]
21. Tonooka, Y.; Liu, J.; Kondou, Y.; Ning, Y.; Fukasawa, O. A survey on energy consumption in rural households in the fringes of Xian city. *Energy and Builds.* **2006**, *38*, 1335–1342. [CrossRef]
22. Mastrucci, A.; Rao, N.D. Bridging India's housing gap: Lowering costs and $CO_2$ emissions. *Build. Res. Inf.* **2019**, *47*, 8–23. [CrossRef]
23. Chan, D.; Cameron, M.; Yoon, Y. Key success factors for global application of micro energy grid model. *Sustain. Cities Soc.* **2017**, *28*, 209–224. [CrossRef]
24. Song, Y.; Chen, X.L.; Hao, T.Y.; Liu, Z.N.; Lan, Z.X. Exploring two decades of research on classroom dialogue by using bibliometric analysis. *Comput. Educ.* **2019**, *137*, 12–31. [CrossRef]
25. Röllin, H.B.; Mathee, A.; Bruce, N.; Levin, J.; von Schirnding, Y.E. Comparison of indoor air quality in electrified and un-electrified dwellings in rural South African villages. *Indoor Air* **2004**, *14*, 208–216. [CrossRef]
26. Siddiqui, A.R.; Lee, K.; Bennett, D.; Yang, X.; Brown, K.H.; Bhutta, Z.A.; Gold, E.B. Indoor carbon monoxide and PM2.5 concentrations by cooking fuels in Pakistan. *Indoor Air* **2009**, *19*, 75–82. [CrossRef] [PubMed]
27. Tadevosyan, A.; Mikulski, M.A.; Wallis, A.B.; Rubenstein, L.; Abrahamyan, S.; Arestakesyan, L.; Hovsepyan, M.; Reynolds, S.J.; Fuortes, L.J. Open fire ovens and effects of in-home lavash bread baking on carbon monoxide exposure and carboxyhemoglobin levels among women in rural Armenia. *Indoor Air* **2020**, *30*, 361–369. [CrossRef] [PubMed]
28. Manel, J.; Gabrion, I.; Raspiller, M.F.; Flesch, F.; Krencker, E.; Lambert, H. Epidemiology and prevention of accidental poisoning by carbon monoxide in the East of France. *Indoor Built Environ.* **1999**, *8*, 179–183. [CrossRef]
29. Burford, N.; Thurrott, J.; Pearson, A. Between Country and Town: New Concepts in Sustainable Rural Housing. In *Sustainable Architecture and Urban Development*; Lehmann, S., Al Waer, H., Al-Qawasmi, J., Eds.; CSAAR Press: Irvine, CA, USA, 2010; Volume 4, pp. 153–168.
30. Burford, N.; Robertson, C. New Housing Models for an Autarkic Rural Community. Architecture_MPS. 2015. Available online: http://architecturemps.com/housing-critical-perspective/ (accessed on 1 August 2023).
31. Di Tommaso, M.; Bordonzotti, I. $NO_x$ adsorption, fire resistance and $CO_2$ sequestration of high performance, high durability concrete containing activated carbon. In Proceedings of the II International Conference on Concrete Sustainability, Madrid, Spain, 13–15 June 2016; Volume 192.
32. Wang, J.B.; Fu, X.L.; Ni, H.W.; Zhou, X.L. Effects of Nitrogen Deposition on Ecosystem Carbon Fluxes in Sanjiang Plain Marsh of Northeastern China. *Appl. Mech. Mater.* **2011**, *71*, 2957–2961. [CrossRef]

33. Ge, J.; Luo, X.; Lu, J. Evaluation system and case study for carbon emission of villages in Yangtze River Delta region of China. *J. Clean. Prod.* **2017**, *153*, 220–229. [CrossRef]
34. Wu, G.; Liu, T.; Tang, M. Analysis of household energy consumption and related $CO_2$ emissions in the disregarded villages of Lijiang City, China. *Int. J. Sustain. Dev. World Ecol.* **2012**, *19*, 500–505. [CrossRef]
35. Xiao, L.S.; Li, X.H.; Wang, R. Integrating climate change adaptation and mitigation into sustainable development planning for Lijiang City. *Int. J. Sustain. Dev. World Ecol.* **2011**, *18*, 515–522. [CrossRef]
36. Siudek, A.; Klepacka, A.M.; Florkowski, W.J.; Gradziuk, P. Renewable Energy Utilization in Rural Residential Housing: Economic and Environmental Facets. *Energies* **2020**, *13*, 6637. [CrossRef]
37. Paolo, M.; Roberto, B.; Roberto, I.; Stefano, M. Solar Energy System in A Small Town Constituted of A Network of Photovoltaic Collectors to Produce Electricity for Homes and Hydrogen for Transport Services of Municipality. *Energy Procedia* **2015**, *78*, 824–829. [CrossRef]
38. Yang, T.; Zhang, X. Benchmarking the building energy consumption and solar energy trade-offs of residential neighborhoods on Chongming Eco-Island, China. *Appl. Energy* **2016**, *180*, 792–799. [CrossRef]
39. Skowronski, P.; Wisniewski, G. Potential of solar domestic hot water systems in rural areas for greenhouse gas emission reduction in Poland. *Environ. Manag.* **1996**, *20* (Suppl. S1), S95–S100. [CrossRef]
40. Fan, J.L.; Yu, H.; Wei, Y.M. Residential energy-related carbon emissions in urban and rural China during 1996–2012: From the perspective of five end-use activities. *Energy Build.* **2015**, *96*, 201–209. [CrossRef]
41. Tavakoli, R.; Najafi, M.; Tabesh, A.; Ashoori, T. Comparison of carbon footprint of trenchless and open-cut methods for underground freight transportation. In *Pipelines 2017*; American Society of Civil Engineers: Reston, VA, USA, 2017; pp. 45–55.
42. Kaya, Y. *Impact of Carbon Dioxide Emission Control on GNP Growth: Interpretation of Proposed Scenarios*; IPCC Energy and Industry Subgroup, Response Strategies Working Group: Paris, Italy, 1990.
43. Xu, D.M.; Zhao, T.Y.; Cheng, W. Study on Low-Carbon and Energy-Saving New Rural Countryside Construction in the Northeast of China. In *Advanced Materials Research*; Trans Tech Publications Ltd.: Stafa-Zurich, Switzerland, 2011; Volume 280, pp. 179–182.
44. Gupta, R.; Gregg, M. Local energy mapping for enabling area-based energy reductions. In Proceedings of the 16th IBPSA Conference, Rome, Italy, 2–4 September 2019; BSO2019; pp. 3690–3697.
45. Shan, M.; Yang, M.; Yang, X. Renewable energy utilization in Chinese buildings: Technology development and demonstration projects. *ASHRAE Trans.* **2012**, *118*, 508–514.
46. Deng, M.; Nie, Y.; Lu, F.; Ma, R.; Yuan, Y.; Shan, M.; Yang, X. Pollutant emission performances of improved solid fuel heating stoves and future implications in rural China. *Energy Build.* **2022**, *257*, 111810. [CrossRef]
47. Lai, A.M.; Clark, S.; Carter, E.; Shan, M.; Ni, K.; Yang, X.; Baumgartner, J.; Schauer, J.J. Impacts of stove/fuel use and outdoor air pollution on chemical composition of household particulate matter. *Indoor Air* **2020**, *30*, 294–305. [CrossRef]
48. Gupta, R.; Barnfield, L.; Hipwood, T. Impacts of community-led energy retrofitting of owner-occupied dwellings. *Build. Res. Inf.* **2014**, *42*, 446–461. [CrossRef]
49. Baborska-Narozny, M.; Stevenson, F.; Chatterton, P. A Social Learning Tool–barriers and opportunities for collective occupant learning in low carbon housing. *Energy Procedia* **2014**, *62*, 492–501. [CrossRef]
50. Ma, L.; Liu, S.N.; Ding, X.H.; Ma, W. Seasonal dynamics of soil microorganism in Zhalong reed wetland. *Appl. Mech. Mater.* **2011**, *71*, 2992–2998. [CrossRef]
51. Xu, Z. Effects of Building Layout on air temperature in community. *Procedia Eng.* **2017**, *205*, 4022–4029. [CrossRef]
52. Jebaraj, S.; Iniyan, S. An Optimal Electricity Allocation Model for the Effective Utilisation of Energy Sources in India with Focus on Carbon Tax. In Proceedings of the ISES World Congress 2007 (Vol. I–Vol. V) Solar Energy and Human Settlement, Beijing, China, 18–21 September 2007; Springer: Berlin/Heidelberg, Germany, 2008; pp. 2935–2939.
53. Zhou, B.; Wang, X.; Yan, D.; Xu, J.; Kang, X.; Chen, Z.; Hao, T. A novel stochastic model for hourly electricity load profile analysis of rural districts in Fujian, China. *Sci. Technol. Built Environ.* **2022**, *28*, 1166–1183. [CrossRef]
54. Deng, M.; Ma, R.; Lu, F.; Nie, Y.; Li, P.; Ding, X.; Yuan, Y.; Shan, M.; Yang, X. Techno-economic performances of clean heating solutions to replace raw coal for heating in Northern rural China. *Energy Build.* **2021**, *240*, 110881. [CrossRef]
55. Li, Y.L.; Han, M.Y.; Liu, S.Y.; Chen, G.Q. Energy consumption and greenhouse gas emissions by buildings: A multi-scale perspective. *Build. Environ.* **2019**, *151*, 240–250. [CrossRef]
56. Jiang, L.; Xing, R.; Chen, X.; Xue, B. A survey-based investigation of greenhouse gas and pollutant emissions from household energy consumption in the Qinghai-Tibet Plateau of China. *Energy Build.* **2021**, *235*, 110753. [CrossRef]
57. Zhou, Z.; Anwar, G.A.; Dong, Y. Performance-based bi-objective retrofit optimization of building portfolios considering uncertainties and environmental impacts. *Buildings* **2022**, *12*, 85. [CrossRef]
58. Cao, J.J.; Lee, S.C.; Chow, J.C.; Cheng, Y.; Ho, K.F.; Fung, K.; Liu, S.X.; Watson, J.G. Indoor/outdoor relationships for PM2.5 and associated carbonaceous pollutants at residential homes in Hong Kong-case study. *Indoor Air* **2005**, *15*, 197–204. [CrossRef]
59. Devine-Wright, P.; Wrapson, W.; Henshaw, V.; Guy, S. Low carbon heating and older adults: Comfort, cosiness and glow. *Build. Res. Inf.* **2014**, *42*, 288–299. [CrossRef]
60. Bondar, D.; Lynsdale, C.J.; Ramezanianpour, A.A.; Milestone, N.B. Alkali-activation of natural pozzolan for geopolymer cement production. In *Sustainable Construction Materials and Technologies*; CRC Press: Boca Raton, FL, USA, 2020; pp. 313–317.
61. Bret, S.; William, M.; Bench, G.; Stewart, F.; Charles, M.; Judith, C.; John, W. Fossil and contemporary fine particulate carbon fractions at 12 rural and urban sites in the United States. *J. Geophys. Res.* **2008**, *D113*, 1–20. [CrossRef]

62. Sadegh, M.; AghaKouchak, A.; Mallakpour, I.; Huning, L.S.; Mazdiyasni, O.; Niknejad, M.; Foufoula-Georgiou, E.; Moore, F.C.; Brouwer, J.; Farid, A.; et al. Data and analysis toolbox for modeling the nexus of food, energy, and water. *Sustain. Cities Soc.* **2020**, *61*, 102281.
63. Kim, Y.J. Moment-Shear Interaction Mechanism for Carbon Fiber-Reinforced Polymer-Strengthened Reinforced Concrete Beams in Flexure. *ACI Struct. J.* **2014**, *111*, 967. [CrossRef]
64. Zhang, P.; He, J.; Hong, X.; Zhang, W.; Qin, C.; Pang, B.; Li, Y.; Liu, Y. Carbon sources/sinks analysis of land use changes in China based on data envelopment analysis. *J. Clean. Prod.* **2018**, *204*, 702–711. [CrossRef]
65. Ahmed, S.; Mahmood, A.; Hasan, A.; Sidhu, G.A.S.; Butt, M.F.U. A comparative review of China, India and Pakistan renewable energy sectors and sharing opportunities. *Renew. Sustain. Energy Rev.* **2016**, *57*, 216–225. [CrossRef]
66. Huang, Z.; Liu, Y.; Gao, J.; Peng, Z. Approach for Village Carbon Emissions Index and Planning Strategies Generation Based on Two-Stage Optimization Models. *Land* **2022**, *11*, 648. [CrossRef]
67. Smith, P.; Martino, D.; Cai, Z.; Gwary, D.; Janzen, H.; Kumar, P.; McCarl, B.; Ogle, S.; O'Mara, F.; Rice, C.; et al. Greenhouse gas mitigation in agriculture. *Philos. Trans. R. Soc. B Biol. Sci.* **2008**, *363*, 789–813. [CrossRef] [PubMed]
68. Lal, R. Soil carbon sequestration to mitigate climate change. *Geoderma* **2004**, *123*, 1–22. [CrossRef]
69. Smith, P.; Powlson, D.S.; Glendining, M.J.; Smith, J.U. Potential for carbon sequestration in European soils: Preliminary estimates for five scenarios using results from long-term experiments. *Glob. Chang. Biol.* **1997**, *3*, 67–69. [CrossRef]
70. Lal, R. Soil Carbon Sequestration Impacts on Global Climate Change and Food Security. *Science* **2004**, *304*, 1623–1627. [CrossRef]
71. Kotsila, D.; Polychronidou, P. Determinants of household electricity consumption in Greece: A statistical analysis. *J. Innov. Entrep.* **2021**, *10*, 19. [CrossRef]
72. Jana, C.; Bhattacharya, S.C. Sustainable cooking energy options for rural poor people in India: An empirical study. *Environ. Dev. Sustain.* **2017**, *19*, 921–937. [CrossRef]
73. Tang, K.; Liu, Y.; Zhou, D.; Qiu, Y. Urban carbon emission intensity under emission trading system in a developing economy: Evidence from 273 Chinese cities. *Environ. Sci. Pollut. Res.* **2021**, *28*, 5168–5179. [CrossRef]
74. Abbood, K.; Mészáros, F. Carbon and energy footprint analysis of Hungarian transportation activities using a multi-region input-output model. *Curr. Res. Environ. Sustain.* **2023**, *5*, 100208. [CrossRef]
75. Feng, H.; Wang, R.; Zhang, H. Research on Carbon Emission Characteristics of Rural Buildings Based on LMDI-LEAP Model. *Energies* **2022**, *15*, 9269. [CrossRef]
76. Zhang, Z.; Wang, B. Research On the Life-Cycle $CO_2$ Emission of China's Construction Sector. *Energy Build.* **2016**, *112*, 244–255. [CrossRef]
77. Zhu, W.; Feng, W.; Li, X.; Zhang, Z. Analysis of the Embodied Carbon Dioxide in the Building Sector: A Case of China. *J. Clean Prod.* **2020**, *269*, 122438.
78. Anand, C.K.; Amor, B. Recent Developments, Future Challenges and New Research Directions in LCA of Buildings: A Critical Review. *Renew. Sustain. Energy Rev.* **2017**, *67*, 408–416. [CrossRef]
79. Chau, C.K.; Leung, T.M.; Ng, W.Y. A Review On Life Cycle Assessment, Life Cycle Energy Assessment and Life Cycle Carbon Emissions Assessment On Buildings. *Appl. Energy* **2015**, *143*, 395–413. [CrossRef]
80. Cang, Y.; Yang, L.; Luo, Z.; Zhang, N. Prediction of Embodied Carbon Emissions From Residential Buildings with Different Structural Forms. *Sustain. Cities Soc.* **2020**, *54*, 101946. [CrossRef]
81. Lotteau, M.; Loubet, P.; Pousse, M.; Dufrasnes, E.; Sonnemann, G. Critical Review of Life Cycle Assessment (LCA) for the Built Environment at the Neighborhood Scale. *Build. Environ.* **2015**, *93*, 165–178. [CrossRef]
82. Zhang, G.; Liu, M.; Gao, X. Dynamic Characteristic Analysis of Indirect Carbon Emissions Caused by Chinese Urban and Rural Residential Consumption Based on Time Series Input-Output Tables from 2002 to 2011. *Math. Probl. Eng.* **2014**, *2014*, 297637. [CrossRef]
83. Tomás, M.; López, L.A.; Monsalve, F. Carbon footprint, municipality size and rurality in Spain: Inequality and carbon taxation. *J. Clean. Prod.* **2020**, *266*, 121798. [CrossRef]
84. Xu, W.; Yin, Y.; Zhou, S. Social and economic impacts of carbon sequestration and land use change on peasant households in rural China: A case study of Liping, Guizhou Province. *J. Environ. Manag.* **2007**, *85*, 736–745. [CrossRef] [PubMed]
85. Xu, Y.; Wang, T.; Liu, W.; Zhang, R.; Hu, Y.; Gao, W.; Chen, Y. Rural system sustainability evaluation based on emergy analysis: An empirical study of 321 villages in China. *J. Clean. Prod.* **2023**, *389*, 136088. [CrossRef]
86. Hao, A.; Hou, Y.; Tan, J. How does digital village construction influences carbon emission? The case of China. *PLoS ONE* **2022**, *17*, e0278533. [CrossRef]
87. Sun, C.; Zhen, M.; Shao, Y. Research on the Thermal Environment of Northeast China's Rural Residences. *Open House Int.* **2017**, *42*, 52–57. [CrossRef]

MDPI AG
Grosspeteranlage 5
4052 Basel
Switzerland
Tel.: +41 61 683 77 34

*Sustainability* Editorial Office
E-mail: sustainability@mdpi.com
www.mdpi.com/journal/sustainability